AF148503

Advanced Nuclear Systems Consuming Excess Plutonium

NATO ASI Series

Advanced Science Institutes Series

A Series presenting the results of activities sponsored by the NATO Science Committee, which aims at the dissemination of advanced scientific and technological knowledge, with a view to strengthening links between scientific communities.

The Series is published by an international board of publishers in conjunction with the NATO Scientific Affairs Division

A Life Sciences **B Physics**	Plenum Publishing Corporation London and New York
C Mathematical and Physical Sciences **D Behavioural and Social Sciences** **E Applied Sciences**	Kluwer Academic Publishers Dordrecht, Boston and London
F Computer and Systems Sciences **G Ecological Sciences** **H Cell Biology** **I Global Environmental Change**	Springer-Verlag Berlin, Heidelberg, New York, London, Paris and Tokyo

PARTNERSHIP SUB-SERIES

1. Disarmament Technologies	Kluwer Academic Publishers
2. Environment	Springer-Verlag / Kluwer Academic Publishers
3. High Technology	Kluwer Academic Publishers
4. Science and Technology Policy	Kluwer Academic Publishers
5. Computer Networking	Kluwer Academic Publishers

The Partnership Sub-Series incorporates activities undertaken in collaboration with NATO's Cooperation Partners, the countries of the CIS and Central and Eastern Europe, in Priority Areas of concern to those countries.

NATO-PCO-DATA BASE

The electronic index to the NATO ASI Series provides full bibliographical references (with keywords and/or abstracts) to more than 50000 contributions from international scientists published in all sections of the NATO ASI Series.
Access to the NATO-PCO-DATA BASE is possible in two ways:

– via online FILE 128 (NATO-PCO-DATA BASE) hosted by ESRIN, Via Galileo Galilei, I-00044 Frascati, Italy.

– via CD-ROM "NATO-PCO-DATA BASE" with user-friendly retrieval software in English, French and German (© WTV GmbH and DATAWARE Technologies Inc. 1989).

The CD-ROM can be ordered through any member of the Board of Publishers or through NATO-PCO, Overijse, Belgium.

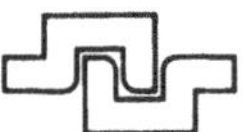

Series 1: Disarmament Technologies – Vol. 15

Advanced Nuclear Systems Consuming Excess Plutonium

edited by

Erich R. Merz
Forschungszentrum Jülich,
KFA,
Jülich, Germany

and

Carl E. Walter
Lawrence Livermore National Laboratory,
University of California,
Livermore, California, U.S.A.

Springer Science+Business Media, LLC

Proceedings of the NATO Advanced Research Workshop on
Advanced Nuclear Systems Consuming Excess Plutonium
Moscow, Russia
13–16 October 1996

A C.I.P. Catalogue record for this book is available from the Library of Congress

ISBN 978-0-7923-4650-0 ISBN 978-94-007-0860-0 (eBook)
DOI 10.1007/978-94-007-0860-0

Printed on acid-free paper

All Rights Reserved
© 1997 Springer Science+Business Media New York
Originally published by Kluwer Academic Publishers in 1997

No part of the material protected by this copyright notice may be reproduced or utilized in any form or by any means, electronic or mechanical, including photocopying, recording or by any information storage and retrieval system, without written permission from the copyright owner.

TABLE OF CONTENTS

PREFACE

This volume summarizes the materials presented at the NATO Advanced Research Workshop on Advanced Systems Consuming Excess Plutonium, held in Moscow, Russia, 13-16 October 1996. The conference was sponsored by the NATO Division of Scientific and Environmental Affairs in the framework of its outreach program to develop co-operation between NATO member countries and the Cooperation Partner Countries in the area of disarmament technologies.

This workshop was the first attempt to address unconventional approaches for maximum plutonium burning both from the point of view of reactor type and from the point of view of fuel cycle including fuel type, since they have not been reviewed and evaluated in depth elsewhere.

It is common understanding that in order to degrade the ex-weapon plutonium security risk to an acceptable level, any disposition option should meet what it calls the "spent fuel standard". This means making the excess plutonium roughly as inaccessible for weapons use as the much larger growing quantity of plutonium in spent fuel from commercial nuclear-power reactors. However, if one intends to go beyond the spent fuel standard to reduce the proliferation risk further and to satisfy the needs created by the entire global stock of plutonium and not only to protect it due to the allocated radiation barrier, an advanced approach is required.

Although weapon plutonium represented the focal essence of the workshop content, the addressed "advanced nuclear systems" apply to all sorts of plutonium as well (e.g. reactor grade Pu), and thus the more generally formulated workshop title seems to be justified.

As a follow-up to NATO`s three preceding Advanced Research Workshops,

- Managing the Plutonium Surplus, ARW in London, U.K., January 1994;
- Mixed Oxide Fuel Exploration and Destruction in Power Reactors, ARW in Obninsk, Russia, October 1994;
- Disposition of Weapon Plutonium - Approaches and Prospects, ARW in St.Petersburg, Russia, May 1995;

dealing with the totality of various topics of ex-weapon plutonium disposition stemming from dismantled warheads, this final workshop in series was hosted by the Nuclear Safety Institute, Russian Academy of Sciences. This far-sighted contemplation includes the exploration and evaluation of all conceivable pathways, even rather futuristic ones.

The conference`s concept envisaged consideration of the various options such as burning the plutonium more or less totally in reactors by special fuel management and repeated fuel reprocessing (partitioning and transmutation) or accelerator-driven subcritical systems. The conference demonstrated the availability, in different countries, of substantial relevant knowledge, experience, technologies, and other resources that can be applied to solve the predetermined goal.

The results and conclusions thus gathered now provides an excellent overview of the state of the art and should help decision makers to initiate reasonable problem solutions. A broad range of methods and processes are available to get rid of the unwanted weapon plutonium.

A thorough examination of all aspects was highlighted at the meeting and the pros and cons discussed. Specific future challenges posed include the safe and secure storage of surplus plutonium until the most suitable disposition option becomes workable. In the meantime, the creation of increased transparency is required to ensure mutual monitoring processes. International safeguarding measures are of paramount importance.

The issue treated on a comprehensive scientific basis at the workshop is a particularly multi-faceted problem. It represents besides important safety and security features a complicated technological challenge, and requires economic assessment as well.

The workshop concluded that there are several alternative solutions at hand for application and some others may be made available in the future. What is needed now is to promote the use of appropriate tools in comparative assessment studies.

Prof.Dr. Erich R. Merz
Co-Director, NATO Advanced
Research Workshop Programme

ACKNOWLEDGEMENTS

The sponsorship and the financial support of NATO is gratefully acknowledged. The editors are most grateful to the Disarmament Panel of the North Atlantic Treaty Organization Science Committee for their support.
The workshop was hosted and supported by the Nuclear Safety Institute of the Russian Academy of Sciences (IBRAE). Thanks to the efforts of many individuals from IBRAE in producing both a technically challenging and a perfect organized workshop.
The organizing committee and the editors would like to express their thanks to the speakers, commentators, and all others who attended the workshop and contributed to the discussions; they all made this publication possible.
We would like to thank Mrs. Annelies Kersbergen for editorial assistance.

LIST OF SPEAKERS AND PARTICIPANTS

Afanasiev, Anatoli A.
Russian Academy of Sciences
Nuclear Safety Institute (IBRAE)
52, Bolshaya Tulskaya
113191 Moscow
Russia

Alberstein, David
General Atomics
3550 General Atomics Court
San Diego, California
USA

Alekseev, Pavel N.
Russian Scientific Centre
Kurchatov Institute
1, Kurchatov square
123182 Moscow
Russia

Arutyunyan, Rafael V.
Russian Academy of Sciences
Nuclear Safety Institute (IBRAE)
52, Bolshaya Tulskaya
113191 Moscow
Russia

Boczar, Peter G.
Atomic Energy of Canada Limited
Chalk River Laboratories
Chalk River, Ontario KOJ 1JO
Canada

Bolshov, Leonid A.
Russian Academy of Sciences
Nuclear Safety Institute (IBRAE)
52, Bolshaya Tulskaya
113191 Moscow
Russia

Broeders, Cornelius
Forschungszentrum Karlsruhe
Institut für Neutronenphysik und
Reaktortechnik
Postfach 3640
D-76021 Karlsruhe
Germany

Carron, Igor
Texas A&M University
Department of Nuclear Engineering
College Station, Texas 77843-3133
USA

Chebeskov, Alexander N.
State Research Centre
Institute of Physics and
Power Engineering
1, Bondarenko square
249020 Obninsk
Russia

Didenko, Andrei N.
Russian Academy of Sciences
14, Leninsky prospect
117915 Moscow
Russia

Dmitriev, Alexander M.
Gosatomnadzor of the RF
34, Taganskaya str.
109147 Moscow
Russia

Doolen, Garry G.
Los Alamos National Laboratory
P.O.Box 1663, H 854
Los Alamos, New Mexico 87545
USA

Efimov, Evgenii I.
State Research Centre
Institute of Physics and Power
Engineering
1, Bondarenko square
249020 Obninsk
Russia

Egorov, Nikolai N.
Vice Minister
MINATOM of the Russian Federation
24/26, Bolshaya Ordynka str.
101000 Moscow
Russia

Ermakov, Nikolai I.
MINATOM of the Russian Federation
24/26, Bolshaya Ordynka str.
101000 Moscow
Russia

Gordeev, Vadim N.
Science, Technique, and
Disarmament Advisor of the
Embassy of Ukraine in Moscow
18, Leont`evskii per.
103009 Moscow
Russia

Gudowski, Waclaw
The Royal Institute of
Technology
Lindstedtsvagen 30
S-10044 Stockholm
Sweden

Hannum, William H.
Argonne National Laboratory
P.O.Box 2528
Idaho Falls, Idaho 83403-2528
USA

Hattori, Sadao
Central Research Institute of
Electric Power Industry, CRIEPI
1-6-1, Ohtemachi, Chiyoda-ku
Tokyo 100
Japan

Hesketh, Kevin W.
British Nucler Fuel
BNFL Springfields
Springfields, Preston, Lancs PR4 OXJ
United Kingdom

Hicken, Enno F.
Institute of Safety Research and
Reactor Technology, ISR-1
Forschungszentrum Jülich
P.O.Box 1913
D-52425 Jülich
Germany

Hyunin, Vladimir G.
State Research Centre
Institute of Physics and Power
Engineering
1, Bondarenko square
249020 Obninsk
Russia

Ipatov, Anatolii P.
Institute of Power Problems
Belarus Academy of Sciences
Sosny, 220109 Minsk
Belarus

Ivanov, Valentin B.
State Research Centre
Institute of Nuclear Reactors
Ulyanovsk region
433510 Dimitrovgrad-10
Russia

Krechetov, Sergei V.
Nuclear Power Agency
Republic of Kazakhstan
13, Respubliki square
480013 Almaty
Kazakhstan

Korobeinikov, Valerii V.
State Research Centre
Institute of Physics and Power
Engineering
1, Bondarenko square
249020 Obninsk
Russia

Kudryavtsev, Evgenii G.
MINATOM of the Russian Federation
24/26 Bolshaya Ordynka str.
101000 Moscow
Russia

Lefevre, Jean
Saclay Research Center
DCC/DIR, CE Saclay
F-91191 Gif sur Yvette, CEDEX
France

Maershin, Alexander A.
State Research Centre
Institute of Nuclear Reactors
Ulyanovsk region
433510 Dimitrovgrad-10
Russia

Marshalkin, Vasilii E.
Russian Research Center VNIIEF
37, Mira pr.
Nizhnii Novgorod region
607200 Arzamas-16
Russia

Matveev, Vyacheslav I.
State Research Centre
Institute of Physics and Power Engineering
1, Bondarenko square
249020 Obninsk
Russia

Medved, Yuri I.
Russian Academy of Sciences
Nuclear Safety Institute (IBRAE)
52, Bolshaya Tulskaya
113191 Moscow
Russia

Merz, Erich R.
Forschungszentrum Jülich
KFA-ISR-3
P.O.Box 1913
D-52425 Jülich
Germany

Minato, Akio
Central Research Institute of Electric Power Industry, CRIEPI
c/o IAEA Vienna, Windhabergasse 22/3
A-Vienna
Austria

Ogorodnik, Stanislav S.
SSTC for Nuclear and Radiation Safety
17, Kharkovskoe shosse
253100 Kiev
Ukraine

Orlov, Victor V.
Research and Development Institute of Power Engineering, a/ya 78
101000 Moscow
Russia

Pankov, Vladimir D.
Russian Academy of Sciences
Nuclear Safety Institute (IBRAE)
52, Bolshaya Tulskaya
113191 Moscow
Russia

Pavlovichev, Alexander M.
Russian Scientific Centre
Kuchatov Institute, Nuclear Reactors
1, Kurchatov square
123182 Moscow
Russia

Phlippen, Peter-W.
Institute of Safety Research and Reactor Technology, ISR-2
P.O.Box 1913
D-52425 Jülich
Germany

Polyakov, Anatoli S.
State Research Centre
Institute of Inorganic Materials
6a, Cosmonaut Volkov str.
125167 Moscow
Russia

Ponomarev-Stepnoi, Nikolai N.
Russian Scientific Centre
1, Kurchatov square
123182 Moscow
Russia

Prusakov, Vladimir N.
Russian Scientific Centre
1, Kurchatov square
123182 Moscow
Russia

Rief, Herbert W.
Commission of the European Union
Nuclear Safety Institute
Joint Research Centre Ispra
I-21020 Ispra, Varese
Italy

Robotnov, Nikolai S.
State Research Centre
Institute of Physics and Power Engineering
1, Bondarenko square
249020 Obninsk
Russia

Rozhdestvenskii, Mikhail I.
Research and Development Institute
of Power Engineering, a/ya 78
101000 Moscow
Russia

Rubbia Carlo
European Organization for
Nuclear Research, CERN
9, Chemin des Tulipiers
CH-1208 Geneva
Switzerland

Salvatores, Max
French Commission (CEA)
Nuclear Reactor Directorate
CEA Cadarache Bld.707
F-13108 Saint-Paul-Lez-Durance,CEDEX
France

Schulte, Nancy
Programme Director NATO
Science Affairs Division
Disarmament Technologies
B-1110 Brussels
Belgium

Sekimoto, Hiroshi
Tokyo Institute of Technology
O-okayama, Meguro-ku
Tokyo 152
Japan

Skiba, Oleg V.
State Research Centre
Institute of Nuclear Reactors
Ulyanovsk region
433510 Dimitrovgrad-10
Russia

Smirnov, Valerii S.
Research and Development Institute
of Power Engineering, a/ya 78
101000 Moscow
Russia

Stenbok, Igor A.
Research and Development Institute
of Power Engineering, a/ya 78
101000 Moscow
Russia

Stukalov, Vladimir A.
Russian Scientific Centre
Kurchatov Institute
1, Kurchatov square
123182 Moscow
Russia

Subbotin, Stanislav A.
Russian Scientific Centre
Kurchatov Institute
1, Kurchatov square
123182 Moscow
Russia

Troyanov, Mikhail F.
State Research Centre
Institute of Physics and Power
Engineering
1, Bondarenko square
249020 Obninsk
Russia

Trutnev, Yuri A.
Russian Nuclear Center VNIIEF
37, Mira pr.,Nizhnii Novgorod region
607200 Arzamas-16
Russia

Usanov, Vladimir I.
State Research Centre
Institute of Physics and Power
Engineering
1, Bondarenko square
249020 Obninsk
Russia

Tsurikov, Dmitrii F.
Russian Scientific Centre
Kurchatov Institute
1, Kurchatov square
123182 Moscow
Russia

Vasiliev, Sergei I.
State Research Centre
Institute of Nuclear Reactors
Moscow representative
2, Treyakovskii prozed
103012 Moscow
Russia

Velikhov, Evgenii P.
Russian Academy of Sciences
14, Leninsky prospect
117915 Moscow
Russia

Venneri, Francesco
Los Alamos National Laboratory
Accelerator Driven Transmutation Technology
P.O.Box 1663, MS H854
Los Alamos, New Mexico 87545
USA

Gol`din, Vladimir Ya.
Institute for Mathematical Modeling
Russian Academy of Sciences
14, Leninsky prospect
117915 Moscow
Russia

Walter, Carl E.
Lawrence Livermore National Laboratory
P.O.Box 808, MS I-125
Livermore, California 94551
USA

Zakharin, Boris S.
State Research Centre
Institute of Inorganic Materials
6a, Cosmonaut Volkov str.
125167 Moscow
Russia

INTRODUCTION TO NATO ADVANCED RESEARCH WORKSHOP

"Advanced Nuclear Systems Consuming Excess Plutonium"

Erich R. Merz
Research Center, KFA-ISR-3, D-52425 Jülich, Germany

The link between nuclear power and weapons capability has been recognized since the end of World War II, as has the potential for nuclear proliferation and its attendant political and security risk. Plutonium produced by the irradiation of uranium and then separated in pure form via chemical reprocesssing may be used by mankind for good or evil depending on the predetermined intention. The misfortune of peaceful nuclear power utilization stems from this entanglement of an identical processing step together with the fact that reactor plutonium can also be misused as weapons material.

The lower content of the plutonium isotope Pu-239 in energy-grade plutonium, as well as the presence of a comparatively large amount of isotope Pu-240, make energy-grade plutonium a less preferable material for nuclear munitions production than weapon-grade. Nevertheless, energy-grade plutonium can be directly used for making a nuclear explosive device. All things being equal, the device will only be larger if it was made of weapons-grade plutonium.

The prevention of nuclear war is of greatest importance to all mankind. A major component of global efforts to prevent proliferation is an efficient control of plutonium production and utilization. However, there is a growing world surplus of separated plutonium (civil and military). The plutonium problem is enhanced as a consequence of nuclear weapons disarmament under the reciprocal reduction agreed upon by the governments of Russia and the United States of America. The nuclear warheads that are to be dismantled without replacement contain approximately 150 - 200 tons of plutonium. At the same time, production of new weapons plutonium has come to an end in the U.S., and has also slowed down drastically in Russia, and will come to a total stop by the year 2001.

Surplus plutonium from both sources requires suitable methods for safe disposition, with the disposition of weapon plutonium exhibiting the highest priority. Unless plutonium is completely fissioned, a large fraction will remain for a long time in one form of storage or another, whether as separated plutonium, in spent fuel, or fixed in another matrix such as high-level waste glass or inert ceramics. Storage in spent fuel, or fixed together with high-level waste in another stable matrix, would make the storage form self-protecting to a significant degree due to its intense gamma radiation.

If one intends to go beyond the so-called spent fuel standard to reduce the proliferation risk further and to satisfy the needs created by the entire global stock of plutonium and not only to protect it by the allocated radiation barrier, an advanced approach is required. Such an elimination would place the plutonium almost completely beyond human access. This includes options such as burning it more or less totally in reactors by special fuel management and repeated fuel reprocessing or accelerator-driven subcritical systems (ADS).

E. R. Merz and C. E. Walter (eds.), Advanced Nuclear Systems Consuming Excess Plutonium, 1–3.
© 1997 *Kluwer Academic Publishers.*

Achievement of elimination is more costly, complex, time-consuming, and risky then minimizing accessibility to meet the spent fuel standard. Nevertheless, it seems worthwhile to treat all relevant aspects with regard to the potential and promise of the various approaches in this Advanced Research Workshop.

Topics to be covered

In the course of consuming plutonium, one should take advantage of its energy potential by the application of the various advanced atomic fissioning systems, e.g.

- advanced LWRs with inherent safety features
- CANDUs using plutonium- and thorium-based fuel
- fast reactors comprising sodium and lead cooling systems, including thorium fuel utilization
- high temperature gas-cooled reactor systems using plutonium- and thorium-type fuels
- molten salt reactor systems using plutonium- and thorium-type fuels
- accelerator-driven subcritical systems with different fuel cycles.

The objective of this Advanced Research Workshop is:

- identify reactor systems which really consume considerable amounts of plutonium
- assess the amount of radioactive products in the plant
- assess the increase in safety with these systems as compared to future innovative LWRs and fast reactors
- identify the required R&D and compare the advantages and disadvantages of fuel cycles for the considered systems taking into account all stages: fuel fabrication, interim storage, reprocessing (if necessary), and final disposal of wastes
- assess the time needed up to the operation of a plant with sufficient plutonium consumption (~1 ton/year)
- assess the amount and quality of the waste from the individual processes.

An in-depth treatment of these items is desired, stressing the discussion on technologies of modern core designs for advanced LWRs (full MOX cores) or Sodium Cooled Fast Reactors (without breeding blankets) with the utilization of traditional MOX fuel, and on the development of new types of fuel (for example, fuel with an inert diluent instead of uranium, or cermet fuel, etc.) as well as on the use of new reactor systems.

The output of this workshop will be a report citing the potential utility of each approach for consuming weapons plutonium, the key unresolved issues of each approach, and the key unresolved issues of each fuel cycle technology.

Sufficient knowledge is already available today to give evidence that large-scale transmutation of plutonium is technically feasible. However, plutonium destruction fractions greater than 80 percent only appear attainable with the help of fuel reprocessing and plutonium recycling. With such repeated reprocessing and reuse, virtually any type of reactor could in principle be used to consume more plutonium than it produces.

One must keep in mind that reactors with a thermal neutron spectrum, such as LWRs, can only fission some isotopes of plutonium (e.g. Pu-239, Pu-241, Pu-243), whereas fast-neutron reactors can fission all plutonium isotopes. However, to do this on a technical scale requires a tremendous expenditure of funds and development time.

The time required for the actual transmutation is a complex function of the percentage of plutonium consumed in each reactor cycle, the function of the plutonium in the fuel cycle that is actually in the reactor where it can be consumed, and the amount of plutonium lost to waste in chemical processing.

Introduction of plutonium transmutation will not eliminate the need for radioactive waste disposal and thus the availability of a geologic underground repository. But reducing the potential long-term risks can make it easier to make decisions about site location, site licensing, and knowledge of how to build a geologic repository. But new problems may arise regarding the siting and safety problems of the necessary reprocessing and transmutation facilities.

Nevertheless, future steps should be taken to reduce the proliferation risks posed by all of the world`s plutonium stocks, military and civilian, separated and unseparated. Options for near-total elimination of plutonium are at hand, but research on defining and exploring these options should be continued at a conceptual level. These options, however, can only be realistically considered in the broader context of the future of nuclear electricity generation, including the minimization of security and risk.

Implementation of partitioning and transmutation is more likely in case of the continuation and expansion of nuclear power than otherwise. There is an ongoing debate about whether the expansion of nuclear power with the partitioning/transmutation strategy should be a valid goal, or whether such a strategy should be evaluated independently of nuclear power expansion. Obviously, there are such small incentives to introduce partitioning and transmutation in the current dominating once-through nuclear fuel cycle, that partitioning/transmutation should be introduced in a longer term decision about using new nuclear power technology as a future energy source.

If prompt measures for weapons plutonium disposition are indispensable then only the well-known approaches applying the "spent fuel standard" are recommendable.

View on The Problem of the Plutonium Utilization in Russia.

A.M.Dmitriev

Gosatomnadzor of the Russian Federation, Moscow, Russia

On behalf of Russian Regulatory Body I am greeting you on the such representative meeting. Our meeting, information exchange and discussion will be doubtless to promote the best understanding of the problem concerned the excessive plutonium consumption.

The nuclear community faces new realities which were not foreseen two or three decades ago.

Nuclear power generation has grown at a far slower rate than expected.

There has been limited interest in fast reactors and delay in their commercialization where they are being developed.

The adoption of a closed nuclear fuel cycle, where chosen, is only partially achieved through burning of mixed oxide fuels.

These new realities have resulted in the accumulation of plutonium in civilian programmes. In addition, as a result of the end of the cold war, there may soon be a large amount of plutonium from dismantled warheads.

From the all possible ways of the excessive plutonium management the purpose of our working meeting is a consideration of the problems concerned the excessive plutonium burning of in the appropriate nuclear installations.

In any case the excessive plutonium storage is inevitable and very expensive. In Russia the plutonium storage problem relates to the weapon grade plutonium as well as to the plutonium extracted from the spent fuel of nuclear power plants (NPPs).

Except the plutonium burning of and it's temporary storage there are several technologies of the plutonium management under the consideration in the world community today. Plutonium immobilization is one of such technologies. It implies the plutonium replacement inside the glass surrounding (vitrification) or synthetic rocks (synrocks) with the further replacement in the

E. R. Merz and C. E. Walter (eds.), Advanced Nuclear Systems Consuming Excess Plutonium, 5–10.
© 1997 *Kluwer Academic Publishers.*

deep geological structures. The Russian point of view, that is common for the organizations using an atomic energy as well as for the nuclear safety regulatory body, consists that the best way of the excessive plutonium elimination will be in it's burning of in the appropriate nuclear installations. In spite on the common approach to this problem solution the initial motives are principally different for Minatom and for Gosatomnadzor of Russia.

The Minaton of Russia representatives declare, that plutonium is valuable power raw material, and consequently it's burning of is the most rational way of it's handling from the economics point of view. Gosatomnadzor of Russia supports such way of the plutonium management because of the plutonium vitrification results (at least in Russia and at least at the present moment of time) in the technological problems which do not guarantee long term preservation of the vitrified structures. It can entail rather fast - during several decades - the occurrence of a problem concerned the destroyed vitrified substanses contained plutonium handling. Today Russia has not the industrial technology of manufacturing the synrocks containing plutonium. It has to be stressed from the Gosatomnadzor of Russia point of view that the immobilization basically permits to extract plutonium, though it is rather difficult problem. It is undesirable from the point of view of nuclear non-prolifiration. As to the plutonium disposal in deep wells, it is rather difficult to guarantee the underground plutonium migration absence during the tens of thousands years, because of the geochemical processes in the depth of earth crust are very complex. Much more easy to deal with the fission products because there are already exist rather wide experience on such radioactive materials handling and the decay time of the most dangerous radioactive nuclides are much more shorter than that of plutonium.

Thus, the statement of our workshop problem restricting the excessive plutonium handling problem by plutonium using in the special nuclear installations is more closest to the Russia and causes maximum interest.

I shall not discuss the principal circuits of nuclear installations for the plutonium burning of which will require for their development and practical realization so large time that the plutonium storage problem will become very sharp. First of all I have in view of very beautiful circuits with using the accelerators engineering. The expectation during large period of time beautiful, safe and economical future technologies will put us before the problem of large expenses on the radiochemically extracted plutonium storage will be required.

There is no basis to think, that use of such technologies can affect the plutonium balance during nearest 30-40 years.

The attitude to plutonium as to the valuable energetic material for usual nuclear reactors requires a detailed estimation of perspectives. It is necessary to realize what we intend to do with the spent MOX fuel. If we decide to develop radiochemical plutonium and uranium extraction from the spent MOX fuel, we should clearly understand that uranium irradiated even the first time in usual pressurised water reactors (PWRs) contains such amount of U-232 that the safety limits for the fuel manufacturing from this extracted and supplementary enriched uranium will much exceed limits established for the usual processes of the fuel manufacturing for the PWRs.

We should also take into account, that the MOX fuel cost can be higher than this one for the usual fuel, even at zero plutonium cost. All these factors force us to define a position - either we shall accept the plutonium as the main fuel for nuclear reactors with it's breeding, that means we shall pass to the plutonium epoch, or select other way at which plutonium will play a smaller role.

It is necessary to agree that the energetics of mankind in 30-60 years will face the serious problems.

Prompt growth of the population, increase of an energy consumption level in developing countries, natural resources exhausting and pollution of an environment force us to look at the atomic energetics development as on the one of few possible ways of mankind surviving.

Russian stocks of organic fuel on period more than 50 years really consist only from coal. Coal burning of is made so far with the release of a plenty of toxic substances, sulphur, dust, carbon dioxide, radioactive substances. Any estimations of the equipment cost and operational expenses on the effective cleaning the releases into atmosphere and the appropriate ashes management result in so large figures that it is very difficult to rely on their wide use in future.

Thus the future of energetics can be largely determined by perfection of nuclear technologies.

There were the plans of fast reactors wide using in Russia about 20 years ago. It was supposed that these reactors will work on mixed uranium-plutonium fuel. The practical experience of MOX fuel use in Russia exists only with regard to the fast reactors. In reality these plans have been strongly changed as regards to the reactors number and to the reactor fuel. Only one sodium cooled commercial fast reactor works in Russia today. It is BN-600 type reactor of Beloyarskaya NPP (nuclear power plant). This reactor has only 7% of its reactore core loaded with MOX fuel. The fuel does not reach the design burn up level on the everage. High expenses which are necessary for the fast reactors construction and operation and the complex economic situation in Russia has resulted in that now there are only two applications for two sodium cooled fast reactors construction - one for the Beloyarskaya NPP and one for the South-Urail NPP.

The financing of these reactors construction was not practically carried out during the last time.

The existing capacities on the MOX fuel manufacturing are not sufficient even for the fuel manufacturing for one reactor having reactor core fully loaded with the MOX fuel. It means that the real rate of plutonium burning of in the fast reactors in the nearest future will be much lower than the rate of its chemical extraction.

The orientation on the plutonium burning of in the fast reactors has resulted in that Russia has not installations for the MOX fuel preparation for WWERs (Russian PWRs).

Accordingly, there is no experience on the MOX fuel burning of in WWERs and in the other reactors of such types advanced for MOX fuel burning of.

The safety of the reactors should be increased while MOX fuel used.

The analysis of the advanced PWR reactors projects in the world shows that their design is being constantly complicated. Decreasing heavy incidents probability and reducing consequences of possible heavy incidents with the reactore core melting require increasing the number and complexity of safety systems. In result, the capital (common) expenses grow so much that the cost of electric power received from the reactors with water becomes comparable with that for the fast reactors. At the same time the decreasing of heavy incidents probability and reducing their consequences does not exclude them completely. Accordingly, the wide development of atomic energy based on the pressurised water reactors (PWR) should result in significant growth of number of reactors in operation. It will also result in the incidents with heavy consequences probability increasing practically proportionally to number of reactors. It will be difficult for mankind to reconcile with such perspective.

The wide development of nuclear energy in the world is possible only on the basis of nuclear reactors which do not lead to the heavy consequences at any accident initiating event (Catastrophe-free nuclear technology).

High temperature reactors with the fuel in form of microsphere (indestructible fuel elements under any severe incident) seem to meet this requirement. The other peculiarity of these reactors are the flexible fuel cycle and good neutron economy.

There are rather good technologies today for the microfuel corn (kernel) manufacturing directly from solutions of any nuclear materials - uranium, plutonium and thorium. There is also the opportunity of MOX fuel manufacturing on the basis of these materials.

The opportunity to receive a high burn up level determines the good profitability of fuel elements under the single using without their further chemical reprocessing.

Such technology was developed previously in Germany and USA. This experience should be used while designing new reactors on the basis of microfuel elements.

It seems to be that the plutonium and thorium burning up in the HTR's fuel may ensure the essential increase of future nuclear energetics safety and better meet the requirements of nuclear non-prolifiration.

The quantity of plutonium in the fuel should be enough to ensure initial period of the fuel work - before the essentional amount of U-233 will appear from thorium in the fuel.

Then the subsequent work of fuel elements up to the highest burn up levels should result in the almost complete plutonium burning up and accumulation in the formed U-233 a significant impurity of U-232.

The absence of plutonium causes the toxicality decreasing of the spent fuel. The high level of gamma-radiation caused by the U-232 decay products strongly complicates possible attempts of U-233 radiochemical extraction. Besides, the thorium dioxide low solubility serves as the additional barrier complicating attempts of U-233 radiochemical extraction.

Refusal from the spent fuel radiochemical reprocessing and high levels of the fuel burning of provide serious reduction of toxical releases as well as the significant reduction of the wastes in nuclear energetics. Besides, the amount of operational radioactive wastes from the high-temperature reactors (HTRs) usually makes 1-2% from the appropriate wastes of water reactors (PWRs and BWRs).

Thus, combination of the high-temperature reactors with the plutonium-thorium fuel manufacturing in kind of microspheres, permits in principal:

- To increase sharply the safety of nuclear energetics;
- To eliminate the weapon-grade plutonium stocks;
- To consume plutonium from water reactors in accordance with its radiochemical extraction;
- To proceed hereafter on the reactors feeding by thorium at the expense of U-233 stocks in kind of spent fuel. The reloading technology of these reactors should allow partial replacement of fuel;
- To increase the level of nuclear non-prolifiration at the expense of U-233 radiochemical extraction difficulties;
- To lower the radiotoxine releases in nuclear energetics at the expense of the low level of radioactive wastes generation in the reactors based on microfuel using and refusal from the radiochemical reprocessing of nuclear fuel.

Such decision assumes a refusal from the idea of "plutonium era" in nuclear energetics but assigns to plutonium an important role in initiating the wide development of more safe nuclear energetics of future.

SAFETY ASPECTS OF ADVANCED NUCLEAR SYSTEMS CONSUMING PLUTONIUM

E.F. HICKEN

Forschungszentrum Jülich GmbH

Institut für Sicherheitsforschung und Reaktortechnik

D-52428 Jülich

1. Introduction

The current situation of nuclear energy in the world is fundamentally influenced on the one hand by the need for electrical energy and on the other hand by concerns resulting from the accidents in TMI and Chernobyl. As a result of the latter in some (mostly Western) countries no more nuclear power plants are built - at least for some time. In countries with less public concerns nuclear power plants are commissioned.

Various constructors of nuclear steam supply systems are developing new products with a higher safety level as compared with existing systems. In addition, higher safety requirements have been defined by licensing bodies.

E. R. Merz and C. E. Walter (eds.), Advanced Nuclear Systems Consuming Excess Plutonium, 11–19.
© 1997 *Kluwer Academic Publishers.*

There has been some need expressed to use weapons-grade plutonium for an efficient energy production. Reactor plutonium is used mostly as MOX in LWRs - in principle with a low efficiency with regard to fully take advantage of its energy content. Only Fast Breeders would use the plutonium efficiently; however, there are concerns regarding the safety of these reactors.

A threat from a final disposal site of radioactive waste is dependent on the amount and species of long-lived actinides. A reduction of the amount and radioactive content of the disposed fission products would be beneficial.

Therefore, it should be studied if advanced systems would have advantages regarding the safety during operation and from a disposal site compared with the next generation nuclear reactors (e.g. EPR, AP 600, etc.).

2. Concerns by the Public

The "public" seems to be most concerned about

- the threat to health and environment resulting from an accidental release of radioactive products from operating nuclear installations (power-, reprocessing and fuel fabrication plants). The releases from normal operation seem to be of less importance.
- a threat resulting from the transport and storage of radioactive or fissile material - especially in case of plutonium.
- a threat from final disposal sites above ground as well as in deep geological formations.

3. General Objectives Regarding Safety

Only some general objectives are given below.

3.1 ENVIRONMENTAL LOADS

For operational states and Design Basis Accidents the permissible loads outside the fence of a nuclear installation are specified by national bodies or the limits as defined by ICRP are used. These limits are specified in a way that no undue health effect exist.

For radioactive loads from accidents being less probable than Design Basis Accidents and with a major release of fission products inside the containment the requirements are quite different in different countries. The toughest requirements seem to exist in Germany and France - to be applied for the EPR design. A law in Germany does require, now, that also in case of Severe Accidents (e.g. core melts for LWRs) no evacuation and relocation of people outside the fence should be necessary. The limits for evacuation and relocation are up to now specified nationally; it is expected that they will be replaced soon by limits proposed by IAEA agreed internationally.

As a consequence these limits - at least - have to be met by advanced systems. If well designed, advanced systems may better and cheaper meet these requirements.

3.2 DEFENCE-IN-DEPTH PRINCIPLE

The defence-in-depth principle will remain the fundamental principle of safety for nuclear power plants. It has to be demonstrated that the three basic safety functions

- reactivity control
- cooling the fuel
- confining radioactive substances

are correctly ensured.

The defence-in-depth principle is applied twofold.

1) The quality of design, construction and operation as well as appropriate measures to cope with incidents, accidents (up to beyond - design-basis-accidents) have to be ensured.

2) Barriers against a release of radioactive substances have to be existing to cope with internal and external hazards. I strongly believe that a leaktight containment - being capable to cope with loads from internal and external hazards - is mandatory for any advanced power plant.

3.3 EXPERIENCES FEEDBACK

40 years experience with nuclear installations has shown that experience in design, commissioning and operation has been beneficial. As a consequence, if advanced nuclear systems are not of "evolutionary" type pilot and demonstration plants creating operating experience are mandatory.

3.4 COST CONSIDERATIONS

When assessing costs the necessary support for R&D, pilot and demonstration plants will not be considered here.

It should be indicated, here, that

- the advanced reactors should have a lifetime up to 60 years and availability factors of 85 %+ as it is planned for the next generation reactors.
- the costs for power production and waste disposal must be competitive with other (allowed) power producers.

4. Some Safety-Related Aspects

The author must confess that the items listed below are mainly influenced by safety considerations for LWRs. However - because they are common practice - they should be considered also for advanced reactors.

4.1 INTEGRITY OF THE PRIMARY CIRCUIT

Independent of any specific design, the integrity of the primary circuit (containing the fuel) is of very high importance.

Because the replacement of major components will result in an outage time of at least a few weeks (example: steam generators of PWRs) a replacement of those components should not be necessary more than once or twice during lifetime.

4.2 REDUCTION OF THE FREQUENCY OF INITIATING EVENTS

The threat from operating nuclear installations results usually from accidents where related safety systems fail and an inherent safe system behaviour does not exist.

Therefore, the frequency of initiating events should be low during all operating states, including full power, low power and shutdown conditions.

4.3 PLANT TRANSIENT BEHAVIOUR

To the extend possible the plant should show a slow reaction to deviations from operating conditions. In addition, the plant should not be sensitive to operator errors and failures in operational systems. The existing guidelines in some countries require automatic reactions to make operator actions not necessary for at least 30 minutes. This should be the minimum requirement.

4.4 PROBABILISTIC SAFETY ASSESSMENTS (PSAs)

It has been experienced from existing power plants that results from a Probabilistic Safety Assessment can provide a useful guidance to identify weak points in the design or operation and to assist efforts for an optimised system regarding safety. PSAs can be performed at the design stage.

4.5 EXTERNAL HAZARDS

External hazards can affect consecutively or simultaneously different lines of defence of the plants and they are, of course, site-dependent. Therefore, due considerations must be paid to the choice of the sites in order to avoid high costs for related design provisions.

It is possible to design a plant to cope with loads from earthquakes, floods, gas explosions and fires.

Special considerations has been given to loads from airplane crashes and military attacks.

Most of the existing nuclear power plants are not designed to cope with loads from an airplane crash. Some (less than 10 %) can cope with loads from an existing military fighter. Military nuclear power plants are built underground. An underground design will result in a cost increase between 25 and 40 %. It is evident that in about 30 - 50 years practically no plant above ground can cope with the loads from an airplane crash. This is already valid now for military attacks. Therefore, the requirements resulting from airplane crashes and military attacks have to be specified later.

4.6 TRANSPORT AND STORAGE OF RADIOACTIVE MATERIAL

It is obvious that a transport of fissile material can be performed safely; due to security reasons plutonium and highly enriched uranium - or more general, fissile material usable for weapons - should not be transported in weapons-grade form.

Requirements for storages can be derived from those for plants; the security aspect must be added. It is evident that a short storage time would be beneficial.

4.7 FINAL DISPOSAL

It is evident that the total amount of waste and its composition determines the quality of intermediate storage sites as well as the final disposal site.

Therefore, the amount of waste has to be minimised. In addition, to the extend feasible the waste should be separated into long-lived and short-lived radioactive materials and to the extend feasible long-lived radioactive products should be transmuted into products with shorter half-life times.

Although transmutation and separation should not be required both treatments may ease disposal with regard to costs and acceptance.

5. Licensing Aspects

It must be realised that existing licensing requirements are tailored to LWR safety. In addition, licensing bodies and experts are mainly knowledgeable for this reactor type.

From experience (fast breeders, HTRs) difficulties are known when other types have to licensed.

Examples are e.g. the requirements for redundancies, diversities, applications of the single failure criterion to passive components/systems, etc.

Some relief will exist for experimental facilities.

6. Some Final Remarks

It is obvious that the ideas presented are (closely) related to advanced requirements for LWRs. This is because about 350 LWRs are operating and several hundred LWRs will be commissioned before the year 2050; most of them - I believe - designed according to advanced requirements. Therefore, any new proposal has to be compared with "reality", because special requirements cannot be enforced upon a nation. Therefore, new ideas have to be convincing and should be based on a thorough assessment.

PROBLEMS OF EXCESS PLUTONIUM UTILIZATION IN NUCLEAR REACTORS

A.N. Chebeskov, V.S. Kagramanian, A.V. Malenkov
State Research Center, Institute of Physics and Power Engineering, Bondarenko Sq.1, Obninsk, Kaluga region, Russia

Abstract

The great amount of fissile materials being released under nuclear disarmament drew attention again to problems of nuclear fuel cycle and to options of plutonium utilization both civilian and weapons-grade. As Russia keeps the strategy of closed fuel cycle, separated plutonium is supposed to be utilized in nuclear reactors for electricity production. Three scenarios of separated plutonium utilization are presented in the paper. The first basic scenario is the transformation of separated plutonium into spent nuclear fuel using fast reactors. The second scenario is plutonium using in presently operating reactors. And in the third one thermal reactor possibilities are considered. Together with safety, people's health protection, environmental and economic acceptance the problem of non-proliferation regime strengthening is extremely important when various concepts of fuel cycle are under consideration.

1. INTRODUCTION

Accumulated stocks of separated civilian plutonium together with the expected great amount of weapons-grade plutonium call for elaboration of appropriate strategy for management of this material. Safety, environment, economics and non-proliferation problems must be taken into consideration.

Taking into account the amount of separated civilian plutonium ~30 t and the expected amount of weapons-grade plutonium up to 100 t it is necessary to elaborate such plutonium utilization scenario which would be technically realizable within visible period of time. From the first steps of development of nuclear energy in the former Soviet Union the concept of closed fuel cycle had the priority. Russian attitude to this point did not change noticeably. Russian point of view consists in reprocessing of spent fuel and plutonium separating. Then separated plutonium is supposed to be utilized in nuclear reactors. In this connection it is necessary to answer the question: are Russian reactors ready to utilize separated civilian and expected weapons-grade plutonium taking into account that this plutonium has various content of higher isotopes?

During last decades the development of nuclear power has been considerably down-sized and the construction of BN-800 fast reactor on the South-Urals site was "frozen". It is reasonably for time being to store separated plutonium in reliable safe interim storage. However it should not be too long for economic, environmental, political and proliferation risk reasons. The basic concept on the plutonium management (both civilian and weapons-grade) has been developed by the Ministry for Atomic Energy which is based on the following principles [1]:

- Our experience of plutonium management should be maximized;
- Protection against sabotage and uncontrolled plutonium use is the most important criteria for selecting an effective plutonium management option;
- Plutonium management should be economically and environmentally ac-ceptable;
- Separated plutonium management option should serve as a good basis for the development of the optimal fuel cycle for the long-term perspective.

On this basis three possible scenarios of plutonium utilization are considered. The first, basic scenario is the transformation of separated civilian and ex-weapons plutonium into spent fuel using fast reactors. The second one assumes plutonium utilization in operating reactors. And in the third scenario thermal reactor possibilities for this purpose are considered.

It is necessary to stress that these three scenarios are not equivalent and in particular from the point of proliferation risk. Taking into account the features of Russian state that is rather large territory and many inhabitable regions are available, the concept of closed nuclear centers including all necessary facilities: spent fuel reprocessing plant, MOX fuel fabrication facility and nuclear reactors with MOX fuel inventory is the most suitable decision which can solve nonproliferation task together with the problems mentioned above.

E. R. Merz and C. E. Walter (eds.), Advanced Nuclear Systems Consuming Excess Plutonium, 21–26.
© 1997 *Kluwer Academic Publishers.*

2. PLUTONIUM ISOTOPIC COMPOSITION VARIETY

Plutonium which is built-up inevitably in nuclear reactors is characterized by the variety and diversity of its isotopic composition. Plutonium isotopic composition depends on the type of reactor-producer, the value of spent fuel burn-up and many other factors. Plutonium being produced in thermal reactors of the civilian nuclear power plants (NPPs) has as a rule large amount of higher plutonium isotopes which can reach up to 60 % of the total plutonium. At the same time plutonium produced for nuclear weapons has comparably small amount of higher isotopes: 6-7 %. The variety and diversity of plutonium composition leads to some problems if plutonium is supposed to be used in nuclear reactors. In what way these problems are being solved in Russia the next section presents.

3. PROBLEMS OF PLUTONIUM UTILIZATION IN NUCLEAR REACTORS

3.1. Fast Neutron Reactors

Comprehensive studies having been carried out earlier in the IPPE show that fast neutron reactors can consume plutonium of any isotopic composition. The variety of plutonium isotopic composition results in some additional measures to be taken that links namely with MOX fuel fabrication and does not influence noticeably on the concept and design parameters of fast neutron reactors. Nevertheless single additional problems may occur. For example, for MOX fuel core inventory with ex-weapons plutonium the use of neutron source for starting period may be needed in contrast to civilian plutonium core inventory.

Within the process of MOX fuel fabrication it is necessary to provide continuous measuring of plutonium isotopic composition and the amount of plutonium in the mixture (fuel enrichment) should be determined in accordance with the real isotopic composition of the plutonium portion that has been just involved in the technology process. The appropriate method of fuel enrichment correction has been developed in the IPPE and introduced into Complex-300 technology process at the PO "Mayak".

Another important issue of MOX fuel fabrication under variety of plutonium isotopic composition is the nuclear safety provision at all stages of technology line. Depending on isotopic composition the different restrictions should be introduced to limit the maximum mass of plutonium in the technology cycle. For ex-weapons plutonium, for example, the nuclear safety restrictions will be the strongest.

One more important point of considered problem is radioactivity. If glove boxes provide sufficient shielding for personnel handling ex-weapons plutonium, the radiation from higher isotopes and their daughter products means that more shielding and greater precautions including remote-controlled systems must be necessary.

3.2. Thermal Neutron Reactors

The utilization of plutonium in thermal reactors upon the variety of plutonium isotopic composition seems to be more complicated task than it is for fast reactors. This follows partly from the presence of non-fissile isotopes, soft neutron spectrum and the fact that this problem in Russia has been not developed properly yet. So we have not got experimental substantiation of plutonium utilization in thermal reactor designs as well as we have not got facility for MOX fuel fabrication for thermal reactors. Till now there is no thermal reactors in Russia which were designed with possibilities of MOX fuel utilization.

The experience of Western Europe and our calculational studies show that some difficulties exist on this way. These difficulties are connected in particular with the accumulation of plutonium-241 and minor actinides in spent MOX fuel with the amounts that are several times higher than those in spent uranium fuel. This results in the noticeably higher radiotoxicity of spent MOX fuel in comparison with spent uranium fuel.

The specific character of Russia consists in the fact that at present we have not got, in contrast to Western Europe, thermal reactors which are capable to use MOX fuel. We should design and construct such reactors if this scenario gets enough support. However, again in contrast to Western Europe, we have got the BN-800 fast reactor design which meets all up-to-date safety requirements. At present this is the only design of nuclear reactor in Russia that has passed all required examination stages including ecology examination.

Economic issue of the problem should be also mentioned. Plutonium utilization in thermal reactors in Western Europe results from the relative high cost of fast reactors. Such a situation has been in Russia too - the cost of electricity produced by the first commercial fast reactor BN-600 is about 40% higher than that of the VVER-1000 thermal reactors. The use of the BN-600 experience in the development of the BN-800 design resulted in decreasing of specific metal demand for the BN-800 design up to 20%. In addition to this the improvement of the BN-800 fuel cycle economics is also determined by changing over from the uranium fuel which is inefficient in fast reactors to the MOX fuel and by its further increase of burn-up. The strict safety requirements adopted after Three Mile Island and Chernobyl accidents also favoured rapprochement of economic indices for fast and thermal reactors. Some inherent features of fast reactors as well as some additional technical measures introduced into the BN-800 design brought this reactor to a level that meets all up-to-date requirements for new generation of NPPs with improved safety.

The work on the new projects of light water reactors with improved safety (VVER-500 and VPBER-600) has not finished yet. Preliminary estimates show that specific capital investment in these designs are approximately equal to the average value for the South-Urals NPP with three BN-800 units.

4. SCENARIOS OF SEPARATED PLUTONIUM UTILIZATION

4.1. Transformation of Separated Plutonium into Spent Fuel Using Fast Reactors

Utilization of ex-weapons plutonium as a MOX fuel in fast reactors and transformation it into spent fuel also aims at the prevention of its reuse in military purposes. Presumably it is the most effective way for decreasing of ex-weapons plutonium potential hazard. This scenario may be realized in Russian nuclear center PO "Mayak" which includes RT-1 spent fuel reprocessing plant and where the construction of MOX fabrication plant Complex-300 and three units of the BN-800 fast reactor has been planned. The creation of this nuclear center was planned earlier in accordance with the program of nuclear energetic development in Russia.

The RT-1 is operating facility aiming at reprocessing of spent fuel discharged from VVER-440 thermal reactors, BN-350 and BN-600 fast reactors, civilian and military ship reactors, research reactors. The plant capacity is 400 tons/year. Up to now about 3000 tons of spent fuel have been reprocessed in total. The main product of the RT-1 plant is the enriched uranium. The plant can also produce as much as 2.6 tons of civilian plutonium per year as a by-product. The plutonium production during last years dropped to 0.6 tons/year. The total quantity of extracted civilian plutonium stored at the "Mayak" site is about 30 tons. This plutonium is stored reliably in the form of dioxide. Since 1982 there is in operation pilot facility at the plant for high level radioactive waste vitrification.

The construction of the Complex-300 plant to fabricate MOX fuel for fast reactors was not completed and stopped at present. If enough financial subsidy is provided, the Complex-300 plant may be put into operation within ten years, i.e. before the first unit of the BN fast reactor will be constructed. It is supposed that the Complex-300 plant will reach its maximum productivity after three units of the BN-800 are put into operation.

The existing BN-800 reactor design allows annual consumption of weapons-grade plutonium of 1.6 tons. If three or four BN-800 units are put into operation, then both civil and ex-weapons plutonium could be "rendered harmless" by transforming it into spent fuel within first three or four decades of the forthcoming century. It is sufficient for this purpose to put off chemical reprocessing of the BN-800 spent fuel until the time when most amount of already extracted civilian plutonium and released ex-weapons plutonium is transformed into spent fuel.

It is necessary to add that the existing BN-800 design has breeding ratio of about 1. Id est, this reactor transforms plutonium into spent fuel without changing plutonium balance. At present the improved BN-800 reactor designs are being considered in order to increase noticeably plutonium consumption rate. The following steps are under consideration: removal of breeder blankets, increasing of plutonium content in the fresh fuel, using of uranium-free fuel and so on.

4.2. Plutonium Utilization in Operating Reactors

The economic situation existing at present in Russia should be taken into account so the possibility of rapid construction of three or four BN-800 reactor units is doubtful. This factor, together with the successful experience in Western Europe on assimilating MOX fuel in thermal reactors impel to consider another ways of plutonium utilization in particular in presently operating reactors. Presently some studies are in progress aiming at substitution of uranium fuel in the BN-600 fast reactor for the MOX fuel. For the fast reactor this process is quite natural and we will not expect to encounter any principal difficulties. However, the operation life of the reactor is ending and it can consume only 10-12 tons of plutonium.

As to the thermal reactors there are 24 units in Russia under operation. But only some of them can be considered as plutonium users. Because of lack of experience there is no point in considering channel graphite reactor design (RBMK type) for this purpose, although there are 11 RBMK-1000 units in operation in Russia. The nuclear power plants with VVER-440 reactors (6 units) are near the end of their operating life, so they should also be eliminated from the consideration. Therefore, only VVER-1000 reactors and their future improved designs can be considered as the MOX fuel users. Presently in Russia there are 7 such units under operation and only 4 of them at the Bolakovskaya NPP meet up-to-date safety requirements.

European experience shows that in modern PWR the maximum MOX fuel loading is limited to 1/3 of the core. Such restriction results in a relatively low rate of ex-weapons plutonium consumption, that is of about 300 kg/year per 1 GW(e). Transferring this experience to the VVER-1000 reactors one can see that 4 units will consume annually only 20 % more plutonium than one BN-600 reactor. Taking into account the design-life of the considered reactors not more than 25 tons of ex-weapons plutonium can be involved into fuel cycle. This scenario realization is also complicated due to the absence of Russian experience in MOX fuel subassembly fabrication for the VVER type reactors.

4.3. Plutonium Utilization in the VVER-1000 reactors

The third possible scenario assumes plutonium utilization in new constructed VVER-1000 reactors. It was planned earlier to put into operation 7 VVER-1000 units. At present it is difficult to imagine that this plan will be realized in the nearest future. Nevertheless according to estimates 11 VVER-1000 units with MOX fuel inventory (1/3 of the core) will consume annually 3.3 tons of ex-weapons plutonium. So about 30 years will be needed to utilize 100 tons of ex-weapons plutonium.

5. MAINTAINING OF NON-PROLIFERATION REGIME

Measures preventing unauthorized use of plutonium and other fissile materials can be elaborated on the basis of barrier concept making more difficult access to these materials and their more dangerous handling. These barriers can be based on various principles exploiting in particular physics properties of the fissile materials [2].

5.1. Risk of Separated Plutonium Storage

Together with increasing of separated plutonium amounts and broadening of the geography of its transportation the risk that plutonium could fall into unauthorized hands is growing. While plutonium and highly enriched uranium can both be used to make a nuclear explosive device, there is an important difference between them.

Highly enriched uranium can be diluted with natural uranium to make low-enriched uranium which is of no use for explosive device. In order to enrich again diluted uranium requires complex enrichment technology and facility to which most unauthorized people do not have access. In contrast to uranium the plutonium separation from other elements with which it might be mixed requires only chemical processing that can be gained practically using only open literature. Thus, the management of plutonium in any form requires greater security than does the management of uranium.

According to the estimates of specialists from the Institute of Technical Physics virtually any combination of plutonium isotopes and not only the separated plutonium but also the large number of its chemical compounds can be used to make nuclear explosive device [3]. This opinion of Russian scientists share too their American colleagues [4].

Not all combinations of plutonium isotopes, however, are equally convenient for making explosive device or efficient. The most suitable and common isotope is plutonium-239. The presence of higher isotopes as well as plutonium-236 and plutonium-238 complicates creating such a device. This fact can be considered as a barrier to overcome it additional efforts and time are needed.

5.2. Transformation of Plutonium into Spent Fuel

As it mentioned above in existing situation in Russia the transformation of accumulated civilian and released weapons-grade plutonium into spent fuel is one of the effective ways of cutting down the proliferation risk. In order to make this it is necessary first and foremost, to fabricate MOX fuel that is connected with noticeably dilution of plutonium especially in case of weapons-grade plutonium which presents in the most concentrated state. Such dilution can be considered as the first barrier against using plutonium for weaponry. The next barrier is MOX fuel cladding and assembling into subassemblies. Nuclear reactor subassembly is rather bulky and heavy device and it is practically impossible to steal it from the plant and not to be detected. And the irradiation of MOX fuel subassemblies in nuclear reactor is the last step resulting in creation several additional barriers. Among them four basic are worth to be mentioned:

1. The intense radioactivity of the fission products;
2. Chemical compound of plutonium with uranium and fission products;

3. Large amount of plutonium higher isotopes and plutonium-238;
4. Increased difficulties for theft.

The intense radioactivity of spent fuel complicates drastically all stages of this fuel management. However, the proliferation risk posed by the spent fuel grows with time as its radioactivity becomes less intense. So the storage of spent fuel should not be too long.

As to plutonium separation from spent fuel, this task for the potential proliferators might be much more difficult that it may seem at the first sight. All the essential processes are described in the open literature. To separate some amount of plutonium needed for creation one or a few explosive devices potential terrorists could rely on a simple technology and primitive low-cost facilities not paying enough attention to safety and workers' health protection. Such a facility could be built in principle in an ordinary warehouse or garage. All the chemicals involved are widely available since they are used for a variety of other industrial purposes. However, significant engineering skill and experience would be required. The greatest difficulties arise because of strong radioactivity so remotely operated equipment is needed. In addition although the process and technology of reprocessing are unclassified the experience gained in actually operating reprocessing plants is not widely available. So, as it usually occurs with the development of new chemical process, the initial attempts to separate plutonium will encounter unexpected difficulties, to overcome which without substantial testing and practice is not possible. So the time required will extend greatly.

The presence of sufficient amount of higher isotopes and plutonium-238 will complicate the task greatly. However, as it mentioned above these difficulties might be overcome.

As to the last point (4) the following consideration can be taken into account. Spent fuel in reactor subassemblies is stored in reactor cooling ponds at the NPP site. To steal such a subassembly it is necessary to use crane and a truck. Because of strong radioactivity there should be a considerable distance between people and the fuel during the action. It is difficult to imagine that the action would be done without being detected by the personnel and the guard.

5.3. Closed Nuclear Center Concept

The development of closed nuclear centers in Russia allows to solve properly all problems of nuclear energetics, including ecology and economic issues as well as people's health protection and non-proliferation regime reinforcement. Closed nuclear center located far enough from the large towns should include all manufacture plants for the optimal operation of nuclear energetics in concept of closed fuel cycle. As an example of this kind of center the PO "Mayak" site may be considered.

The non-proliferation of fissile materials and especially plutonium is solved in this concept with the most completeness and reliability. First, the area of fissile material transportation is limited by the borders of guarded zone. The transportation of these materials between plants carries out within uninhabited limited area that decreases the risk of accidents and if it happens, excludes poisoning the population. The local transportation of fissile materials within the center under reliable guarding with limited personnel excepts almost at all the risk of theft of fissile materials.

The concept of closed nuclear centers also meets virtually the demand of full exception of diversion of fissile materials for weapons purposes. On the one hand it is obvious that in such a center it is easier to start or renew works on nuclear weapons for all necessary facilities are available. On the other hand the effectiveness of closed nuclear center as barrier preventing the risk of diversion will be defined greatly by the reliability of comprehensive control on all stages of fuel cycle. Such a control can be organized properly if all plutonium activity is concentrated in one or a few nuclear centers. In another situation when nuclear power plants are dispersed all over the country the necessity of fissile material transportation decreases the effectiveness of control and increases the risk of accidents and thefts. At the condition of strict control inside one or a few closed nuclear centers any attempt of diversion of fissile materials for weaponry must be resulted in stopping the International control and so will be inevitably fixed.

6. CONCLUSION

1. Russian concept of plutonium management is based on closed nuclear fuel cycle. No problems exist while plutonium of various isotopic composition including ex-weapons plutonium being utilized in fast power reactors. Presently the BN-800 fast reactor design due to both inherent features and additional technical adjustments meets up-to-date safety requirements for the new generation of NPPs with improved safety. Presently this is the only design in Russia that has passed all required examinations including environmental examination.

2. The short-term plutonium management program in Russia is based upon safe and reliable storage of plutonium before it s utilization in nuclear reactors. Transformation of separated civil and released ex-weapons plutonium into spent fuel by irradiating it in the BN type fast reactors is the basic scenario of plutonium utilization in Russia.

3. Another considered scenarios of plutonium utilization viz. using it in operating reactors or using it in thermal reactors don't provide either its full utilization within visible period of time or encounter great technical and financial difficulties, to overcome them more additional time and money are needed in contrast to the basic scenario realization.

4. Storage of separated civil and released ex-weapons plutonium poses the increasing risk that these materials could fall into unauthorized hands. This risk will be minimized if all plutonium is transformed into spent nuclear fuel. Concept of closed nuclear centers situated in inhabitable regions of Russia and concentrating all plutonium activity is the most reliable barrier preventing any violation of non-proliferation regime and any diversion of fissile material use for unpeaceful purposes.

5. Russia welcomes wide-scale International cooperation aiming at the elaboration of optimal options for plutonium utilization on the basis of solving all related problems: non-proliferation task, environment and people's health protection as well as economic, political and social issues.

REFERENCES

1. Murogov V.M., Kagramanian V.S., Troyanov M.F. et al. (1995) "The management of plutonium in Russia", In Proc. of International Conference on Evaluation of Emerging Nuclear Fuel Cycle Systems" Global 95, p. 946.
2. Murogov V.M., Kagramanian V.S., Troyanov M.F. (1995) "Concept of nuclear fuel cycle resistant to proliferation with the use of plutonium in Russia", Consultancy on development of concept of nuclear fuel cycle resistant to proliferation of nuclear materials, IAEA.
3. Ptitsyna N.N., Chitaikin V.I., Shibarshov L.I. (1995) "Plutonium and its chemical compounds: the problem of nuclear weapon non-proliferation", Proc. of the NATO Advanced Research Workshop on Managing the Plutonium Surplus: Applications and Technical Options.
4. "Management and Disposition of Excess Weapons Plutonium", Committee on Intern. Security and Arms Control, National Academy of Sciences, National academy press, Washington, D.C., 1994.

RECENT NEUTRON PHYSICS INVESTIGATIONS ON THE INCINERATION OF PLUTONIUM AND OTHER TRANSURANIA ELEMENTS

C.H.M. BROEDERS, I. BROEDERS, G. KESSLER and E. KIEFHABER
Forschungszentrum Karlsruhe, Postfach 3640
D76021 Karlsruhe, Germany

ABSTRACT

In early strategic considerations for the production of nuclear energy in fission reactors, plutonium should become the main fissile material in fast breeder reactors (FBR) in order to achieve a high utilization of the uranium resources. The startup of breeder reactors should be realized with available plutonium. An improvement of the startup was considered by enhanced plutonium production rates in high converting light water reactors (HCLWR).

However, political and economical aspects have lead to a significant delay in the market introduction of FBRs. So, instead of enhanced plutonium production in LWRs, the incineration of plutonium of present reactors is now the main objective of strategic studies. Moreover, the considerable amounts of "weapons-grade" (WG) plutonium must be disposed in the most useful way.

The paper will discuss recent FZK neutron physics investigations on plutonium incineration in three types of nuclear systems:

- Pressurized water reactors (PWR),
- liquid metal cooled fast reactors (LMFR) and
- accelerator driven systems (ADS).

All systems need a closed fuel cycle with sufficient capabilities.

The use of mixed uranium and plutonium oxide (MOX) in PWRs is proven technology in several countries. Strategic investigations for plutonium multi-recycling in PWRs will be presented for pools of PWRs with full UOX and full MOX core-loadings. It will be shown, that for a ratio of UOX/MOX≈5/3 after about 80..100 years a near to equilibrium constant plutonium inventory may be reached at a level of about half the value of the case of direct spent fuel disposal. First investigations for the use of WG-plutonium in PWRs also will be discussed.

Plutonium incineration in fast spectrum reactors is studied within the common European CAPRA project. Especially the consequences for the buildup of neptunium and americium will be discussed in some detail.

For the more advanced accelerator driven systems, suitable nuclear data libraries and calculational procedures have been established and validated. First results of studies for plutonium incineration with a fast spectrum ADS will be presented.

1. Introduction

One of the most challenging problems at the end of this century, from the technical, ecological and political point of view, is the back end of the nuclear fuel

E. R. Merz and C. E. Walter (eds.), Advanced Nuclear Systems Consuming Excess Plutonium, 27–42.
© 1997 *Kluwer Academic Publishers.*

cycle. Besides the existing considerable amounts of transurania and fission products from military and commercial applications of nuclear fission, nuclear power plants all over the world are producing nuclear waste, containing large amounts of potential new nuclear fuel. In Germany the Atomic Act prescribed for a long period of time the recycling of spent reactor fuel for the use for further energy production. However, a recent amendment now also allows the direct disposal of spent fuel.

An important objective of a small research and development (R&D) program at the Forschungszentrum Karlsruhe (FZK) consists in maintaining competence and preserving existing capabilities to judge the potential of the various alternative options for the back end of the nuclear fuel cycle. In this contribution recent developments in the neutron physics area will be discussed. These investigations focus on a qualified assessment of the capabilities of a number of different concepts for the incineration of transurania and fission product isotopes in comparison to direct disposal of spent fuel.

In early strategic considerations for the production of nuclear energy in fission reactors, plutonium should become the main fissile material in fast breeder reactors (FBR) in order to achieve a high utilization of the uranium resources. The startup of breeder reactors should be realized with available plutonium. An improvement of the startup was considered by enhanced plutonium production rates in high converting light water reactors (HCLWR).

However, political and economical aspects have lead to a significant delay in the market introduction of FBRs. So, instead of enhanced plutonium production in HCLWRs, the incineration of plutonium is now the main objective of strategic studies. Moreover, the considerable amounts of WG-plutonium must be disposed in the most useful way.

The paper will discuss recent FZK neutron physics investigations on plutonium incineration in three types of nuclear systems:

- Pressurized water reactors (PWR),
- liquid metal cooled fast reactors (LMFR) and
- acellerator driven systems (ADS).

All systems need a closed fuel cycle with sufficient capabilities.

The use of mixed uranium and plutonium oxide (MOX) in PWRs is proven technology in several countries. Strategic investigations for plutonium multi-recycling in PWRs will be presented for pools of PWRs with full UOX and full MOX core-loadings. These investigations are a continuation of earlier activities for tight lattice light water reactors, aiming to preserve uranium fuel resources by enhanced plutonium generation. The applied methods are qualified for fast, epithermal and thermal reactor systems. It will be shown, that for a ratio of UOX/MOX$\approx$5/3 after about 80..100 years a near to equilibrium constant plutonium inventory may be reached. The amount of plutonium at that time is about half the quantity of plutonium to be produced without fuel reprocessing. First investigations for the use of WG-plutonium in PWRs also will be discussed.

The fast reactor investigations are based on the comprehensive activities in the area of fast breeder reactors at the Forschungszentrum Karlsruhe. Within the European CAPRA program the aim is now to incinerate plutonium in future reactors with hard neutron spectra. Especially the consequences for the buildup of neptunium and americium will be discussed in some detail. As an extreme case, the use of uranium-free fuel is considered, plutonium nitride (PuN) being chosen as the most promising candidate of a ceramic compound.

For the more advanced accelerator driven systems, suitable nuclear data libraries and calculational procedures have been been made available and validated, especially by benchmark participation. The application of alternative methods for transport calculations has been studied in some detail. First results of studies for plutonium incineration with a fast spectrum ADS will be presented.

2. Investigations for Pressurized water reactors.

The use of $(UPu)O_2$ mixed oxide (MOX) fuel in PWRs is proven technology in several countries, e.g. France and Germany. Up to about 30% of the fuel assemblies (FA) in existing PWR cores are replaced by MOX instead of UOX assemblies. A main objective during this replacement is to maintain the safety characteristics of the PWRs, especially the coolant density reactivity coefficient (CDRC) and the fuel Doppler coefficient (DC). From earlier studies, e.g. in reference [1], it is well known that high plutonium fractions in the MOX fuel of PWRs may lead to problems with the CDRC. Due to the neutron captures in fissile isotopes, the fraction of non-fissile isotopes in the plutonium increases and its K_∞-value decreases if its irradiation time is increased, e.g. by enlargement of the burnup or by multi-recycling of the spent fuel. As a consequence, the plutonium fissile (and also the total) fraction of MOX fuel with depleted or natural uranium must be increased considerably during plutonium multi-recycling and one must be cautious concerning the CDRC. In a recent study [2, 3] the buildup of transurania in PWRs has been investigated in some detail, both for direct disposal of spent fuel from UOX cores and for plutonium multi-recycling. In the present paper some results for the transurania buildup in UOX cores and in pools of PWRs with a mixing of full UOX and full MOX cores will be discussed. Preliminary results of investigations for the use of WG-plutonium in PWRs also will be presented. The calculations are based on a reactor lattice from benchmark investigations on plutonium multi-recycling, which were performed in collaboration of Forschungszentrum Karlsruhe (FZK) and Electricité de France (EDF) [4, 5]. For the WG-plutonium also a wider, overmoderated, lattice is investigated.

2.1. **Transurania buildup in UOX fuel.** In table 1 a summary is given of the transurania buildup in modern PWRs with UOX fuel in dependence of the U^{235} enrichment for discharge burnups of 33 and 50 gigawattdays per tonne heavy metal (GWD/THM). The weights are normalized to 1 tonne initial heavy metal (TIHM). The burnup calculations are performed on the basis of the realistic power rating of

Material (kg/TIHM)	Burnup							
	33 GWD/THM				50 GWD/THM			
	U^{235} Enrichment				U^{235} Enrichment			
	3.2%	3.5%	4.0%	4.5%	3.2%	3.5%	4.0%	4.5%
Pu	9.680	9.666	9.640	9.612	11.833	11.874	11.946	12.017
Np^{237}	0.4254	0.4306	0.4368	0.4408	0.6453	0.6650	0.6916	0.7119
Am^{241}	0.0370	0.0373	0.0374	0.0371	0.0540	0.0565	0.0604	0.0638
Am^{243}	0.1002	0.0869	0.0694	0.0561	0.3220	0.2914	0.2467	0.2095
Transurania	10.243	10.220	10.184	10.146	12.855	12.886	12.944	13.003

Table 1: Transurania buildup, power rating 164 W/cm

164 W/cm, corresponding to ≈35 W/g. One tonne of spent fuel from modern PWRs contains also about 10..13 kg of transurania isotopes, mainly plutonium. These PWRs have a loading of about 80 tonnes of heavy metal fuel per gigawatt electric (GWe), leading to about 800..1000 kg transurania per GWe per reactor core life or to amounts of around 200 kg discharged plutonium per GWe.a energy production.

U^{235} (%)	Burnup											
	33 GWD/THM						50 GWD/THM					
	Pu isotope (%)					Fiss. (%)	Pu isotope (%)					Fiss. (%)
	238	239	240	241	242		238	239	240	241	242	
3.2	1.5	56.8	22.1	14.0	5.6	70.8	2.9	47.6	24.5	14.8	10.2	62.4
3.5	1.5	58.3	21.3	13.9	5.0	72.2	2.9	48.8	24.0	14.9	9.4	63.7
4.0	1.4	60.6	20.3	13.5	4.2	74.1	2.8	50.8	23.2	15.0	8.2	65.8
4.5	1.3	62.8	19.1	13.1	3.7	75.9	2.7	52.8	22.3	15.0	7.2	67.8

Table 2: Plutonium compositions, power rating 164 W/cm

Table 2 shows the influence of discharge burnup and U^{235} enrichment on the plutonium composition. The fissile fraction decreases significantly if the discharge burnup is increased from 33 to 50 GWD/THM. Higher U^{235} enrichment causes less in situ plutonium burning for the same discharge burnup and leads to higher fissile fractions of the plutonium.

2.2. **Plutonium multi-recycling in PWRs.** The use of fuel assemblies with MOX fuel is proven technology in PWRs, e.g. in France and in Germany. Until now mainly good quality plutonium with high fissile fractions has been utilized for discharge burnups of ≈35 GWD/THM. The potential of plutonium multi-recycling in PWRs has been analyzed in some detail in common benchmark investigations of

FZK and EDF [4, 5]. These investigations were based on simplified lattice calculations. The main result was, that after a small number of recyclings the fissile fraction of the plutonium becomes so unfavourable, that the fissile enrichment of the plutonium has to exceed 6..7% if about 50 GWD/THM discharge burnup should be reached with natural or depleted uranium in the MOX. With such plutonium fractions in the MOX fuel it cannot be guaranteed, that the CDRC remains satisfactory. Starting from this experience, in reference [2] whole core calculations for pools of PWRs with full UOX and full MOX cores have been performed. In figure 1 the principle of this reactor scenario is shown.

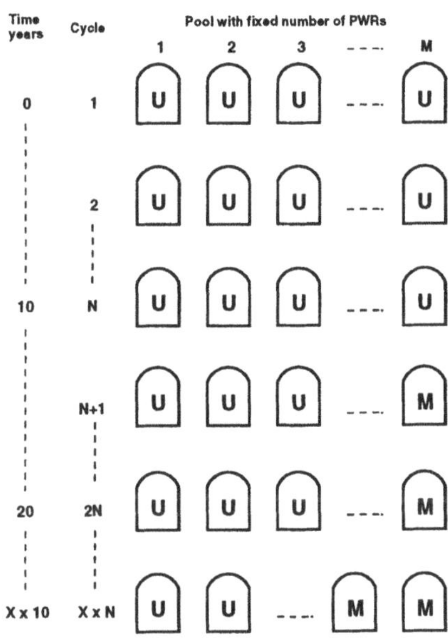

Figure 1: Scenario for plutonium multi-recycling in a pool of PWRs.

To avoid problems with the CDRC, the fissile fraction of the plutonium was limited to ≈6%. In order to get enough excess reactivity at reactor startup to achieve the required discharge burnups, U^{235} enriched uranium is used if necessary.

In table 3 important results are summarized for the near equilibrium situation. For discharge burnups of 33, 40 and 50 GWD/THM the normalized plutonium production in UOX cores and the net plutonium incineration in the MOX cores are given. The production to destruction ratio for plutonium is nearly independent of the discharge burnup: ≈0.6. This means, that the plutonium inventory is nearly constant

Target-burnup (GWD/THM)	Pu-balance (kg/TIHM)		Ratio UOX / MOX	U^{235} enr. (%)
	UOX	MOX		
33	+9.6	-16.2	0.6	0.7..1.0
40	+10.6	-18.5	0.6	2.0..2.5
50	+11.9	-21.0	0.6	3.5..4.0

Table 3: Near equilibrium cycles in pools of PWRs with UOX and MOX fuel.

in a pool of PWRs with a ratio of 3 MOX to 5 UOX cores. In table 3 we also may observe, that the required U^{235} enrichment increases rapidly with the increase of the discharge burnup if the fissile plutonium fraction of the MOX fuel is limited to 6%. In figure 2 the normalized buildup of plutonium for a period of about 100 years is shown. The calculations are based on ex-core times of 7 years (cooling + reprocessing) and 3 years fabrication time. For each discharge burnup a suitable number of reactor burnup cycles during these 10 years ex-core time is applied. We may observe that after 80..100 years the plutonium inventory is nearly constant at a level of about half the value of the case without plutonium multi-recycling.

Pu buildup in PWRs, normalised to installed GWe

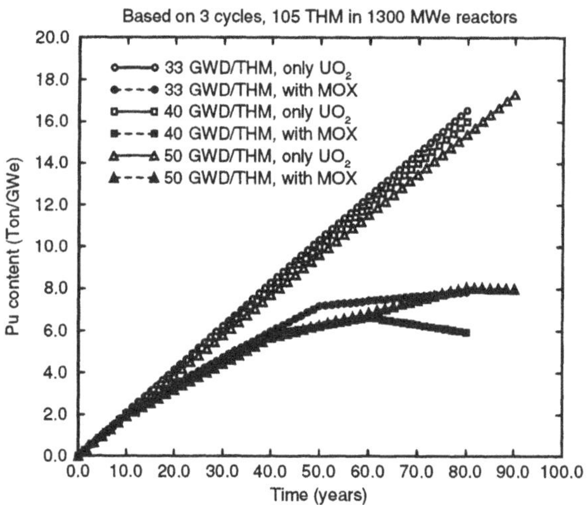

Figure 2: Plutonium buildup in PWRs for burnups of 33, 40 and 50 GWD/THM.

2.3. **Incineration of WG-plutonium in PWRs.** On the basis of full core calculations for equilibrium reactor systems some preliminary studies for the incineration of WG-plutonium in PWRs have been done. The calculations have been performed for a standard and for a wider PWR lattice as proposed in reference [6]. The results are summarized in the tables 4 and 5.

Case	Lattice p/d	Pu-fiss. (%)	Plutonium fiss. fract.		K_{eff}^{EOC}	Burnup (GWD/THM)	Con-vers. Ratio	Plutonium-burnup	
			BOL	EOL				(GWD/THM)	(kg/GWe.a)
Normal	1.3824	4.0	0.94	0.642	1.004	48.6	0.66	13.9	265
Wide	1.5926	3.5	0.94	0.505	0.997	47.8	0.60	18.3	350

Table 4: Burning of weapons-grade plutonium in a 1300 MWe PWR, 510 EFPD, normal and wide lattices.

In table 4 the main core characteristics for a normal and a wider lattice are given. For the WG-plutonium a fraction of 94% of Pu^{239} is assumed. Discharge burnups are slightly below 50 GWD/THM. A wider reactor lattice leads to a decrease of the initial Pu_{fis}-fraction in the MOX from 4.0 to 3.5% and an increase of the Pu incineration rate from 265 to 350 kg/GWe.a. The fissile fraction of the unloaded plutonium is significantly smaller for the wider lattice.

Case	Fraction in weight percent					
	Pu^{238}	Pu^{239}	Pu^{240}	Pu^{241}	Pu^{242}	Pu_{fis}
WG MOX 50/4.0* normal	1.2	46.8	28.3	17.4	6.3	64.2
WG MOX 50/3.5 wide	1.4	33.5	35.7	17.0	12.4	50.5
PWR MOX1⁺ 50/6.0	4.4	42.8	30.2	10.7	11.9	53.5
PWR MOX7 50/6.0‡	5.6	34.6	31.5	9.3	19.0	43.9
UOX 50/4.5	2.7	52.6	22.5	15.0	7.2	67.6
UOX 33/3.2	1.5	56.4	22.3	14.1	5.7	70.5

* xx/yy: xx discharge burnup (GWD/THM), yy initial fissile fraction (%)
⁺ MOXi: PWR plutonium after i recyclings, i=1 from UOX spent fuel
‡ MOX fuel with 6% Pu_{fis} and 3.8% U^{235} enrichment

Table 5: Unloaded plutonium compositions for selected cases.

In table 5 plutonium compositions of unloaded fuel are compared for selected UOX and MOX fuel assemblies.

3. Investigations for fast reactors.

Fast reactors have the well-known advantages of

- low values of $\alpha = \frac{\sigma_c}{\sigma_f}$ for most of the heavy isotopes so that the probability for a neutron absorption reaction of ending up in a fission event rather than in a capture process is considerably increased and, consequently, the production of transurania isotopes and so-called minor actinides (MAs) is significantly reduced as compared to thermal reactors and - due to the low α-values - they offer high values of $\eta = \frac{\nu\sigma_f}{\sigma_a}$ and a resulting high neutron excess which, in former times, was used to breed fresh fissile material and could in a transmutation strategy be favorably used for incinerating MAs and long-lived dangerous fission products and
- low mean cross section values for most of the fission products so that even for fairly high burnups the neutron balance is not too much deteriorated by neutron absorptions in these fission products.

On the other hand, fast reactors suffer from the disadvantages of

- generally rather small average cross sections requiring high concentrations of fissile material and a rather high neutron flux intensity and
- their fairly large fraction of high energy neutrons ($E \geq 0.1MeV$) which are inducing pronounced radiation damage in structural materials.

These pecularities of fast reactors with "hard" neutron spectra are responsible for the limitation of the residence time of fuel assemblies which can withstand damage rates of about 100-200 displacements per atom (DPAs). The corresponding neutron fluences of several $10^{23} n/cm^2$ are usually not sufficient for burning almost completely the initial fissile content of the fuel. Furthermore, due to the conversion of fertile to fissile material, the burnt fuel still contains a large fraction of fissile (and fertile) isotopes so that reprocessing is practically mandatory for fuel irradiated in fast reactors and using of an inert matrix for incinerating MAs can hardly be envisaged in a representative LMFR neutron flux due to the remaining high radiotoxicity of the irradiated material which may be a severe obstacle for its direct disposal (considered to be one of the main advantages of using an inert matrix material).

The incineration capabilities of fast reactors with conventional mixed oxide (MOX) fuel is essentially limited by the solubility of the MOX fuel which becomes much worse when the Pu-content is increased above about 45% (as is well known, there exists a rather strong correlation between the fissile enrichment and the maximum incineration rate which can be achieved). Accepting the above enrichment constraint, an incineration rate of roughly 70 kg(Pu)/TWe.h$\approx$600 kg(Pu)/GWe.a was obtained for a CAPRA-type reactor with MOX fuel. Of course such a burner reactor has no longer any fertile blankets as previously encountered in fast breeder reactor designs. Moreover, so-called diluents were loaded in the core to increase the neutron leakage so that a large fast reactor with a power of 1500 MWe could be loaded with fuel having a significantly larger Pu-content than earlier breeder-related designs such as SUPERPHENIX or EFR. Figure 3 shows the approach to achieve a core with high plutonium content within the fuel. A three-level dilution strategy is applied: small diameter fuel pins with hollow fuel pellets, heterogeneous fuel assemblies containing about 30% "fuel-free" pins and a heterogeneous core layout with a number of "fuel-free" sub-assemblies.

In addition to burning plutonium, special investigations were devoted to the transmutation of MAs. The addition of MAs (especially of Np^{237}) causes a moderate reduction of the burnup reactivity swing but causes a pronounced reduction of the Doppler effect due to the large absorption cross sections of these MAs in that part of the resonance region which mainly contributes to the Doppler effect. This drawback of MA addition can be mitigated or compensated in such a way that the ratio of coolant void effect to Doppler effect is similar to that of a burner reactor without MA addition by replacing diluent or inert material in diluent subassemblies (S/As) or in so-called "empty" pins of fuel S/As by a suitable moderator material like BeO or $B_4^{11}C$ (boron carbide with a practically vanishing content of B^{10}). This enables a fast reactor to incinerate at least its self-generated MAs with a non-negligible capacity to accept an additional amount of MAs e.g. originating from LWRs. Table 6 shows possible combinations of NpO_2-contents in the fuel, of fuel enrichments in the inner and outer core zones and of B_4^{11}-contents in fuel and diluent S/As with nearly constant ratios of the reactivity coefficients $\Delta\rho_{Void}$ and $\Delta\rho_{Doppler}$.

NpO_2 Content in Fuel (%)	Core Enrichments CI (%)	CE (%)	$B_4^{11}C$ Content in Fuel S/As (%)	Diluent S/As (%)	$\Delta\rho_{Void}$ (pcm)	$\Delta\rho_{Doppler}$ (pcm)	$\Delta\rho_{Void}/\Delta\rho_{Doppler}$
0.	40.38	42.06	0.	0.	1558.1	590.8	2.6373
1.	40.60	42.28	2.69	0.	1597.4	605.7	2.6373
3.	41.05	42.73	9.06	1.	1674.5	634.8	2.6378
4.	41.16	42.84	9.06	16.	1651.7	626.3	2.6372
8.	42.71	44.39	9.06	49.	1527.2	579.2	2.6367

Table 6: CAPRA cores 04/94 with NpO_2- and $B_4^{11}C$-addition.

An inevitable disadvantage of the addition of Np^{237} to the MOX fuel is the considerable amount of Pu^{238} produced during irradiation. Depending on the quality of the plutonium of the fresh fuel, the fraction of Np^{237} in the fuel has probably to be kept in the range of 5..10% if the percentage of Pu^{238} in the unloaded fuel should not exceed about 10% of the total plutonium, an amount which might require extra precautions in reprocessing this Pu^{238}-loaded fuel.

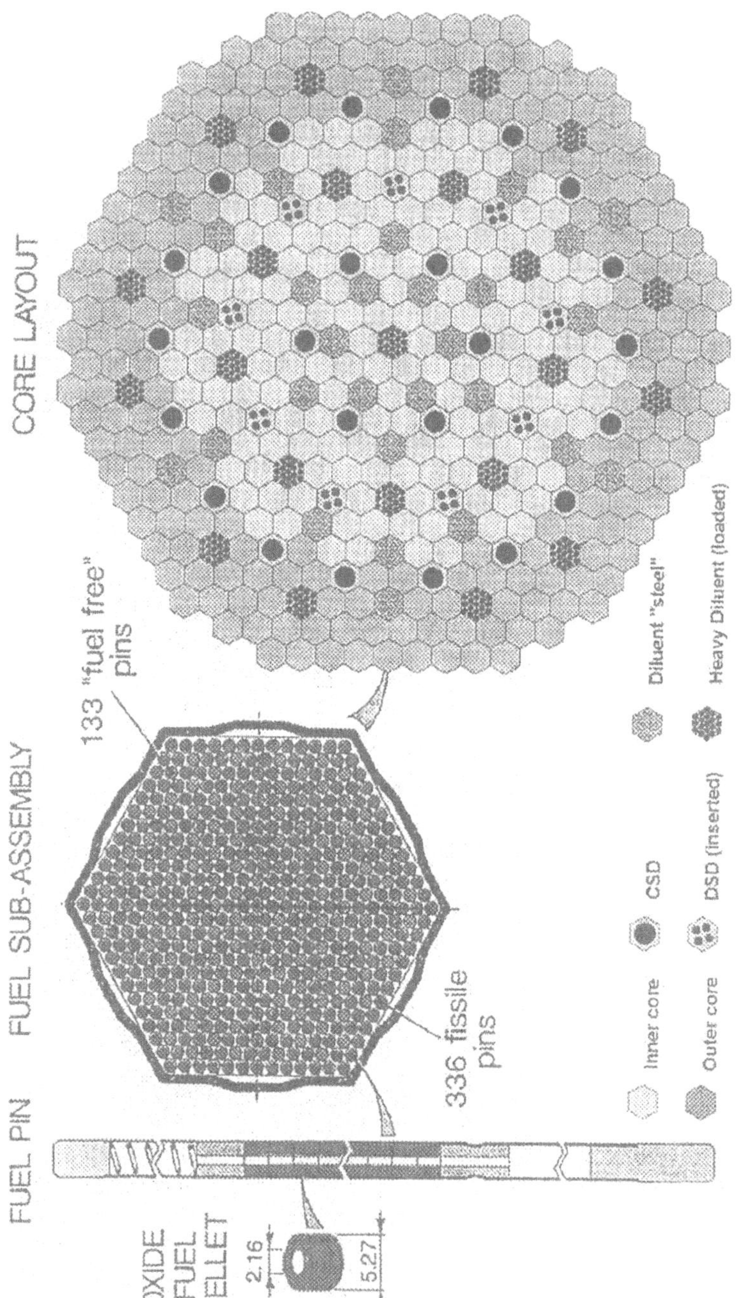

Figure 3: Characteristics of the reference CAPRA 4/94 core.

Having in mind the complications and economic aspects of fabricating and handling pins and S/As loaded with pellets containing significant amounts of americium it seemed appropriate to concentrate such highly radioactive MAs in special S/As, placed in reflector positions of the core. Obviously this leads to only a small degradation of core performance parameters (e.g. safety related reactivity coefficients corresponding to Doppler- and sodium-void-effect) by the addition of MAs (in the reflector) and an operational behaviour of the MA-loaded reflector S/As which is similar to that of conventional radial breeders S/As. By moderating or softening the neutron energy spectrum in these peripheral S/As, the incineration rate might be increased remarkably; nevertheless the irradiated and unloaded fuel still contains considerable amounts of MAs, so that its handling becomes more difficult if reprocessing is regarded as a feasible option. If direct disposal of that spent fuel is envisaged, it might be questionable whether the transmutation effort leads to such a significant reduction of the long-term radiotoxicity and/or the waste volume or the needed repository volume that a long-lasting irradiation period and the associated handling risks are justified.

As a particular design of an extraordinary burner reactor a so-called U-free core was studied. In order to comply with solubility requirements, nitride fuel was chosen and the necessary Doppler feedback was delivered by using a fairly "dirty" Pu isotopic composition with high fractions of the even isotopes Pu^{240} and Pu^{242}. This version of a burner could be considered as adequate to deal with multi-recycled LWR-Pu and, therefore, possibly as representative for the later or last phase of nuclear energy production. It would allow to come close to the theoretical limit of about 120 kg(Pu)/TWe.h≈1000 kg(Pu)/GWe.a.

4. Investigations for accelator driven Systems.

ADS investigations started at FZK about 5 years ago with the installation of parts of the HERMES[7] code version of KFA Jülich. The module HETC for the calculation of intermediate energy processes and the Monte Carlo module MORSE for transport calculations below 20 MeV are operational both on IBM-mainframe computer and on UNIX-workstations. Moreover, interfaces were established between HETC and both the Monte Carlo program MCNP[9] and the discrete ordinates transport code TWODANT[8]. For burnup investigations the program PROSDOR[10] was developed. The latest workstation version consists of a semi-automatic sequence of the codes HETC, MCNP-4 and the depletion code KORIGEN[11]. KORIGEN was developed at FZK from the original ORIGEN code from Oak-Ridge. Some special provisions for the PROSDOR application were made. Recently for the burnup calculations a sequence of HETC, MCNP/TWODANT and the burnup code KARBUS[1] has been established. KARBUS was developed at FZK for more general fission reactor burnup calculations. Its depletion part is also based on ORIGEN formalisms. Much effort has been spent for best estimate determination of one-group cross sections.

4.1. **Validation work.** The first activity after installation of part of the HERMES package was to gain experience with the application of the code. This could be realized by a successful participation to the NEA/NSC international code comparison benchmark for intermediate energy nuclear data for thick targets [12]. Generally the FZK results agree satisfactorily with the other solutions. Further extensive comparisons were performed between the Monte Carlo code MCNP and the discrete ordinates code TWODANT with the objective to replace the time-consuming burnup sequence HETC-MCNP-KORIGEN by the faster sequence HETC-TWODANT- KARBUS [13]. These comparisons are based on non-multiplicative targets and slightly subcritical systems with thorium fuel. As an example, in table 7 one-group data for U^{233} from MCNP and TWODANT calculations for a Th-U^{233} ADS are compared. Except for the threshold cross sections (n,2n) and inelastic scattering, the differences between the calculational methods are small. These threshold processes must be analyzed carefully because of their sensitivity to spectral changes in the results of the calculations. Especially the (n,2n) cross sections may influence significantly the buildup of actinide isotopes, resulting from such processes.

4.2. **Applications.** First results have been published by Segev et. al. [14] for different applications, mainly transmutation of long-lived MAs and fission products. Recently, investigations on the incineration of LWR-plutonium have been performed [13]. As a first application the program sequence HETC-TWODANT-KARBUS has been used to investigate the capabilities of Pu - incineration in an accelerator driven subcritical core of 3000 MWth power with $Th^{232} - Pu$ - fuel and sodium coolant. The radius of the core was 110 cm, its height 220 cm, the primary energy of the protons was 1.6 GeV. Burnup calculations have been carried out for $3 \cdot 365$ days. Within each period of 365 days full power operation has been assumed for $3 \cdot 110$ days and zero power for 35 days. The depletion and buildup of isotopes during $3 \cdot 365$ days is shown in figure 4 for U^{233}, Pu^{239} and Pa^{233}. The dotted and dot-dashed lines show the results, when spectra of MCNP calculations are used instead of TWODANT spectra. The total amount of Pu incinerated within $3 \cdot 365$ days is found to be 1940 kg corresponding to 600 kg/GWe.a (efficiency assumed: 40%). During the same time 1470 kg of U^{233} are build corresponding to 450 kg/GWe.a. Th^{232} is reduced from 19100 kg in the fresh core to 16200 kg at the end of the burnup time discussed. We may observe that the incineration of the plutonium is counterbalanced by the buildup of U^{233}. To judge such a system it will be necessary to handle this U^{233} in a proper way. The K_{eff} of this system varies from $\approx$0.98 to $\approx$0.96. The corresponding proton currents for achieving 3000 MWth are in the range of 18 to 40 mA. Total voiding of the ADS leads to small negative ΔK_{eff} values. The hot to cold reactivity change is less than +1% in K_{eff}, so the ADS remains safely subcritical during shutdown. Figure 5 shows the neutron flux density spectrum in the fresh core and after irradiation. The influence of the plutonium resonances at 0.3 eV in Pu^{239} and 1 eV in Pu^{240} may clearly be recognized, being more pronounced for the fresh core.

	TWODANT 69 groups		MCNP-4A	MCNP-3A	$\frac{TW-MC}{MC}\%$[1]
	k_{eff}	inhomogen. source	k_{eff}	inhomogen. source	inhomogen. source
σ_{tot} $[\frac{1}{cm}]$	9.5640	9.5738	10.0466 (.0016)	10.0793 (.0019)	-5.0
σ_{el}	6.5330	6.5403	7.0542 (.0017)	7.0528 (.0019)	-7.3
$\sigma_{(n,2n)}$	1.5172E-3	2.1227E-3	7.9018E-4 (.0297)	1.6015E-3 (.0421)	32.5
σ_{fiss}	2.2597	2.2622	2.4602 (.0017)	2.4631 (.0208)	-8.2
σ_{inel}	0.5852 [2]	0.5841 [2]	0.3176 (.0017) [3]	0.3143 (.0461) [3]	85.9
$\sigma_{(n,\gamma)}$	0.1847	0.1850	0.2137 (.0021)	0.2141 (.0208)	-13.6
ν $[\frac{neutrons}{fission}]$	2.5393	2.5406	2.5240 (.0014)	2.5266 (.0208)	-0.56
k_{eff}	0.9035	0.9142 [4]	0.9339 (.0016)	0.9510 [4] (.0204)	

[1] TW=TWODANT MC=MCNP

[2] inelastic from discrete levels and from energy continuum

[3] only continuum inelastic

[4] determination of k_{eff} - values for calculations with external neutron source with formula from reference [9]:

$$k_{eff} = \frac{M - R^{(0)}(\sigma_{n,xn}) - 1}{M - \frac{1}{\nu}}$$

with $M = R(\nu\sigma_{fiss}) - R(\sigma_{fiss}) + R(\sigma_{n,xn}) + 1$

where R is the corresponding reaction rate.

$R^{(0)}$ means reaction rate from criticality calculation

Table 7: U-Th ADS, comparison of one-group data of U^{233}

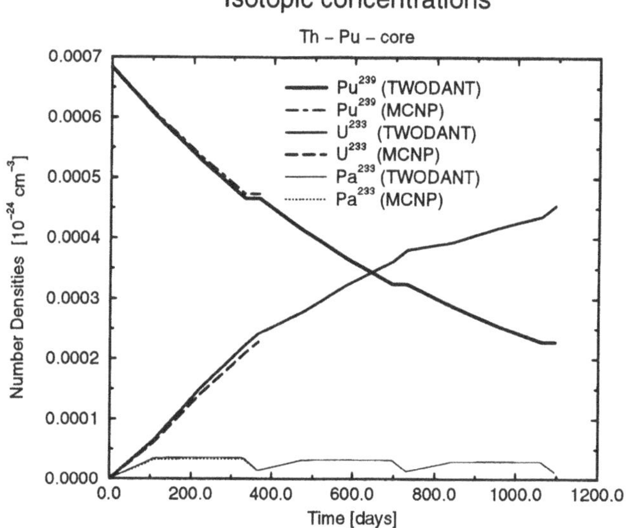

Figure 4: Th-Pu ADS, time dependence of isotopic composition.

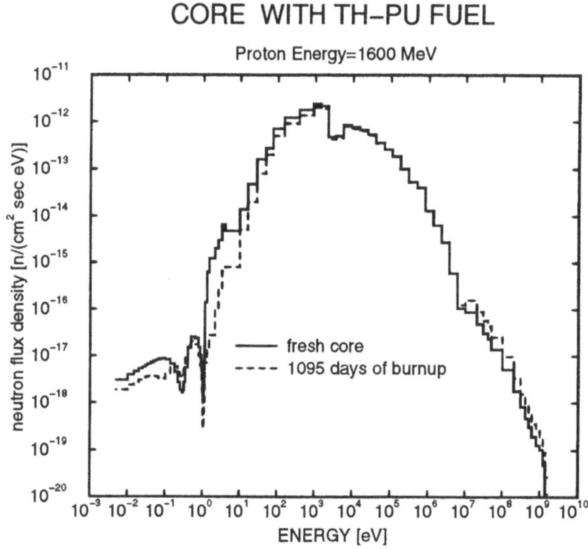

Figure 5: Neutron flux density spectrum for a sodium cooled Th^{232} - Pu core for different burnup states (normalized to $\sum_{g=1}^{G} \Phi_g = 1$, Φ_g = volume integrated flux in energy group g)

5. Summary.

At the Forschungszentrum Karlsruhe (FZK) research and development investigations on the incineration of plutonium and other transurania elements are performed for three types of nuclear systems: pressurized water reactors (PWR), liquid metal fast reactors (LMFR) and accelerator driven systems (ADS). Strategic studies for plutonium multi-recycling in PWRs with full core models indicate that a stabilization of the total plutonium inventory in pools of PWRs with full UOX and full MOX cores may be obtained after 80..100 years at a level of half the inventory at that time when following a reactor strategy without plutonium recycling. The incineration rates for plutonium are about 300..500 kg(Pu)/GWE.a in the MOX cores of the PWRs in these pools. weapons-grade plutonium may be incinerated with similar rates. Widening of the reactor lattice, leading to overmoderated neutron spectra, may improve the incineration rates in PWRs significantly, producing unloaded plutonium with a considarable smaller fissile fraction.

The LMFR investigations are imbedded in the European CAPRA project. The main objective is to optimize a fast reactor to burn plutonium. For this purpose a high plutonium content in the fuel is required. At present the plutonium content is limited to 45% to remain compatible with the existing manufacturing processes. These high plutonium contents require high dilutions of the fuel in the core. These are obtained by a three-level dilution strategy: small diameter fuel pins with hollow fuel pellets, heterogeneous fuel assemblies containig about 30% "fuel-free" pins and a heterogeneous core layout with a number of "fuel-free" sub-assemblies. A main problem for the core-layout is the high burnup reactivity, compared to existing LMFRs. The FZK neutron physics investigations within the CAPRA project especially focus on the incineration of the minor actinides neptunium and americium. Neptunium may be incinerated in LMFRs by admixing it with the MOX fuel. The deterioration of the important safety parameters void- and Doppler-reactivity may be counterbalanced by the insertion of moderator materials, e.g. $B_4^{11}C$, in special sub-assemblies of the core. For americium the use of special sub-assemblies located at reflector positions is proposed. These sub-assemblies then may stay over a longer period (≈10 years) in the core until the maximum allowed material damage is reached. The incineration rates in LMFRs strongly depend on the core-loading (mixing of plutonium, neptunium and americium). The following rates may be achieved: plutonium; 600 kg/GWe.a, Np^{237}; 140 kg/GWe.a and Am^{241}; 80 kg/GWe.a.

The investigations for accelerator driven systems (ADS) started at FZK about 5 years ago. In a first stage adequate calculational procedures have been established and validated. For the calculation of the intermediate energy processes above ≈20 MeV the HETC module of the HERMES code version of KFA Jülich has been implemented. At lower energies coupled transport calculations are possible with the Monte Carlo code MCNP and with the deterministic code TWODANT. These calculational procedures could be qualified by the participation to an international benchmark investigation. Especially the use of faster deterministic solutions for the calculation of the neutron transport in these systems, compared to time-consuming Monte Carlo procedures, has been studied in more detail. Satisfactory agreement could be observed for non-multiplicative targets and for slightly sub-critical systems with a hard neutron spectrum and thorium fuel. As a first application the new procedures have been used to investigate the capabilities of Pu - incineration in an accelerator driven subcritical core of 3000 MWth power with $Th^{232} - Pu$ - fuel and sodium coolant. The total amount of Pu incinerated within $3 \cdot 365$ days is found to be 1940 kg corresponding to 600 kg/GWe.a (efficiency assumed: 40%). During the same time 1470 kg of U^{233} are build corresponding to 450 kg/GWe.a. Th^{232} is reduced from 19100 kg in the fresh core to 16200 kg at the end of the burnup time discussed. We may observe that the incineration of the plutonium is counterbalanced by the buildup of U^{233}. To judge the potential of such a system it will be necessary to deal with this U^{233} in a proper way. Moreover, the consequences of handling other irradiation products like U^{232}, Pa^{231}, Pa^{233} etc. must be investigated in more detail.

References

1. C.H.M. Broeders, KfK 5072 (1992)
2. C.H.M. Broeders, FZKA 5784 (1996)
3. C.H.M. Broeders, ICENES96, Obninsk, Russia (1996)
4. J. Vergnes, C. Broeders, L. Payen, Internal EDF Note (1995)
5. C. Broeders, H. Küsters, L. Payen, J. Vergnes, Draft Proposal NEA/NSC Benchmark (1995)
6. L. Payen, J. Vergnes, C. Broeders, ENS TOPSEAL Conference Stockholm (1996)
7. P. Cloth, D. Filges et. al., Jül-2203, (1988)
8. R.E. Alcouffe, R.S. Baker et. al., LA-12969-M, (1995)
9. Judith F. Briesemeister, Editor, LA-7396-M, Rev.2 Manual, (1986)
10. M. Segev, KfK 5328 (1994)
11. U. Fischer, H.W. Wiese, KfK 3014 (1983)
12. D. Filges, P. Nagel, R.D. Neef, NSC/DOC(95) 2 (1995)
13. I. Broeders, C.H.M. Broeders, ICENES96, Obninsk, Russia (1996)
14. M. Segev, H. Küsters, S. Pelloni, NSE 122, 105-120 (1996)

THE BENEFITS OF AN ADVANCED FAST REACTOR FUEL CYCLE FOR PLUTONIUM MANAGEMENT

W.H. HANNUM, H.F. MCFARLANE, D.C. WADE AND R.N. HILL
Argonne National Laboratory
9700 S. Cass Avenue
Argonne, Illinois 60439 U.S.

Abstract

The United State has no program to investigate advanced nuclear fuel cycles for the large-scale consumption of plutonium from military and civilian sources. The official U.S. position has been to focus on means to bury spent nuclear fuel from civilian reactors and to achieve the spent fuel standard for excess separated plutonium, which is considered by policy makers to be an urgent international priority. Recently, the National Research Council published a long-awaited report on its study of potential separation and transmutation technologies (STATS), which concluded that in the nuclear energy phase-out scenario that they evaluated, transmutation of plutonium and long-lived radioisotopes would not be worth the cost. However, at the American Nuclear Society Annual Meeting in June, 1996, the STATS panelists endorsed further study of partitioning to achieve superior waste forms for burial, and suggested that any further consideration of transmutation should be in the context of energy production, not of waste management.

The U.S. Department of Energy (DOE) has an active program for the short-term disposition of excess fissile material and a "focus area" for safe, secure stabilization, storage and disposition of plutonium, but has no current programs for fast reactor development.

Nevertheless, sufficient data exist to identify the potential advantages of an advanced fast reactor metallic fuel cycle for the long-term management of plutonium. Some of the key advantages are:

1. Tens of tonnes of plutonium could be quickly secured in a single reactor system.
2. Use of a metal alloy fuel would allow economic fuel recycling at any scale to match the energy production requirements.
3. All actinides would remain in the fuel cycle, out of the waste stream.
4. Throughout the fuel cycle, the plutonium would remain in a highly radioactive environment equivalent to the spent fuel standard.
5. The net rate of plutonium consumption could be controlled to meet future energy requirements.

E. R. Merz and C. E. Walter (eds.), Advanced Nuclear Systems Consuming Excess Plutonium, 43–63.
© 1997 *Kluwer Academic Publishers.*

6. Because all actinides fission in the fast spectrum, the more radiotoxic transuranic isotopes would not build up as they do in a thermal spectrum.
7. Specific fission products would be partitioned into the waste forms in which they would be most stable for disposal.

Introduction and Background

It is now widely accepted that rigorous active management of plutonium is a matter of highest priority [(Ref.1)]. Stores of separated plutonium require severe physical protection and accountancy to protect against diversion, and to provide timely warning should there be any attempted diversion. Plutonium which is intimately intermixed with highly radioactive materials ("self-protecting") requires a lesser but still a significant degree of physical protection and accountancy, in that a variety of technologies are available and known by which plutonium can be chemically separated.

Internationally agreed priorities are to assure physical protection and accountancy for separated plutonium on an urgent basis, and to provide a web of protection to detect and deter any attempt at an unauthorized program for the chemical separation of plutonium which is now in a self-protecting condition. These measures are generally considered to be capable of providing protection against diversion for the next few years or perhaps decades. Ultimately, a more permanent means of dealing with world accumulations of plutonium must be developed and deployed [(Ref. 2)].

Given the difficulty in defining an acceptable means of storing plutonium for the millennia necessary for radioactive decay, destruction by fissioning must be considered as an environmentally attractive means for dealing with plutonium. The energy release generated by the fission process creates the prospect for this to be not only environmentally attractive but economically sound. It is this possibility that this paper addresses.

No technology, regardless of its benefits, is without its risks and liabilities, and we must address risks openly and frankly; not just the benefits of various strategies and technologies. For this, we would like to begin by describing the magnitude of the problem as we understand it, and the principal risks and concerns.

Current world holdings of plutonium amount to over 1000 tonnes and are growing at 60 to 80 tonnes per year [(Ref. 3)]. According to some estimates, several hundred tonnes exist as separated plutonium. We will take it as given that plutonium can be stored and used in peaceful applications without endangering either humans or the environment. The precautions are demanding, but are well within the capabilities of existing technologies. It is outside the scope of this paper to discuss the state of technology for environmentally safe and responsible geological disposal of plutonium. The risks under discussion here therefore are those associated with misuse of the material.

The opinions expressed in this paper are the technical opinions of the authors, and do not represent an established position of Argonne National Laboratory, the U. S. Department of Energy, or of the U. S. Government.

Two misuse scenarios can be postulated: dispersal of plutonium by chemical or other means, and use as a nuclear explosive. For the first, even reasonably small quantities are of concern, and purity is not an issue; the dirtier the better. The direct physical consequences of such an event would be modest [Ref. 4], but could result in massive property and psychological damage Given the world inventories of radiological wastes and spent fuel, this is a concern which must be addressed regardless of the process that is ultimately adopted for plutonium disposition or consumption.

The second scenario is more complex, and is the subject of much mis-information. Plutonium can only be used to make a nuclear explosive with an implosion device [Ref. 5]. We consider the technology for implosion type weapons to be beyond the capability of terrorist and sub-national groups. If there were to be a terrorist-developed nuclear explosive device, it would almost certainly be based on highly enriched uranium, not plutonium. Diversion by a sub-national group is of concern only because of the potential for it being used in a dispersion type device.

For a technologically advanced country, it is evident that development or expansion of a nuclear weapons capability can be greatly facilitated by improved access to plutonium. For such countries, there are easier ways to accumulate significant stores of weapons-usable plutonium than to separate it from spent commercial reactor fuel; recent incidents suggest that the route of separation is subject to detection in time to permit diplomatic intervention. Diversion of significant stocks of separated plutonium is most likely to be detected at the source, but until complete and secure protection is provided for all separated plutonium, diversion remains as a plausible route for obtaining or augmenting national stocks of separated plutonium.

Based on this perspective, there are two overriding conditions to be addressed when discussing final and permanent disposal of plutonium: minimizing the exposure of separated plutonium to diversion, and assuring a degree of active management and accountancy for large stocks of non-separated plutonium.

The most effective way to minimize the exposure of separated plutonium to division is to minimize the world inventory of separated plutonium [Ref. 6]. Ideally, plutonium would not be separated until there is an immediate use for it. The only effective long term way to assure active management of non-separated plutonium is to consume the plutonium by way of a fission process; there is then no non-separated plutonium to manage and account for.

External Factors

There are clearly external factors which will ultimately influence choices, and which are in fact likely to be decisive. Dominant among these are the need for energy and the ability of nuclear energy to compete economically with alternatives [Ref. 7]. This in turn will depend at least in part on the availability of nuclear and other energy supply resources. Political prejudices with regard to nuclear power may also be a significant factor in some situations. These considerations are outside the scope of this paper.

Schedule Realities

There is no existing industrial infrastructure available to consume or otherwise permanently dispose of the world's inventory of plutonium. On an energy basis, fissioning of one tonne of any material yields enough energy to produce approximately one GW-year of electricity, so the several hundred tonnes of separated plutonium will involve a significant industrial commitment. While various wartime emergency programs of this general scale have been completed in a few years, it is generally accepted that the development of an appropriate technology, followed by the preparation of the required industrial infrastructure will require several decades.

The U.S. Policy Position

The U.S. policy is addressed at reaching a stable plateau of world inventories of separated plutonium as quickly as possible. To this end, the U.S. has active programs underway to assist in assuring adequate physical protection for existing stocks of separated plutonium, and is actively working to discourage separation (reprocessing) of spent nuclear fuel where there is no established end-use. This priority is based on the recommendations of an august panel of experts [Ref. 2], with detailed programs outlined in a published Draft Environmental Impact Statement [Ref. 8].

The Physics of Plutonium Consumption

The basic physics of plutonium consumption is well understood by specialists, but is frequently not stated clearly when discussing overall strategic alternatives. The basic principles which we consider relevant and significant for this discussion are the following:

- The only way to destroy plutonium is to cause it to fission; all other disposition options amount to storage;
- Transmutation of Pu-239 to a mix of higher plutonium isotopes greatly diminishes attractiveness of the material for weapons use, but does not totally eliminate the potential for misuse[Ref. 9]. Nuclear explosives involve fission by fast neutrons, and all plutonium isotopes are fissionable by fast neutrons. However, material containing mixed isotopes of plutonium are far less attractive because of handling complications associated with the radioactivity of the higher isotopes, and because of their heat generation[Ref. 10]. It may become possible in the future, with advanced techniques, to do an effective isotopic separation, in which case material containing mixed isotopes of plutonium would be, from a safeguards point of view, no better than other means of providing for "self-protection";
- No fission process is complete; there will always be some neutron capture and transmutation to higher mass transuranic elements. As with higher plutonium isotopes,

these higher transuranics are fissionable with fast neutrons. In general, the heat and radiation from these materials provides a high degree of self-protection;

- The fission process in a fast reactor provides sufficient excess neutrons that, with multiple recycle, total consumption of all isotopes and any higher actinides produced can readily be accomplished. Some concepts for plutonium consumption in a thermal spectrum rely on a feed stream of additional thermally fissionable material (i.e. U-235, Pu-239 or Pu-241). Providing excess neutron to subcritical systems, such as spallation sources from impacting an accelerated particle beam onto a heavy material target, is also feasible. The physics of each of these processes is well known and well demonstrated;

- To achieve total destruction, it is not possible to rely on the physical stability of the material being destroyed. In practice, this means the plutonium must either be reformulated (recycled) or be in a form where structural integrity is not a requirement; e.g. a fluid or slurry;

- Any system which consumes plutonium based on neutrons generated by the plutonium fission itself can only asymptotically achieve total consumption;

- In a reactor situation (i.e., no external source of neutrons), any system which burns plutonium without producing new fissile material will require a wide band of reactivity control.

How the Fast Reactor Approach Addresses the Fundamental Disposition Requirements

The goal of elimination of plutonium implies that essentially all transuranic materials be fissioned - none left as waste for disposal. Total consumption can in practice be accomplished only in a fast neutron spectrum, or with an external source of neutrons.

The fission of an actinide atom (i.e., atomic number greater than or equal to 89) gives rise to two fission-product atoms, the release of about 200 MeV of energy (approximately 1 MW-day of heat is released per gram fissioned), and the release of two or three neutrons (average about 2.5). To consume the plutonium without an external supply of neutrons, each transuranic atom fissioned uses one neutron to sustain the chain reaction into the next fission generation. In addition, there is unavoidable neutron capture that causes parasitic transmutation of Pu-239 and higher-mass transuranic isotopes, which build to steady state during long exposure to the neutron flux. Illustrative steady-state transuranic inventories which would result from very long exposure to fast and thermal neutron flux are shown in Table 1. In practical designs, still further neutron losses occur from capture in structural materials and leakage from the system. The greater the burnup per cycle, the more fission products there are which compete for excess neutrons.

Table 1. Equilibrium Distribution of Transuranic Isotopic Masses for Thermal and Fast Neutron Spectra

Isotope	Thermal Neutron Spectrum	Fast Neutron Spectrum
Np-237	5.51	0.75
Pu-238	4.17	0.89
Pu-239	23.03	66.75
Pu-240	10.49	24.48
Pu-241	9.48	2.98
Pu-242	3.89	1.86
Am-241	0.54	0.97
Am-242m	0.02	0.07
Am-243	8.11	0.44
Cm-242	0.18	0.40
Cm-243	0.02	0.03
Cm-244	17.85	0.28
Cm-245	1.27	0.07
Cm-246	11.71	0.03
Cm-247	0.75	2.E-3
Cm-248	2.77	6.E-4
Bk-249	0.05	1.E-5
Cf-249	0.03	4.E-5
Cf-250	0.03	7.E-6
Cf-251	0.02	9.E-7
Cf-252	0.08	4.E-8

Note: All values are atom % of transuranic inventory built up as a result of extended exposure to a neutron flux. (Calculated as the steady-state solution of the depletion-chain equations–independent of criticality considerations.)

The number of neutrons per fission lost to parasitic capture in the transuranics under steady-state conditions can be determined from their capture and fission probabilities; typically, with no fission products present, this is about 0.25 in a fast neutron spectrum and 1.25 in a thermal spectrum. As shown in Fig. 1, for transuranic isotopes the probability of fission relative to parasitic capture is much lower for the thermal neutron spectrum characteristic of a light-water-cooled reactor (LWR) than for the fast neutron spectrum characteristic of a liquid metal cooled reactor (LMR). Indeed, the even-mass-number transuranic isotopes generally do not fission in a thermal neutron spectrum. Thus, in the fast-neutron system a *minimum* release of 1.25 neutrons per fission is required to sustain the steady state, while in a thermal-neutron system at least 2.25 are required. Each actinide fission releases only about 2.5 neutrons, so a modest margin exists for a thermal reactor prior to building of a large inventory of fission products, while in a fast reactor there are sufficient neutrons to generate excess plutonium to accommodate vary high burn up, or to give the designers considerable flexibility.

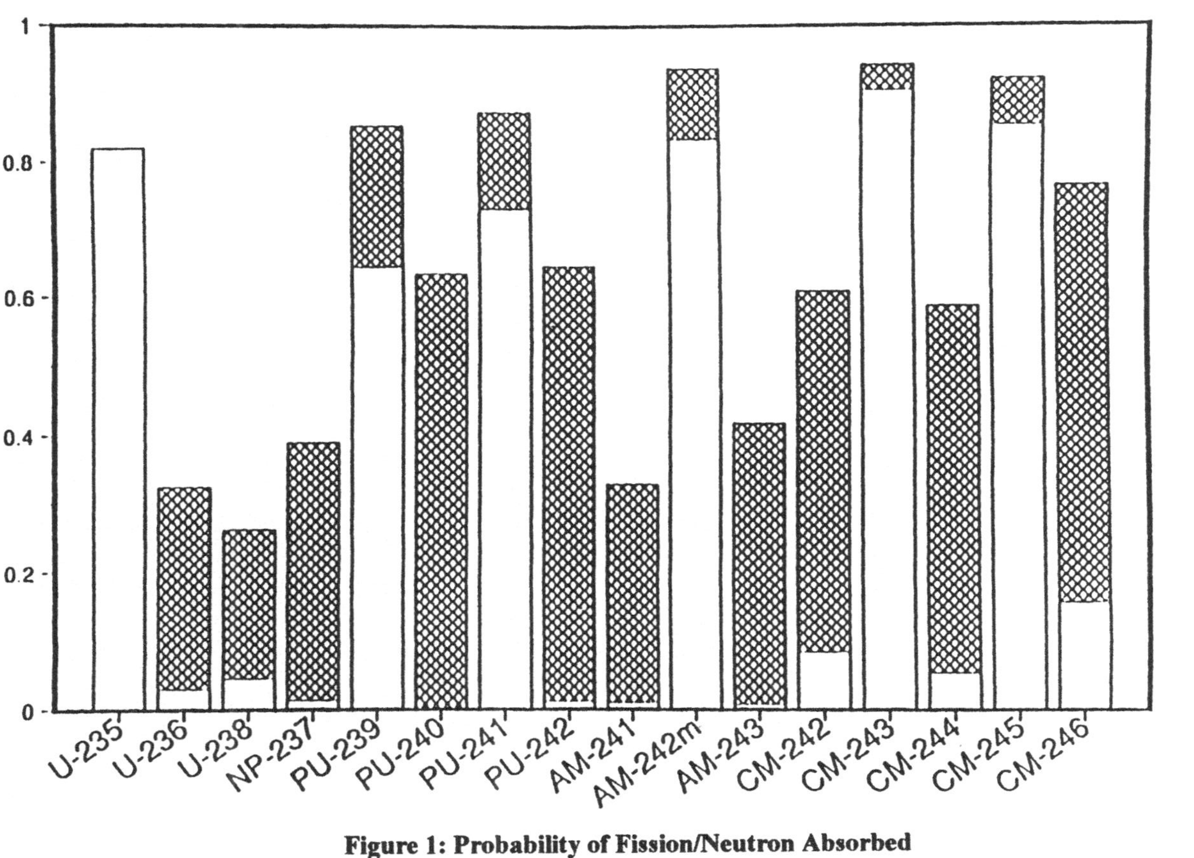

Figure 1: Probability of Fission/Neutron Absorbed

Recycle Considerations

Practical engineering considerations have led reactor designers world-wide to use solid fuel elements. With solid fuel elements, multiple recycle of fuel is required because of irradiation damage to structural materials. To avoid unacceptable losses to wastes, recycle must be essentially complete. Certain accelerator driven concepts have proposed using a molten salt carrier for the plutonium, but the technology for such a system is not yet proven, and various on-line chemical processes which amount to recycle are also generally required.

Recycle technologies enter the discussion in two quite unrelated ways. From a proliferation point of view, the critical questions are: does the recycle process involve separated plutonium that is attractive for weapons use. And how difficult is it to provide an adequate base of safeguards for both intermediate and final products of the recycle stream? These are precisely the two overriding conditions which we identified in the introduction as needing to be addressed when discussing final and permanent disposal of plutonium.

Because of the limited neutronic margins, thermal recycle requires a clean Pu recycle product, which is generally assumed to come from a PUREX type process [Ref. 11]. This clean product plutonium is suitable for use in a broad range of thermal reactors, but falls directly into the class of materials classified as "weapons-usable." The content of higher isotopes, and therefore the degree of attractiveness, depends on the feedstock, and not on the process. Stringent, real-time safeguards are employed to prevent diversion and to detect attempted diversions.

The AIROX process [Ref. 12], involving cleanup of LWR fuel for further use in a CANDU type reactor, avoids this concern but involves only a modest further consumption, and does not approach elimination unless followed by a more traditional recycle process.

The real advantage of the fast reactor system is its tolerance for direct use of self-protecting fuel; that is, a recycle technology can be used which does not remove all fission products. By utilizing a process which emphasizes the total recovery of all transuranics at the expense of having only a moderate capability for fission produce removal, it is possible to reduce waste burdens dramatically, and simultaneously to avoid the added burden of providing rigorous safeguards for direct-use material[Ref. 13].

Managed Inventories

At first blush, the concept of using a fast reactor to consume excess plutonium may sound like an oxymoron; after all, the traditional purpose of the fast reactor was to produce excess plutonium to fuel new reactors in an ever-expanding economy [Ref. 14]. But, in fact, a fast reactor produces less net plutonium than does a light water or gas cooled reactor, since it gets essentially all its power from fission of plutonium, not uranium. A light water reactor typically produces 200 to 250 kg of plutonium per GWe reactor year. The fast reactor closed fuel cycle can be constrained to produce plutonium at the rate at which it is needed for use, but only as there is a *near-term* need for it. In fact, practical designs (see for example Ref.

15) are best suited to operating on a break-even or near-break-even basis, wherein the plutonium acts simply as a catalyst, allowing power production from the backlog of depleted uranium left over from the enrichment process. Given appropriate safeguards, this means that the plutonium inventory can be tailored to the power demand and not, as with the LWR, continually increasing the waste burden.

A fast reactor can be configured to have a conversion ratio anywhere from 0.5 to as high as about 1.3. This large range of performance is precisely what is needed for managing the world's burden of transuranics. Region by geographic region, it allows the amount of plutonium and other transuranics to be held constant (conversion ratio = 1), or to increase (conversion ratio >1) in response to energy requirements. Similarly, existing transuranic inventories could be reduced in an ecologically responsible manner by using core loadings that consume (conversion ration <1) the working inventories of decommissioned sibling units. In this way, the fast reactor fuel cycle allows the transuranic working inventory to be matched to power demand, so that essentially no transuranics are consigned to long-term storage or waste [(Ref. 16, 17)], and the plutonium is always tied up in the power-producing working inventory.

Figure 2 shows the expected growth in world inventory of plutonium, based on official projections of future nuclear capacity through 2030 [(Ref. 18)], extended with an assumed linear growth in nuclear capacity from 2030 to 2045 equal to that between 2020 and 2030, and zero growth beyond 2045. The upper two curves show that there is some benefit (reduction in plutonium surplus) from thermal reactor recycle as is being practiced in France [(Ref. 19)] and as is proposed for Russia [(Ref. 20)]. The third curve in Fig. 2 is for strategy in which fast reactors are introduced at the following rate:

2010 to 2015:	1 GWe/y
2015 to 2020:	2 GWe/y
2020 and beyond:	all new nuclear power plant construction

The nuclear power capacity is the same as the LWR case. By 2045 all plutonium is in use. The figure demonstrates the fundamental advantages of the fast reactor fuel cycle relative to a throw-away or thermal recycle strategy:

- The world inventory of plutonium can all be put in use in working inventory within a credible planning horizon;
- the inventory can be realistically adjusted to match demand on a near-real-time basis;
- the inventory in all cases is less than that of a throw-away or thermal recycle strategy;
- inventory reduction is a feasible alternative if preferable energy sources become practicable.

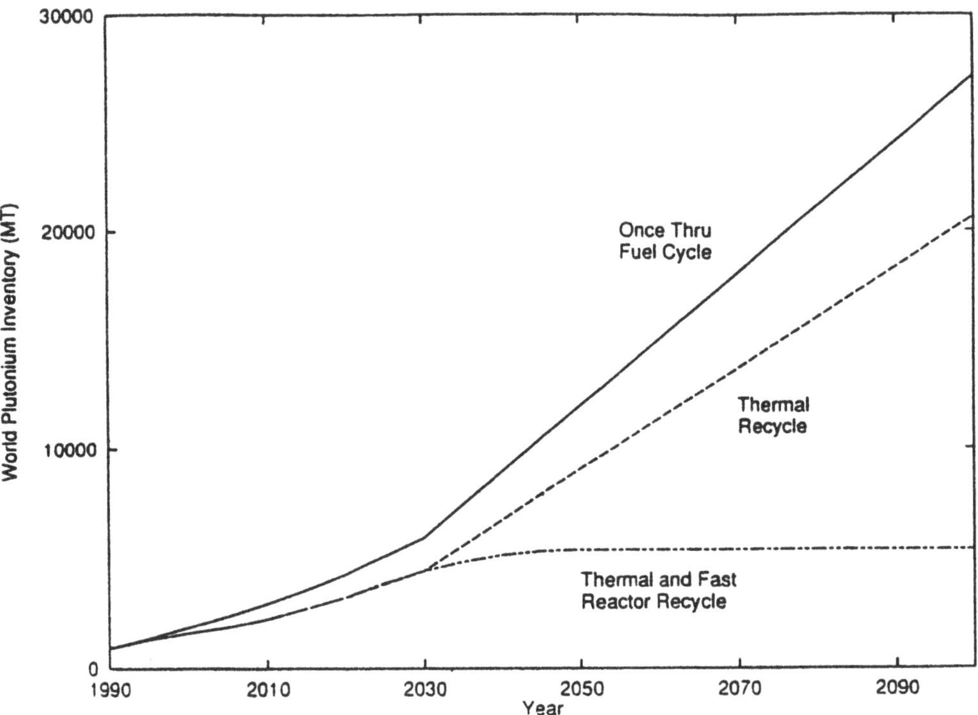

Figure 2: The Effect of Recycle on World Inventories of Plutonium

In short, for all fast reactor scenarios and at all times, transuranic inventory consigned to waste is maintained at essentially zero, and the worldwide transuranic inventory is contained entirely in the working inventories of power-producing fast reactor plants, which ebb and flow with time according to local needs. The revenue stream provides an income to support rigorous safeguards. Thus, rather than steadily growing with no further energy benefit, as is happening today with the once-through LWR throw-away cycle, the world's inventory of transuranics would grow or shrink along with the need for power, and future generations will not have to safeguard a burden of transuranics in spent-fuel storage pools or repositories as a legacy of their predecessors' energy policies.

From a proliferation point of view, the objective should be to leave a residue in the waste stream of no more weapons attractiveness than natural uranium; that is, some small fraction of a percent plutonium. It appears that electrorefining will do far better than this, not only preventing the waste stream from being a long term proliferation concern, but also greatly simplifying the waste disposal concerns. The electrometallurgical recycle technology is designed to discharge less than on part per thousand of the recycled transuranic fuel into the waste -- thereby eliminating any significant actinide contribution to long-term waste radiotoxicity hazard.

Unlike conventional reprocessing using PUREX technology, which emphasizes multi-step separation of plutonium to a purity of about one part-per million, electrorefining recycling technology simply concentrates plutonium in a metal alloy of uranium, transuranics, and rare earth fission products. An alloy containing no more than 70% plutonium is the practical limit of product purity. Subsequent process steps partition the fission products into the waste stream while recycling the residual actinides in the electrorefining process fluids. Because there is no need for a pure product for fast reactor fuel, emphasis can be economically placed on producing a relatively pure (actinide-free) waste stream. Because the process fluids are chloride salts and molten metal instead of organic solvents, there is no radiation damage limit to their recycling.

At some level, most of the metal fuel cycle has been demonstrated. Excellent performance of uranium alloy fuel has been demonstrated, although additional testing of U/Pu/Zr alloy would be required before it could be licensed. Uranium electrorefining of spent metallic fuel is currently being demonstrated, although transuranic partitioning has been dropped. An integrated program to qualify the electrorefining waste forms for geologic disposal is currently under way. It would take approximately five additional years to complete development of the metal fuel cycle for large-scale destruction of plutonium.

Radiotoxicity Legacy

The radiotoxicity hazard of the short-lived fission products decays to below that of uranium ore within 500 years -- a time span comparable to demonstrated longevity of human engineered structures and societal institutions. Sequestering the radiotoxicity hazard of the short-lived fission products in a storage repository for 500 years is well within the realm of demonstrated human achievement.

The several-cycle MOX recycle option in thermal reactors in no way would preclude subsequent further recycle to total fission consumption in fast burner reactors. The choice between starting with thermal MOX recycle with subsequent fast burner reactor deployment versus starting immediately with fast burner reactors can be made country-by-country on the basis of existing infrastructure, timelines, and cost effectiveness. Whichever choice is made, in the end the long-term radiotoxicity hazard from actinides -- in contrast to being consigned to the stewardship of future generations -- is instead held in the working inventory of the fast burner reactor and its associated recycle equipment.

The case considered presumes a continued deployment of nuclear power. The situation is different if we assume an early, universal phase-out of the use of nuclear power. In this instance, the case for the fast reactor is less clear. The U.S. National Academy of Sciences has recently concluded a study on the potential benefits of separation and transmutation technologies for dealing not only with excess plutonium, but also the transuranics and long-lived radioisotopes in spent nuclear fuel under the assumption of a prompt phase-out of nuclear power [Ref. 21]. The study discussed fast reactor systems, thermal reactor systems, and accelerator transmutation devices. The conclusion was that transmutation could not be justified on the basis of improvement in performance of the first repository, primarily because the risk of any fuel cycle, including the once-through, is so minuscule that substantial investment to introduce new technology simply for waste management could not be justified. However, the report emphasized the potential benefit of advanced separations technologies for the purpose of making better waste forms for the long-lived radioisotopes, and accordingly encouraged a focused research and development program with emphasis on improved separations processes.

Description of a Fast Reactor System for Plutonium Management

Any fission reactor "burns" plutonium at the same rate, roughly 1 gram of plutonium fissioned per MWt-day of energy produced. However, all conventional reactor systems utilize fuel which is primarily uranium; thus, the destruction of plutonium is at least partially compensated by in-situ production of Pu-239. The available range of destruction/production characteristics in fast reactor cores provides for flexibility in plutonium inventory management strategy. Conventional fast reactors maintain or can even increase the plutonium inventory (conversion ratio of 1.0-1.3). Alternately, by removing fertile material and/or altering the neutron balance, the transuranic inventory can readily be reduced.

Plutonium management characteristics have been evaluated for a wide variety of fast reactor core configurations and fuel cycle strategies. Traditionally, the focus was on the plutonium breeding potential of the system based on a perceived need for a quickly expanded energy economy. However, in recent years there have been numerous studies addressing the design and implementation of fast reactor burner systems. These studies include evaluations of the basic physics of higher actinide transmutation [(Ref. 22, 23)], the design for core flexibility within the same reactor system[(Ref. 15)], the development of burner designs which optimize a specific performance parameter (e.g., low sodium void worth) [(Ref. 24, 25)], the impact of weapons-plutonium introduction[(Ref. 26)], and radiotoxicity evaluations of closed cycle fast reactor systems [(Ref. 27)].

To illustrate the capabilities of an advanced fast reactor fuel cycle to operate in a plutonium destruction mode, the performance characteristics of a typical fast reactor burner design will be reviewed. By using a core design with a conversion ratio less than one, the inventory is gradually consumed through repeated recycle; and an external feed of makeup fissile material is required. For this paper, the moderate burner design developed in Ref. 25 will be utilized. These moderate burner core designs are also referred to as "conventional burner" designs because they utilize conventional fuel enrichments. The minimal conversion ratio of conventional burners is roughly 0.5; further reductions in the conversion ratio would require higher enrichment levels. Pure burner designs (where all uranium is removed) were also investigated in Ref. 26; however, unfavorable changes in the safety behavior were observed for such systems.

The moderate burner core design has a power rating of 840 MWt as used in the Advanced Liquid Metal Reactor (ALMR) U. S. design project [(Ref. 15)]. Radial and axial blanket zones are eliminated to avoid Pu-239 production. The resulting homogeneous core layout is shown in Fig. 3; this core consists of 354 driver assemblies (two enrichment zones), 28 control assemblies, 12 special control elements and 3 alternate shutdown assemblies. A geometry with a short core height (46 cm) compared to a large core diameter (4.44 m) enhances the axial leakage of neutrons which reduces the conversion ratio. An operating cycle length of 12 months (at an assumed capacity factor of 85%) is applied, with a seven batch refueling strategy.

In Ref. 25, the nuclear and safety performance characteristics of this moderate burner design were evaluated in detail for three alternative feed streams (weapons-plutonium, recycled fast reactor transuranics, and recycled thermal reactor transuranics) and two different fuel cycle scenarios (startup and repeated recycle). For the illustrative purposes of this paper, the case with closed recycle within the fast reactor system and an external makeup feed of recycle LWR transuranics is chosen. The performance characteristics are summarized in Table 2.

The 840 MWt fast reactor burner design produces 260,000 MWt-days of energy each year. The core operates at a conversion ratio of 0.5 making it a net consumer of 124 kg/y of transuranics. Thus, an external feed of 124 kg/y of recycled LWR or weapons transuranics is required in addition to the 543 kg/y of transuranics recycled within the closed fuel cycle. The in-core inventory is 4.3 MT of transuranics. The fuel enrichment is at the upper limit of current fast reactor experience, namely 30-35% transuranics/heavy metal. Most of the neutronic performance parameters (power density, discharge burnup, etc.) are similar to current fast power reactor operating conditions; however, the burnup reactivity loss is significantly higher than conventional designs since the internal blankets have been removed.

The burner designs has a large burnup reactivity loss (>$7) and correspondingly large control rod worth. However, the moderate burners has been designed to significantly decrease the sodium void worth and enhance the radial expansion feedback as compared to conventional systems. The net result is that transient response will be similar to conventional systems where the passive feedback is quite favorable. Thus, the detailed evaluation in Ref. 25 concludes that moderate burner fast reactor core designs have performance characteristics which are comparable to conventional fast reactor cores. In addition, it was shown that alternative feed materials (e.g., weapons plutonium) can be utilized in conventional burners without adversely impacting their performance.

In summary, moderate burner designs (with a conversion ratio near 0.5) will exhibit performance characteristics similar to conventional fast reactor core designs. A 840 MWt moderate burner produces 261 GWt-days of energy each year and consumes 125 kg of transuranics. The in-core inventory of transuranics is 4.3 MT with a similar amount retained in the closed fuel cycle processing facilities.

Table 2: Illustrative Reactor Characteristics

Conversion Ratio[a]	0.490
Net TRU Consumption Rate, kg/y	124
Enrichment, wt.%TRU/Heavy Metal	29/36
Heavy Metal Inventory, MT	
Transuranics	4.28
Pu-239	1.71
Total Heavy Metal	13.90
Equilibrium Loading, kg/y	
Make-up TRU	124
Recycled TRU	543
Total Heavy Metal	2,110
Burnup Reactivity Loss, %Δk	2.37
Peak Linear Power, W/cm	270
Ave. Discharge Burnup, Mwd/kg	118

[a] Ratio of time-integrated transuranic production rate to transuranic destruction rate

Safeguarding Materials in an Electrorefining Fuel Cycle

Safeguarding is always composed of a combination of institutional barriers and technical barriers. Safeguards measures include physical protection, careful accounting, and on-site inspections by the IAEA. There is no basis for treating these requirements differently for a fast reactor fuel cycle than for any other nuclear fuel cycle. *Institutional barriers* include international and intranational agreements, primarily the NPT, and the response of the community of nations to evidence of diversion.

Technical barriers are intrinsic susceptibility to detection of diversion, and materials of intrinsically low attractiveness. The effectiveness of a technical barrier depends both on the difficulty of accessing and transporting the material and on the probable time delay before the diversion is detected and announced to the community of nations. Attractiveness depends both on the difficulty of the process steps required to convert the material, once diverted, into weapons usable form, and on the isotopic purity of the output material (degree of contamination by isotopes that undergo spontaneous fission or emit significant heat, or do not fission readily on absorbing a neutron).

Safeguards discussions generally distinguish between active inventories and waste and scrap inventories. For the electrorefining-based fast reactor fuel cycle, this is somewhat meaningless, in that the fast reactor is designed to work on a continual recycle with no significant release of plutonium (or other transuranics) to wastes, and inventories are matched to power requirements. For the foreseeable future, currently envisioned fast reactor fuel cycles would be deployed to reduce the present excess of plutonium; a premature end

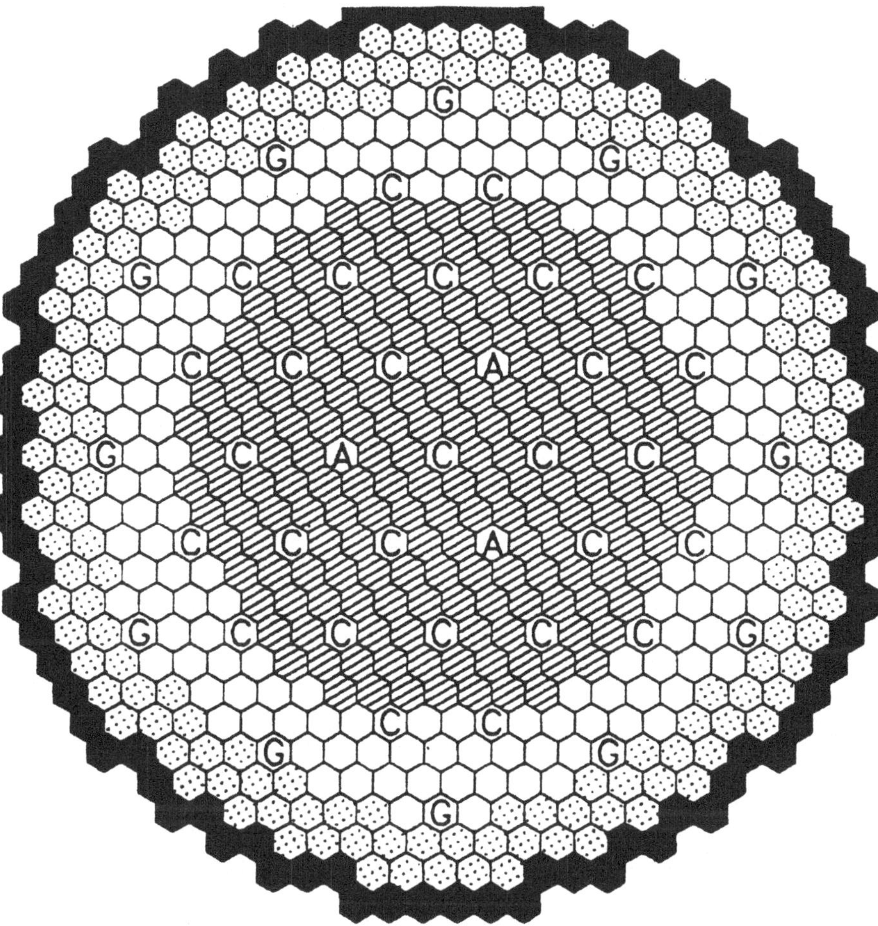

- High Enr. Driver (162)
- Low Enr. Driver (192)
- Reflector (162)
- Shield (90)
- Ⓒ Control (28)
- Ⓐ Alternate Shutdown (3)
- Ⓖ GEM (12)

Figure 3:

to the use of this option, with remaining inventories declared to be waste, would not be worse than if the fast reactor deployment had never been used.

In the following discussion, the safeguards implications of the electrorefining-based fast reactor fuel cycle is compared with two major fuel cycle alternatives that are already deployed: once-through thermal fuel cycle, and recycle using PUREX reprocessing.

The electrorefining process was developed specifically to yield a plutonium product that is inherently commingled with minor actinides (americium, curium, neptunium), uranium, and fission products [(Ref. 28)]. The minor actinides provide substantial decay heat and contamination with alpha, beta, gamma, and neutron emitters.

The electrorefining chemistry inherently limits fission-product decontamination to a factor no greater than about 1000. A typical product composition is compared with a typical PUREX product from the reprocessing of LWR fuel in Table 3. Table 3 also shows the intrinsic heat deposition rates in the transuranics-bearing materials, which is due mostly to alpha decay of the minor actinides. The heating rate per gram of heavy metal (including uranium) is five times that of the unprocessed LWR fuel and about 2.6 times higher as a processed product. Even with radioactive decay, the heating rate per gram never falls substantially below the rate for the heavy metal in LWR spent fuel. Table 3 also shows that the spontaneous neutron emission rates (neutron/s) per gram of heavy metal in the fast reactor spent fuel is three times more for heavy metal from LWR spent fuel.

From the heavy metal alone, the decay heat and spontaneous neutron emission rates are much higher in the electrorefining case. In addition to this, the presence of the residual fission products causes the transuranic-containing materials, at every step of the cycle, to be radioactive enough to be self-protecting due to the gamma radiation from the lanthanides. The radiation level of the material at each step of the pyroprocess easily meets the self-protection criterion of one Si/h at 1 m for the batch quantities of recycle fuels. The PUREX product for LWR recycle is necessarily very low activity.

After examining these factors, U.S. weapon designers have concluded that IFR fuel and recycle materials could not be used to make a nuclear weapon without significant further processing [(Ref. 9)].

Table 3 further shows that, even if electrorefined material were diverted (from any stage of the cycle) and processed in an unsafeguarded PUREX plant, the pure plutonium from PUREX processing of the diverted material would have spontaneous neutron emission rates and heating rates essentially as large (withing 30%) as those in the pure plutonium that comes from PUREX processing of spent LWR fuel. For weapons purposes, there is no particular significance to the somewhat higher fissile content of the electrorefined plutonium, since the yield, yield uncertainty, and manufacturing difficulty are comparable for the two materials. In both cases, further *isotopic* separation would be needed in order to make highly reliable, efficient nuclear weapons [(Ref. 29)].

Table 3: LWR and IFR Spent Fuel: Composition, Decay Heat and Spontaneous Neutron Source Levels

	Relative Isotopic Mass (g/kg HM)		Decay Heat (W/kg HM)		Spontaneous Neutrons (neutrons/s/kg HM)	
	LWR	**Fast Reactor**	**LWR**	**Fast Reactor**	**LWR**	**Fast Reactor**
Spent Fuel at Discharge (Normalized to 1 kg HM basis)						
Total Pu	11.23	219.9	0.10	1.43	3.38e+03	4.75e+04
Other Ac	1.12	3.74	2.20	10.4	1.18e+06	3.64e+06
Total TRU	12.35	223.7	2.30	11.8	1.19e+06	3.79e+06
Total U	987.7	776.3	1.48e-03	8.73e-05	1.23e+02	4.18e+00
Total HM	1000.0	1000.0	2.30	11.8	1.19e+06	3.79e+06
Normal Process Products PUREX for LWR and Electrorefining for the Fast Reactor						
Total Pu	1000.0	219.9	9.62	4.30	301e+05	1.42e+05
Other AC		3.74		21.01		9.22e+06
Total TRU		223.7		25.31		9.36e+06
Total U		776.3		1.08e-05		5.17e-01
Total HM	1000.0	1000.	9.62	25.31	3.01e+05	9.36e+06
Pure Pu Product After PUREX Processing of Diverted Materials						
	1000.0	1000.0	9.62	6.56	3.01e+05	2.17e+05

Summary

Plutonium is a fact. World inventories currently exceed 1000 tonnes, and are increasing at 60 to 80 tonnes per year. This can be considered a valuable energy resource or a political and environmental burden. The best approach is that which will maximize the benefits and minimize the burden. A closed fast reactor fuel cycle using an advanced recycle technology provides such an option by using plutonium as a catalyst to extract the full energy content from the world's uranium reserves, while eliminating excess inventories of plutonium and of other long lived transuranic byproducts. Such a system is fully compatible with rigorous safeguards, and in fact presents few safeguard challenges beyond those which are associated with the once-thorough fuel cycle.

The most important long-term contribution of the fast reactor approach to safeguards and prevention of proliferation is that it provides a positive means of managing the overall size of the world's plutonium and transuranic inventory [Ref. 30]. With a fuel cycle management strategy driven by economics, the fast reactor can readily absorb excess plutonium stocks, leaving the world inventory sequestered in plants producing useful energy.

References

1. Shapar, H. K.; The Policy of the United States with Respect to the Reprocessing of Spent Fuel, *Global '95, International Conference on Evaluation of Emerging Nuclear Fuel Cycle Systems*, September 11-14, 1995 Palais des Congres, Versailles, France.

2. Committee on International Security and Arms Control, Management and Disposition of Excess Weapons Plutonium, *National Academy of Sciences*; Washington, DC 1994.

3. Albright, D., Berkhout , F. and Walker; William; (1992). World Inventory of Plutonium and Highly Enriched Uranium, Oxford University Press, 1993.

4. Sutcliffe, W.G., et al; A Perspective on the Dangers of Plutonium, *UCRL-JC-11825*, April, 1995.

5. Carson, M.; The Explosive Properties of Reactor-Grade Plutonium, *Science and Global Security*, 4, 1993, pp. 111-128.

6. Barre, B.; Perspectives for Long-Term Plutonium Utilization, *Global '95, International Conference on Evaluation of Emerging Nuclear Fuel Cycle Systems*, September 11-14, 1995 Palais des Congres, Versailles, France.

7. Starr, C.; Global Energy and Electricity Futures, *In Energy*, 18(3), 1993, pp. 225-237.

8. U.S. Department of Energy; (1996). Storage and Disposition of Weapons-Usable Fissile Materials Draft Programmatic Environmental Impact Statement, February; *DOE/EIS-0229-D*.

9. DeVolpi, A.; (1986). Fissile Materials and Nuclear Weapons Proliferation. *Annual Review of Nuclear Particle Science*, 36, 83.

10. Goldman, D.J.; (1994). Some Implications of Using IFR High-Transuranic Plutonium in a Proliferant Nuclear Weapons Program, *Lawrence Livermore National Laboratory*, Document COTDU-94-0199.

11. Laurent, J. and Giraud, J.; Plutonium Recycling, a Mature Civilian Industry and a Key Contribution to the Weapons–Plutonium Inventory Disposition Issue; *Global '95, International Conference on Evaluation of Emerging Nuclear Fuel Cycle Systems*, September 11-14, 1995 Palais des Congres, Versailles, France.

12. Feinroth, G. J. and Majumdar, D.; An Overview of the AIROX Process and Its Potential for Nuclear Fuel Recycle, *Proceedings of the International Conference and Technology Exhibition on Future Nuclear Systems: Emerging Fuel Cycles and Waste Disposal Options*, p. 709 Seattle, WA, 12-17 September.

13. Hannum, W.H., Wade, D.C., McFarlane, H.F. and Hill, R.N.; Nonproliferation and Safeguards Aspects of the IFR; *Progress in Nuclear Energy*, Vol. 31, No. 1,2 pp. 203-217, 1997.

14. U.S. AEC (1967). Civilian Nuclear Power: A Report to the President, *U.S. Atomic Energy Commission and Civilian Nuclear Power*, the 1967 Supplement to the 1962 Report to the President.

15. Quinn, J. E., Magee, P. M., Thompson, M. L., and Wu, T.; ALMR Fuel Cycle Flexibility Proc. Physics Studies of Weapons Plutonium Disposition in the Integral Fast Reactor Closed Fuel Cycle, *American Power Conf.*, 55, 1079 (1993).

16. Hannum, W.H. and Wade, D.C. ; (1993). Incorporation of Excess Weapons Material into the IFR Fuel Cycle. *Proceedings of the International Conference on Future Nuclear systems; Emerging Fuel Cycles and Waste Disposal Options*, p. 812, Seattle, WA, 12-17 September.

17. Hannum, W.H. and Lineberry, M.J.; (1993). Flexible Plutonium Management with IFR Technology. *Transactions of the American Nuclear Society* 69, 91.

18. Uranium: Resources Production and Demand (Paris: *PECD/NEA*, 1993)

19. DeGalassus, B. ; (1994). Perspectives on U.S. Plutonium Policy. *International Policy Forum: Management and Disposition of Nuclear Weapons Materials*, Leesburg, VA, 8-11 March.

20. Murogov, V.M.; (1994). Beyond Weapons–A Plan to Convert Weapons Grade Fissile Materials to Commercial Use, *International Policy Forum: Management and Disposition of Nuclear Weapons Materials*, Leesburg, VA, 8-11 March.

21. Nuclear Wastes: Technologies for Separations and Transmutation; Committee on Separations Technology and Transmutation Systems; Board on Radioactive Waste Management; Commission on Geosciences, Environment, and Resources; *National Research Council.*

22. Hill, R. N., Wade, D. C., Fujita, E. K. and Khalil, H.; Physics Studies of Higher Actinide Consumption in an LMR, *Int. Conf. on the Physics of Reactors*, Marseille, France, April 23-27, 1990, p. I-83.

23. Cockey, C. L., Wu, T. A., Lipps, J. and Hill, R. N.; Higher Actinide Transmutation in the ALMR, *Proc. Conf. on Future Nuclear Systems*, Seattle, Washington, September 12-17, 1993, p.123.

24. Hill, R. N.; LMR Design Concepts for Transuranic Management in Low Sodium Void Worth Cores; *Proc. Int. Conf. on Fast Reactors and Related Fuel Cycles*, Kyoto, Japan, October 28 - November 1, 1991, p. 19.1-1.

25. Hill, R. N., Kawashima, M., Arie, K., and Suzuki, M.; Calculational Benchmark Comparisons for a Low Sodium Void Worth Actinide Burner Core Design, *Proc. Mtg. on Advances in Reactor Physics*, Charleston, South Carolina, March 8-11, 1992, p. 1-313.

26. Hill, R.N., Wade, D.C., Liaw, J.R. and Fujita, E.K.; Physics Studies of Weapons Plutonium Disposition in the Integral Fast Reactor Closed Fuel Cycle, *Nucl. Sci. Eng.*, 121, 17 (1995).

27. Grimm, K. N., Hill, R. N. and Wade, D. C.; An Evaluation of Waste Radiotoxicity Reduction for a Fast Burner Reactor Closed Fuel Cycle - NEA Benchmark Results; *Proc. Int. Conf. on Evaluation of Emerging Nuclear Fuel Cycle Systems*, Versailles, France, September 11-14, 1995, Vol. I, p. 172.

28. Hannum, W.H., Wade, D. and Stanford, G; Self-Protection in Dry Recycle Technologies; UCRL-ID-124105; *Global '95, International Conference on Evaluation of Emerging Nuclear Fuel Cycle Systems*, September 11-14, 1995 Palais des Congres, Versailles, France.

29. Wymer, R.G., Bengelsdorf, H.D., Choppin, G.R., Coops, M.S., Guon, J., Pillary, K.K.S. and Williams, J.D.; (1992). An Assessment of the Proliferation Potential and International Implications of the Integral Fast Reactor, *Martin Marietta*, Report K/ITP-511.

30. Walter, C. E.; A Strategy For an Advanced Nuclear-Electric Sector—Proliferation-Resistant, Environmentally Sound, Economical; *Global '95, International Conference on Evaluation of Emerging Nuclear Fuel Cycle Systems*, September 11-14, 1995 Palais des Congres, Versailles, France.

The submitted manuscript has been authored by a contractor of the U.S. Government under contract No. W-31-109-ENG-38. Accordingly, the U.S. Government retains a nonexclusive, royalty-free license to publish or reproduce the published form of this contribution, or allow others to do so, for U.S. Government purposes.

EXPERIENCE OF SSC RF RINR ON USAGE OF URANIUM-PLUTONIUM OXIDE FUEL IN NUCLEAR REACTORS

Ivanov V.B., Skiba O.V., Mayorshin A.A., Porodnov P.T., Bychkov A.V. and Kisly V.A.
SSC RF "Research Institute of Nuclear Reactors"

Works on Pu involvement in the fast reactors fuel cycle have been carried out at RIAR since mid 70s. Development of production technologies for mixed fuel, fuel elements and fuel assemblies is the first stage of organizing the closed U-Pu fuel cycle of nuclear reactors. In contrast to traditional fuel cycle technologies, based on aqueous methods of fuel reprocessing and fuel elements fabrication with compact pelletized rod, the fuel cycle developed at RIAR is based on non-traditional technologies.

Up to now the main technical problems of the RBN closed fuel cycle both in the field of fuel production and in the field of creating on its basis the new type fuel elements are solved. In total, the base principles of the closed fuel cycle built on the mutual compatibility of technologies for U-Pu fuel regeneration and fuel elements and fuel assemblies refabrication were experimentally substantiated.

These principles include:

- "dry" pyroelectrochemical processes for irradiated fuel reprocessing and wastes utilization;
- production during reprocessing of fuel in the granulated form with high particle density;
- production of fuel elements from granulated fuel by vibropacking;
- usage of remotely-controlled automated equipment during fuel reprocessing, production and monitoring of fuel elements and fuel assemblies.

The RBN closed fuel cycle scheme with oxide fuel is presented in Fig.1.

The pyroelectrochemical method of fuel reprocessing in the U-Pu fuel cycle allows reprocessing the fuel with any burnup and exposure time for a small number of technological stages. Media in which these processes are conducted, namely molten chloride salts are radiation resistant ion liquids without moderators. This enables handling high concentrations of fissile and radioactive materials and not use reagents that are destroyed under radiation. That is why "dry" processes are highly productive, compact and give few wastes.

The principal flow sheet of the pyroelectrochemical oxide fuel reprocessing is presented in Fig.2. The initial material for the pyrochemical process can be:

- uranium and plutonium in any chemical form (oxide, metal, carbide, nitride) including weapon alloys;
- irradiated fuel of nuclear reactors.

At the first stage fuel is dissolved in chlorides melt. Fuel from this melt is isolated by electrolysis or precipitation methods in the form of crystalline oxides, e.g. UO_2, PuO_2, $(U,Pu)O_2$, $(U,Pu,Np)O_2$. After salts separation from crystalline products the polydispersed composition graanulate is produced with particles, no less than 10.7 g/cm^3 by density and no more than 1.0 mm in size. Such fuel is ready for vibropacking to fuel lement. Pyroelectrochemical process allows fuel purification from fission products with the purification coefficient more than 100, which is sufficient from the viewpoint of reactor physics. On the other hand, this excludes usage of the product for military purposes, as it continues to be within the frames of the "irradiated fuel" standard. The characteristic feature of the processes is the fact that practically all fission products are concentrated into a solid phase. Therefore, the volume of high-active wastes is minimal.

In total about 4500 kg of oxide fuel were produced at RINR which enabled production of more than 28,000 fuel elements of different reactors.

Technological process of fuel vibropacking is simple, it can be easily automated and high productivity can be provided.

Simple apparatus arrangement allows the process to be realized under remote conditions.

To form the vobropac fuel element rod use is made of two types of the initial fuel: mechanical mixture of granulated UO_2 and PuO_2 and coprecipitated fuel $(U,Pu)O_2$. The comparative reactor and post-reactor investigations did not reveal any difference in fuel elements behaviour. The princi-

E. R. Merz and C. E. Walter (eds.), Advanced Nuclear Systems Consuming Excess Plutonium, 65–67.
© 1997 *Kluwer Academic Publishers.*

pal scheme of vibropac fuel elements fabrication is presented in Fig.3 and it is universal for any fuel rod composition.

Direct usage of fuel, produced after pyroelectrochemical reprocessing, immediately for fuel elements fabrication is the main distinguishing feature of this approach to fuel cycle.

Since 1981 up to now the BOR-60 reactor has been operating using vibropac U-Pu oxide fuel.

Reactor tests and material science investigations of the basic design fuel elements, containing a getter made of metallic U powder in the fuel composition, confirmed their high serviceability up to burnup 26% achieved on standard fuel assemblies and burnup 30% on experimental fuel elements. No thermomechanical and physicochemical fuel-cladding interaction was observed in any of the analyzed cross-sections.

Analysis of radiation characteristics of the vobropac fuel elements serviceability showed the following:

- usage of the fuel composition with the added getter ($UPuO_2$ + U) allows full exclusion of corrosion processes stipulated by availability of Cs and halogens as well as possible technological impurities, which practically eliminates the burnup limit because of the physicochemical fuel-cladding interaction;
- stressed state of claddings in transient processes is many times lower, and stresses relaxation comes considerably faster than in fuel elements with pellet fuel;
- increased smear density of the fuel rod (> 9.0 g/cm^3) allows enough temperature reserve up to the melting temperature;
- increased Cs migration to the low-temperature field of the fuel element and absence of the fuel-cladding gap lead to the fact that during unsealing of fuel elements claddings the number of volatile products entering the coolant is less than from fuel elements with pellets, and the fuel is not practically washed out.

To confirm the vibropac fuel elements serviceability under power reactor conditions, their tests were performed in the BN350 and BN-600 reactors. According to the test results of fuel elements with vibropac $UPuO_2$ fuel in the Bn-350 reactor additional information was obtained and changes were made to the fuel element design. These improvements were realized during fabrication of the BN-600 fuel assemblies that had been successfully tested up to burnup 9.6%.

A complex of reactor and post-reactor investigations of vibropac fuel elements for thermal-neutron reactors was performed. Fuel elements were irradiated in the SM-2 and MIR reactors. Burnup 40 MW d/kg U was achieved in the MIR reactor in both the base operating mode and power manoeuvring mode. These investigations showed that fuel elements features are also fully revealed in thermal reactors.

One of the advantages of the pyroelectrochemical method of nuclear fuel reprocessing and vibropacking technology is the possibility of fuel composition fabrication in the form of the components mechanical mixture. This feature was realized during experimental fuel elements fabrication for study of minoractinides (Np, Am, Cm) transmutation. In particular, tests of fuel compositions (U, Np)O2+U are being completed and (U,Pu,Np)O_2 and fuel with additions Am_2O_3 are prepared for tests.

Tests of 2 experimental fuel assemblies with PuO_2 content - 30 and 40% were performed for study of radiation aspects of Pu utilization.

For complex inspection of the main stages of the RBN fuel cycle a pilot complex was created including the following:

- the remotely-controlled facility for production of granulated oxide U, Pu and mixed U-Pu fuel;
- the automated, remotely-controlled facility "Oryol" for production and monitoring of the BOR-60 reactor fuel elements and fuel assemblies;
- the pilot facility for developing the regeneration technology for irradiated oxide U-Pu fuel and its production in the form of granules.

In terms of the equipment composition and auxiliary systems the pilot facility had all basic components for fuel cycle demonstration. Technological equipment is placed in hot cells. In spite of the fact that the process was developed on the "energy" Pu, the equipment design, control, repairs and maintenance systems were maximally approximated to the irradiated fuel operating conditions.

Operation of the complex from 1976 till 1986 allowed solution of a number of principal fuel cycle problems. The following are among them:

- for full technological cycle there is made the first BOR-60 reactor core from U-Pu oxide granulated fuel and constant operation of this reactor on U-Pu fuel was provided since 1982; - serviceability of the equipment and auxiliary systems under long operating conditions was checked. The reliability level of the equipment was achieved on the whole for the complex 0.95, and for separate equipment 0.98;
- full material balance of fissile materials of the whole cycle is made up. It is shown that direct fuel yield to the ready product makes up 96.6%, and with allowance for secondary products recycle - 99.8%;
- composition, properties and technological processing methods of all radioactive wastes categories are determined;
- radiation situation at all stages of Pu processing is studied. Within the frames of this works stage the equipment was developed abd the facility for development of the technological process with Pu recycle was created. The experiments were conducted, the pilot batch of regenerated PuO2 from spent fuel assemblies of the BN-350 and BOR-60 reactors was obtained. Operations, developed on "energy" Pu, are entirely acceptable for irradiated fuel, and physico-mechanical properties of fuel are identical.

At the end of 80s the developments used industrial mastering of the technology for fuel reprocessing, fuel elements and fuel assemblies fabrication. The experimental base for this stage was the pilot-research complex (PRC) created at the expense of the first pilot facility redesign. This complex was developed as an industrial module prototype and designed for development of the main technical solutions of the closed fuel cycle. In the PRC structure there are all attributes of the industrial technological line, though its productivity makes up 200 fuel assemblies/year for the BN-600 reactor, including for fuel - 1.5 t/year, for fuel elements - 10 fuel elements/h. PRC was put into operation in 1990 and in under mastering now. 26 fuel assemblies of the BN-600 type were produced in the period of pilot operation. Rates of mastering and improving the complex are determined by the financing level and demand in fuel.

Recently the investigations have been performed with the aim of estimation of the possibility to produce mixed fuel from weapon Pu by the pyroelectrochemical method within the frameworks of French-Russian cooperation.

The flow sheet for weapon Pu conversion to the MOX fuel includes the following operations:

- dissolving Pu alloy in melts;
- electrochemical or precipitating production of PuO_2 and/or UO_2-PuO_2;
- preparation of the produced fuel for fabrication of pellets or vibropac fuel elements.

As the result of pyrochemical conversion of Pu alloy to the MOX fuel, its purification against alloying additions (Ga) and Am occurs.

For vibrofuel production such a scheme is applicable without amendments. A series of preliminary experiments on technological procedures inspection was performed.

The main results:

- the metallic Pu dissolving rate makes up 200 mg/cm^2 min with Cl_2 blowing and flowrate 5 l/h; presence of a passivation layer (e.g. PuO_2) does not affect the metallic Pu dissolving rate;
- gallium violates as $GaCl_3$ (boiling temperature T=201 °C);
- Pu oxide contains 0.002% Ga by mass; size of PuO_2 grains - from 5 to 50 mm;
- characteristics of the produced oxide powder do not depend on the starting Pu form (PuO_2 or Pu-Ga, etc.), but are determined by the precipitation mode;
- the possibility of applying the pyrochemical technology for pellet fuel fabrication is shown. This is proved by the experiment with pyrochemical UO_2.

These investigations demonstrate additional possibilities of the pyrochemical technology while solving the Pu utilization associated problems. Thus, the aggregate of "dry" technologies of fuel reprocessing and fuel vibropacking processes combined with unique properties of vibropac fuel allows realization of a new complex approach to the fuel cycle problem using the high-active Pu fuel. Wide tests of vibropac U-Pu oxide fuel elements in the BOR-60 reactor together with successful tests of fuel elements in the BN-600 reactor, reliable operation of the pilot-research complex facilities allow the conclusion about reaal possibility of safe, cost-beneficial U-Pu fuel cycle on the basis of the enumerated technologies as well as about the possibility of both energy and weapon Pu utilization.

Energy Source for the Human Demand

Sadao Hattori
Central Research Institute of Electric Power Industry
1-6-1, Ohtemachi, Chiyoda-ku, Tokyo 100 Japan

1. Introduction

Lack of water and food, extinction of plants and animals, human catastrophe and other serious environmental crises are occurring as a result of the large increase in the population of mankind. Yet, the Asian countries are experiencing strong technical and economical development.

Special data from the United Nations shows us that strong technical and economic development will result in stabilization of the population. There is clear evidence that population growth rate diminishes with the increase in GNP.

Many of the problems presently facing human society can be alleviated through the development and construction of an adequate supply of clean energy. This clean energy source is nuclear power. Utilization of plutonium to provide this clean energy is essential in meeting the significant energy demand requirement for the future.

Technology development and the initiation of effective management systems against nuclear proliferation are currently being implemented to control the use of this excellent nuclear fuel. In this context, weapon Plutonium from major countries is extremely important to fuel the cores of fast breeder reactor. This is an important opportunity for the prosperity of mankind brought about through the epoch-making "Plowshare" policy.

E. R. Merz and C. E. Walter (eds.), Advanced Nuclear Systems Consuming Excess Plutonium, 69–77.
© 1997 *Kluwer Academic Publishers.*

2. Design Concept of 4S-Reactor

The nuclear reactor in future shall be simple and inherently safe. An example of the reactor concept with a philosophy of "Super Safe Small and Simple"[1-7] is introduced here. The inherent safety of the reactor allows for a response where the reactor power decrease and goes to the reactor shutdown for all hypothetical accidents such as sodium boiling or loss of coolant without scram. Figure 1 shows the reactor assembly.

The core of the reactor is slender (less than 1 meter diameter and 4 meters height) and has a cylindrical reflector with partial height of the core (1.5 meters height). This reflector can sustain the criticality of the core. The axial power distribution of the core has its peak at the part surrounded by the reflector and gets down gradually at the upper side and the lower side.

In general, the small core would lose the reactivity rapidly and require the frequent refueling. The reflector is moved up very slowly around the long core connected fuel assembly. Then, this concept with reflector control could bring a non-refueling reactor more than ten to thirty years.

A cylindrical electro-magnetic pump is located above the core and the hot sodium getting out of the core flows up to the intermediate heat exchanger. The sodium which exhcnge heat to secondary sodium flows down as cold sodium along the inside surface of the reactor vessel and gets into the cylindrical flow pass of the electro-magnetic pump to be driven by electro-magnetic force.

The drive mechanism of the reflector needs very small force during the normal operation because the cylindrical reflector receives the upward force by sodium up flow. The small drive components above the reflector stick to the core barrel and move the reflector up slowly by electro-magnetic force.

The reactor power following the load change of the steam turbine is naturally controlled by the negative reactivity coefficient of the sodium temperature in this reactor.

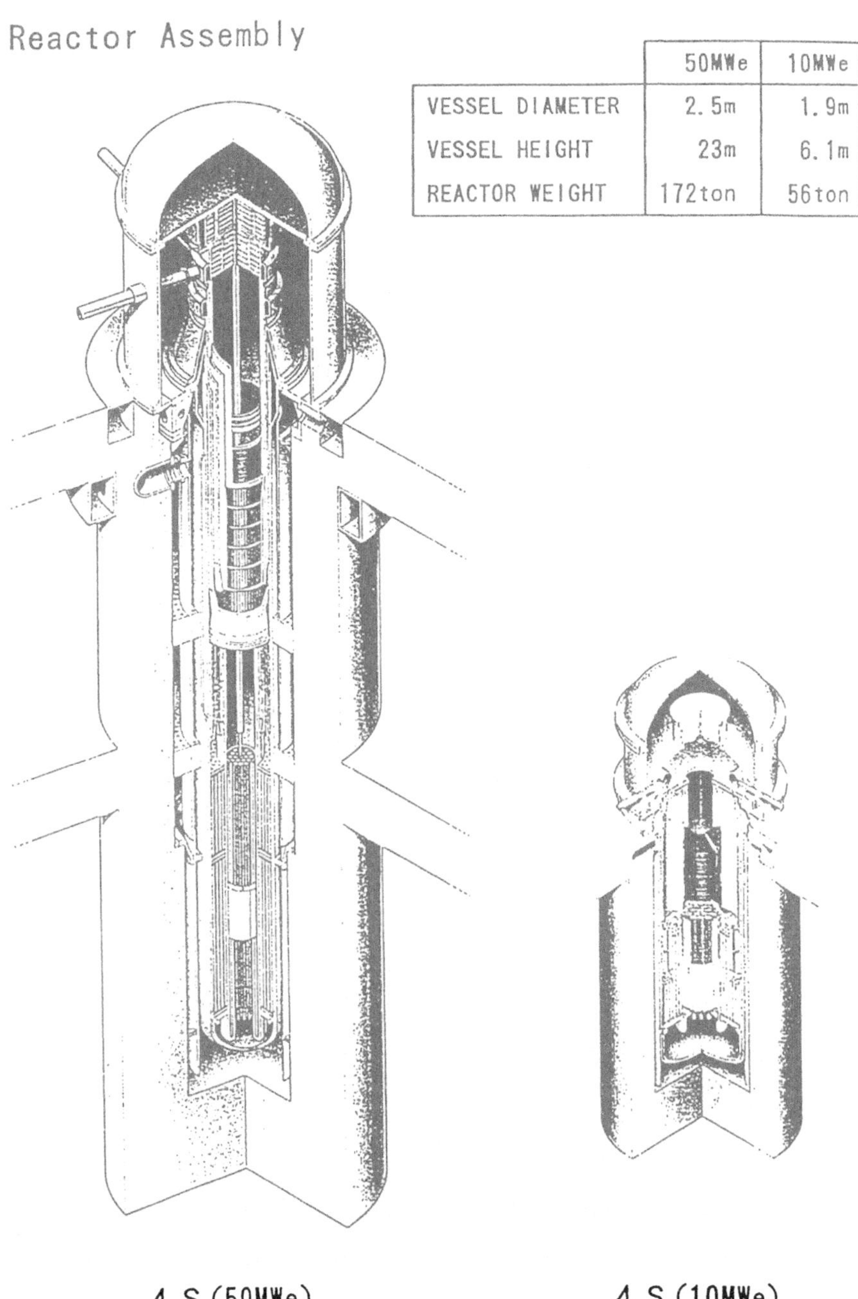

	50MWe	10MWe
VESSEL DIAMETER	2.5m	1.9m
VESSEL HEIGHT	23m	6.1m
REACTOR WEIGHT	172ton	56ton

Fig. 1 Reactor Assembly of 4S (50 and 10 MWe)

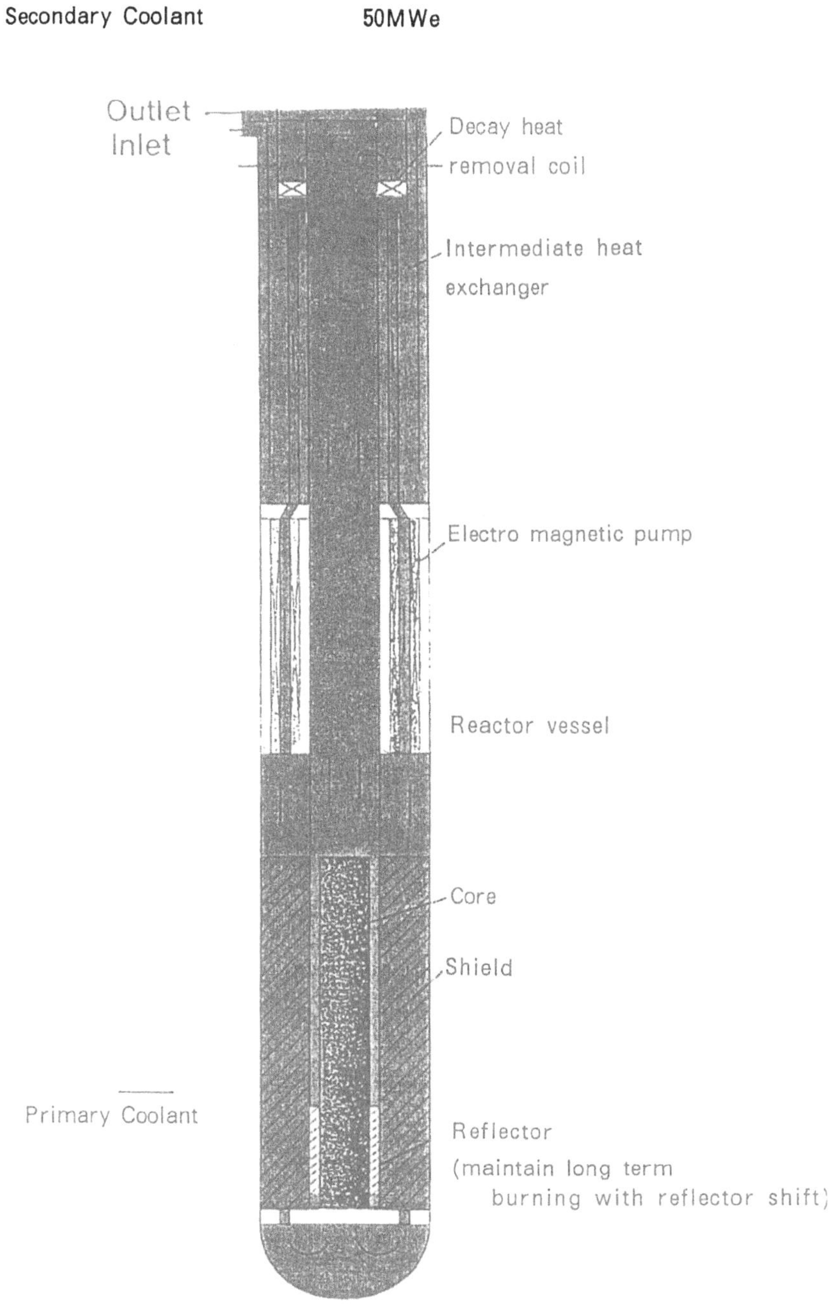

Secondary Coolant

50MWe

Outlet

Inlet

Decay heat removal coil

Intermediate heat exchanger

Electro magnetic pump

Reactor vessel

Core

Shield

Primary Coolant

Reflector (maintain long term burning with reflector shift)

Reactor concept of 4S.

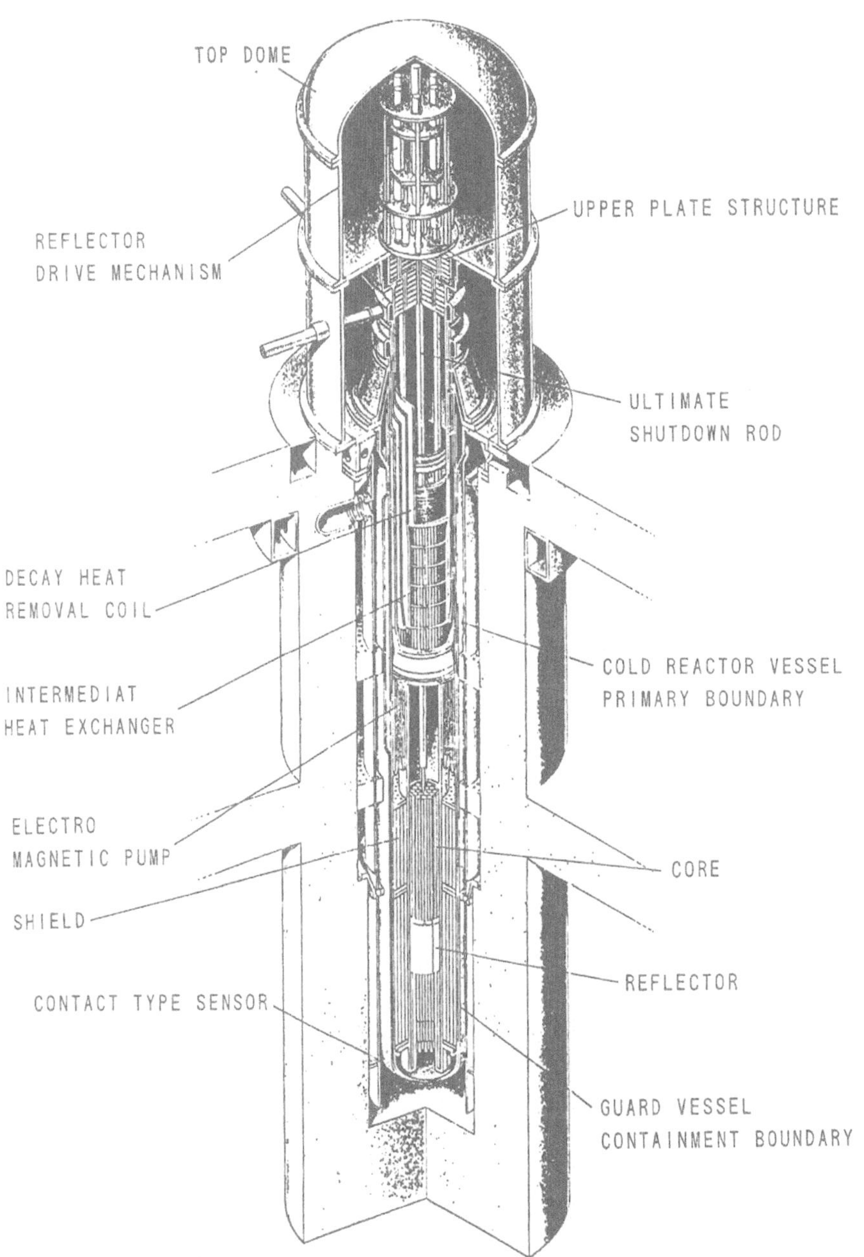

Reactor internal structure of 4S.

CORE PERFORMANCE DATA

ITEM \ FUEL	METAL	MOX
Core Life Time(year) (100% load)	24	18
Avg.Burnup(GWd/t)	108.4	81.2
Max.Burnup(GWd/t)	135.9	103.2
Heat Rate(w/cm)		
Average	80	80
Maximum	377	399
Na Void Reactivity(% Δ K)	-1.4	-1.5
Doppler Coeff.(Tdk/dt)	-6.3×10^{-3}	-8.9×10^{-3}
Pu Inventory(BOL/EOL)		
^{239}Pu(kg)	1698/1222	1612/1167
^{240}Pu(kg)	109/ 238	103/239
^{241}Pu(kg)	9/ 16	9/ 18
^{242}Pu(kg)	0/ 2	0/ 2
^{total}Pu(kg)	1816/1478	1724/1426
^{fiss}Pu(kg)	1707/1238	1621/1185

CORE DESIGN PARAMETERS

(50MWe)

ITEM \ FUEL	METAL	MOX
Power Output (MWt)	125	125
Fuel	U+Pu+Zr	UO_2+PuO_2 (Vibropack)
Smear Density(%)	75	85
Pin Diameter(cm)	1.0	1.0
p/d	1.15	1.15
Pins per Assembly	217	217
Assembly Pitch(cm)	18.2	18.2
No. of Assemblies	18	18
Pu Avg.Enrichment(%)	18.3	22.3
Pu_{fiss} Avg. Enrichment(%)	17.2	21.0
Core		
Diameter(cm)	83	83
Height(cm)	400	400
Volume Fraction		
Fuel	45.0	45.0
Coolant	33.0	33.0
Steel	22.0	22.0
Reflector(Carbon)		
Length(cm)	150	150
Thickness(cm)	15	15
Coolant (Sodium)		
Temp.(In/Out) (℃)	355/510	355/510
Pressure Drop(kg/cm²)	2.4	2.4

Comparison of Total R&D Cost for Commercialization

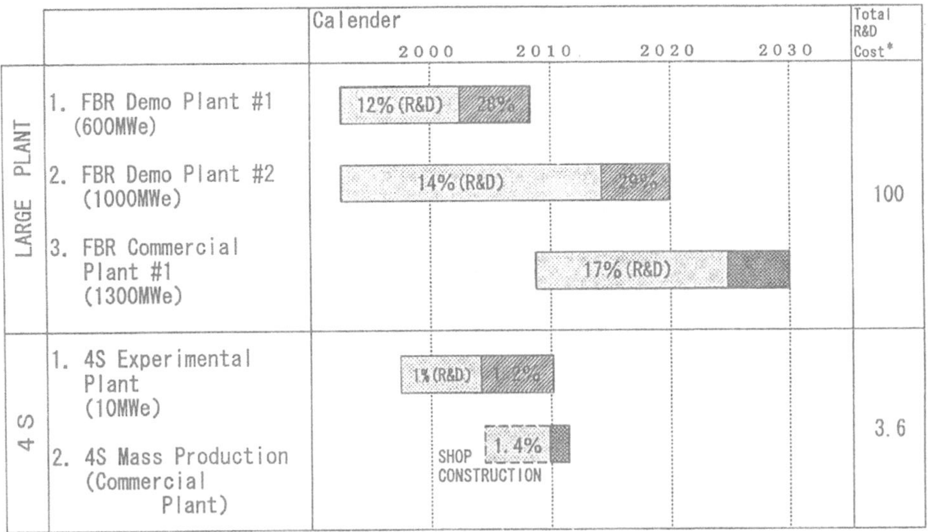

* Total R&D Cost : Cost of R&D and Demo Plant Construction for Commercialization

Ultimate Safety of Metal Fuel

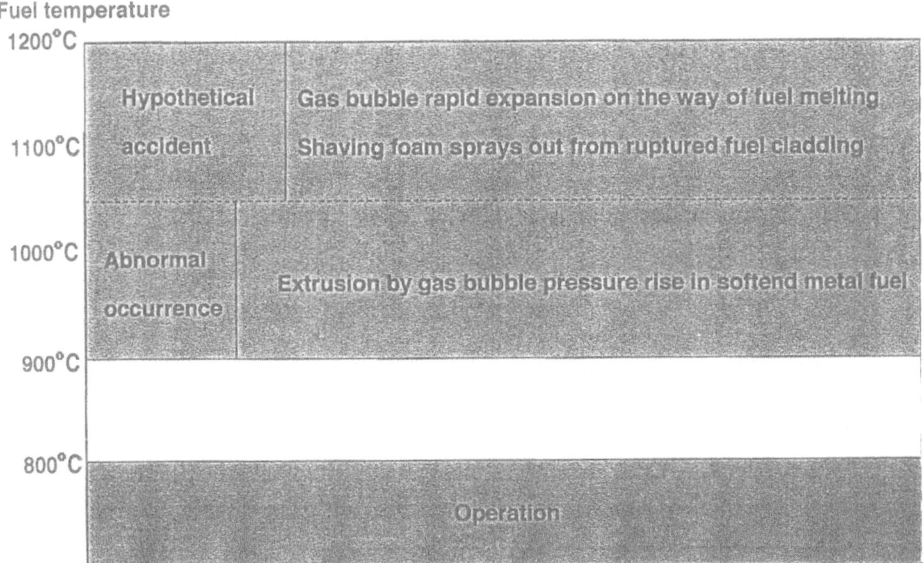

3. Special Features of the Reactor

(1) No refueling is needed for more than ten to thirty years. It would extremely promote and almost complete the IAEA nuclear materials control.

(2) No reactor operator would be needed, because nothing is to be under human observation and no component is to be under human operation.

(3) No components are needed to be positively operated, under any abnormal events, however, conservative design provides the back up shut down mechanism by dropping of the neutron absorber rod into the center of the core.

(4) No structural damage of the reactor and of the core fuel can be considered by any assumptions of hypothetical accident referring conventional safety evaluation conditions in world wide FBR activities.

(5) Extremely low cost and very short time is expected for research and development phase before deployment of this reactor because of its small size.

4. Risk Reduction

More than ten years ago, we obtained a clear rationale on why small reactor strategies came out after many tens of years developmental activities of large nuclear power plants. If you reduce the capacity of the reactor unit from 1000 MWe to 100 MWe, you can obtain the risk reduction of three orders of magnitudes (1/1000) by not only reduction of the numbers of components but also obtaining passive system design concept and reduction of the amount of the radio-active materials. The risk reduction especially by this reactor concept (50 MWe, totally passive) is surely more than above discussion.

5. Economy

When it comes to consider the cost, most important point is how much money we need before achieving the commercial deployment.

An relative explanation compared to the monolithic large FBR development is here advisable. The "Commercial Deployment" means electric power production by similar cost to conventional methods of power production. Time and cost of "Commercial Deployment" are as follows:

	Large monolithic FBR development	Small & simple modular concept
Achievement	2030	2010
Cost estimate	100	3.6

6. Conclusions

Effort have been focused in recent years on developing small and medium sized power reactor with inherent, passive safety characteristics. Successful development of such reactors will lead to a wider range of application of nuclear power.

This reactor as a new concept of fast reactor designed to meet the goals of nuclear power and offers many attractive advantages. Commercial operation of this reactor is expected to solve a number of problems that humans will encounter in the 21st century.

References

1. Hattori, S. and Handa, N. (1989) Trans. Am. Nucl. Soc.,60, 437
2. Ueda, N., Minato, A., Handa, N. and Hattori, S. (1991) Proc. of Int. Conf. on Fast Reactors and Related Fuel Cycle, Tokyo, 5-7
3. Hattori, S. Minato, A. and Handa, N. (1991) Int. Specialist Mtg. on Potential of Small Nucl. Reactors for Future Clean and Safe Energy Sources
4. Hattori, A. and Minato, A. (1992) Proc. of Int. Conf. Design and Safety of Advanced Nuclear Power Plants.
5. An, S and Minato, A. (1993) 4th Annual Scientific & Technical Conference of the Nuclear Society, Russia
6. Hattori, S. and Minato, A. (1993) Cof. on Nuclear Engineering (ICON-2), Vol.2, 611
7. Hattori, S., Minato, A. and Ikemoto, I. (1995) International Topical Meeting on Safety Culture in Nuclear Installations, Vienna

PLUTONIUM-BASED NUCLEAR POWER AND NONPROLIFERATION

V.V. Orlov, RDIPE, Russia

The development of the nuclear power got its start in the 50ies based on 235 U thermal reactors of all the types, which had been earlier developed for production of arm plutonium and tritium and for submarines. It was assumed that as thermal reactors accumulated Pu and fast breeders were being mastered (their first power units were commissioned in 1972-1975 in the USSR, France, UK), the next stage of the plutonium-based nuclear power would be implemented with reaching, in this centenary already, scales of several thousand GW (e). However, such development of the nuclear power in this centenary turned out to be both unclaimed and unprepared: the world market of conventional fuels was stabilised after the oil crisis; severe accidents at TMI and Chernobyl required to upgrade NPPs, which resulted in the rise in their cost and aggravation of the antinuclear opposition in the society; the first fast reactors turned out to be much more expensive than thermal reactors, and they were not developed in practice; disposal of large quantities of radiowastes gives rise to objection in the society, and the closed fuel cycle with Pu extraction contradicts the interests of nuclear weapon non-proliferation. The world nuclear power reached a level of ~ 330 GW (e) based on the reactors of the first stage, it produces ~ 18 % of electricity and replaces about 5 % of conventional fuels in terms of total expenses.

Although some fuel-deficit countries developed (France) or aspire to develop (Japan, Korea) the nuclear power which could provide their main demands in electricity, the demand for NPPs dropped in the world, on the whole, and, according to predictions, their contribution into the world electricity production will keep the current level or will be cut in the following decades.

At this level, within the framework of cheap uranium resources of somewhat higher than 10^7 t, the nuclear power will be able to develop till the middle of the century based on 235 U thermal reactors of conventional types.

At the same time predictions indicate on the doubling in fuel and trebling in electricity world demand by the middle of the century. It is likely that global problems of cheap hydrocarbonic fuel resources exhaustion and hazardous climatic changes due to release of combustion products will arise with aggravation of international relations.

Stabilisation of chemical fuel consumption becomes one of the crucial issues, and during their large-scale replacement nuclear fission has no alternative which could be considered now as realistic. The nuclear power growing by the middle of the century by an order of magnitude from the current level, i.e. to somewhat higher than 3 thousand GW (e), and till the end of the century - by a factor of 2-3, could decrease the accretion of the world consumption of chemical fuels, let's say, by a factor of 2.

A similar scenario can not be implemented by conventional 235 U reactors. It requires to return to the concept of the second stage of the breeding-based nuclear power, but certainly at a new level, learning all the lessons of the semi-centennial experience.

The 235 U reactors will produce ~ 10^4 t of fissionable Pu, and the necessity to utilise it together with Pu from the cut nuclear weapons which is actually 100 times less predetermines the choice for a further stage of the U-Pu fuel cycle and, therefore, fast reactors possessing in this cycle apparent advantages as compared to thermal reactors and any reactors in the Th-U cycle. The advantages of the fast reactors with liquid-metal cooling are high as applied to large NPPs for centralised production of electricity which is the major direction of development of the power engineering.

But, further, providing fast reactors with a small thorium blanket will enable also to produce 233 U for thermal reactors of different types which are preferable in small nuclear power plants for providing local needs in heat and electricity and for which the Th-U fuel cycle turns out to be preferable in terms of economics and safety.

The analysis shows that regardless of the existing stereotype, the liquid metal fast reactor in the U-Pu cycle simpler that LWR with respect to the physical and technical principals of design and control, and, taking into account more efficient use of fuel and heat, it must be cheaper.

For a higher cost of the first generation fast reactors there are no other reasons but the complication of the cooling systems, design, control, refuelling, NPP constructions when using Na. The

E. R. Merz and C. E. Walter (eds.), Advanced Nuclear Systems Consuming Excess Plutonium, 79–80.
© 1997 *Kluwer Academic Publishers.*

chemical activity of Na and small boiling margin, as well as low density and heat conductivity of oxide fuel do not allow to realise the safety potential of the fast reactor which is the most close to the ideal of deterministic safety with respect to its principles. The fast reactor allows to burn (to fission) all actinides which creates conditions for acceptable solution of the problem of high-active long-living radiowastes [2].

Study of the fast reactor with dense and heat-conductive mononitride fuel UN-PuN without U-blanket, with BR = CBR ~1, cooled by chemically passive high-boiling liquid lead (experience of naval reactors with Pb-Bi) allows to count on realisation of the fast reactors potential with respect to safety and economical aspects [1]. Decrease in the growth rates of the power engineering as compared to the post war decades and accumulation of high quantities of Pu by thermal reactors allow to refuse now the requirements of short doubling times of Pu, which earlier served one of the major causes of the choice of Na.

Danger of "crawling" the nuclear weapon during spreading in the world the up-to-date processing technology of irradiated fuel with Pu extraction is today one of the most serious barriers to a large-scale development of the nuclear power on the basis of breeding.

To avoid the use of the power technology for military purposes, a new technology of the closed fuel cycle will be required. This technology does not demand separation of the fuel components but for cleaning against fission products, and it is not capable of such separation. Another requirement is physical protection of fuel against thefts due to return of all actinoides in the composition of the main fuel into the reactor for their transmutation into fission products and during not very deep cleaning against FPs.

The fast reactors considered in [1] with CBR ~ 1 and without uranium blanket do not require division of U and Pu in any chains of the fuel cycle, and they create conditions for use of such simplified technology. It is likely that one will have to transfer from the chemical processing methods to another ones based on the difference in atomic weights or another properties of actinoides and FPs. However, if one could extract Cm, this would facilitate radiation conditions during fabrication of fuel.

During search and development of such technology of the fuel cycle, one can see the most important and nearest task of nuclear developments. The future of the nuclear power depends on this task. Its implementation could be much more productive than the current invention of various special "burners" of plutonium.

It is obvious that no new technology can prevent non-legal use of the current technology of Pu extraction from the irradiated fuel, as well as isotope enrichment of U for obtaining arm materials. This danger exists now, and it will rise in the next century due to production of ~ 10^4 t of Pu by the reactors of the first stage. Very small part of this Pu is enough to produce nuclear weapons. This danger can be avoided only by improvement of the security measures and political regime of non-proliferation. It is necessary to require from the new technology that it could not have been used for arm purposes, and its spread would not give use to the current danger of "crawling" the nuclear weapon.

What's more, involvement of Pu into the power engineering based on the fast reactors with the breeding ratio ~ 1 and with closing the fuel cycle at NPP will result in the decrease in the quantity of Pu being under processing and in storages with simplification of monitoring and reliable physical protection against thefts. The reactors themselves are the most protective "storages" of plutonium.

However, we can not avoid Pu extraction from fuel of thermal reactors, which is accumulated in storages, for production of fuel for first loading the fast reactors. This can be made in plants of fuel re-processing and mixed fuel fabrication (these plants exist in several nuclear countries), perhaps, in case of their expansion and upgrading.

Another more complicated way requires international agreements, foundation of centres under international jurisdiction. These centres would perform this work, as well as storage of excessive amount of Pu transferred to the centres by countries.

REFERENCES

1. E. Adamov, V. Tsikunov, A. Filin, V. Leonov, V. Orlov, V. Smirnov, A. Sila-Novitsky. Liquid Lead Cooled Fast Reactor Concept, Proc. of Int. Topical Meeting ARS'94, v. 1, p. 502. Pittsburgh, USA, 1994.
2. E. Adamov, I. Ganev, V. Orlov. Attainment of Radiation Equivalency in Nuclear Power Radioactive Product Management, Nuclear Technology, v. 104, N 2, 1993

The Cost of Weapons Plutonium Disposition Through MOX Utilisation in Existing Commercial LWRs

Kevin Hesketh

LWR Business Unit
BNFL Fuel Business Group
Springfields,
Preston,
Lancs,
PR4 OXJ,
England

NATO Advanced Research Workshop on
Advanced Nuclear Systems Consuming Excess Plutonium
October 13-16, 1996
Moscow

1. INTRODUCTION

The disposition of weapons grade plutonium has been a topical area for several years now. A number of reactor and accelerator based systems are potentially capable of removing weapons grade plutonium from stock. Depending on the system, the plutonium may be made less accessible, degraded so as to be unsuitable for weapons use, or simply destroyed. Political views differ, with the US wanting to achieve the "spent fuel standard" for weapons plutonium dispositioning. This is a concept to make weapons plutonium as unattractive and inaccessible as the plutonium in spent fuel from commercial reactors. Russia, however, sees weapons plutonium as an assest which should be utilised for power generation.

In many ways the simplest means of meeting both the above objectives is to use weapons plutonium in Mixed Oxide (MOX) fuel in Light Water Reactors (LWRs). This technology is already well established, with known and reasonable costs and there are enough LWRs world-wide that, in theory at least, 100 to 200 tons of weapons plutonium could be treated within one or two decades. There are, however, some practical limitations which would make such a rate of consumption difficult to achieve, not least of which is the lack of sufficient MOX fabrication capacity, and political constraints which might rule out a good fraction of the world's LWRs from consideration.

Advanced, fast flux reactors and accelerator systems have been proposed for weapons plutonium disposition. The theoretical advantage of these systems is that all plutonium isotopes are fissile in the fast spectrum, whereas only the odd-numbered isotopes are fissile in a thermal spectrum. Thus, either destruction or utilisation (whichever is

E. R. Merz and C. E. Walter (eds.), Advanced Nuclear Systems Consuming Excess Plutonium, 81–87.
© 1997 *Kluwer Academic Publishers.*

desired) can be achieved more efficiently. However, it is not clear that costs will be competitive and a lot of development is required. Developers must ensure that, in spite of this, their systems will be attractive compared with the LWR MOX base option.

This paper considers the cost of weapons plutonium disposition via LWR MOX as a baseline to guide the advanced system developers.

2. MOX STRATEGY

This section defines a MOX scenario to be used as a base against which the economics of advanced systems can be compared :

For the sake of discussion, let us define a target to aim for of removing 100 t of weapons plutonium from stocks by a date of 15 years from now, 2011, using MOX in existing thermal reactors. Even with full political commitment to achieve this goal, it would take at least 5 years before the necessary MOX fabrication capacity was available and the various LWR utilities were ready to accept it. Plutonium disposition cannot therefore begin at the necessary rate until at least 2001, and we should have to treat 10 t of plutonium per year thereafter. At current LWR fuel burnups (in the mid-40's GWd/t), MOX fuel with weapons plutonium would need to have about 4.5 w/o total plutonium content. This implies that the MOX fabrication and irradiation capacity would need to be about 220 t/year.

This amount of MOX, at mid-40's GWd/t burnup would generate roughly 10 GWy(e). However, because of the limitations of current Pressurised Water Reactors (PWRs) (they typically have a maximum MOX loading of around 30%), around 30 GWy(e) of reactor capacity would need to be made available for the weapons grade plutonium programme. PWRs are considered here because they are the only thermal system in which commercial scale MOX usage is demonstrated. Boiling Water Reactors (BWRs) may have theoretical advantages for MOX applications and may even be able to utilise 100% MOX cores, but are discounted here because there is much less experience of MOX irradiation in them, and this effectively reduces the total available capacity. Of course, a number of PWRs are already committed to using MOX derived from civil plutonium and there are plans for BWRs to do the same, which reduces the reactor capacity potentially available for a weapons plutonium disposition programme. Even so, a more than sufficient number remain available to meet any reasonable requirement.

The world installed capacity of PWRs amount to more than 225 GW(e). Not all of these LWRs would be suitable for consideration as part of a plutonium consumption programme; on the timescale under consideration some of the existing reactors would have reached the end of their useful lives. Nevertheless, the capacity required to achieve the 100 t Pu target by 2011 is only a modest fraction of that potentially available. Additionally, the Russian designed VVER-1000's are also suitable technically for MOX burning which further increases the capacity potentially available. Setting aside the possibly difficult political issues, the strategy of utilising existing LWRs to consume weapons plutonium would therefore be perfectly feasible logistically and technically.

3. ECONOMIC BASELINE

If the above MOX strategy was to be pursued, it would be necessary to obtain the agreement of all the utilities involved and it would be reasonable for them to expect financial compensation in respect of any extra fuel, operating, design and licensing costs that might be incurred over and above the costs of operating their plants with conventional UO_2 fuel. Additional guarantees may also be necessary to cover any financial risks associated with commercial irradiation of ex-weapons plutonium. The purpose of this section is to estimate the level of compensation might be required in order to make participation in the programme acceptable to a utility which would otherwise have no direct interest. This will define an economic base against which any advanced plutonium burning systems should be compared. It is not necessarily the case that advanced systems should match this base cost, but any advanced systems with an estimated cost considerably higher than the base should perhaps be considered unrealistic unless they can demonstrate very clear technical or commercial advantages.

The economic assessments which follow are mainly expressed in terms of the costs for a single utility with an annual output of 1 GWy(e). The costs for a larger utility might be lower, because there might be shared items such as design and licensing that might not need to be duplicated at each plant. The overall costs of the programme are later obtained by multiplying the single utility costs. Some of the economic data presented are necessarily subjective ; the issues involved are sometimes commercially sensitive and it is not always possible to be too specific or too precise. Nevertheless, it is hoped that the overall estimate is a useful reference point for this Workshop.

There are six main areas where costs would be incurred :

3.1 Plutonium Acquisition Costs

It may not be appropriate to assume that the weapons plutonium will be available free issue, as is generally assumed to be the case for reactor grade plutonium (since reactor grade plutonium generally belongs to the utility). Certainly Russia considers weapons plutonium to be a valuable strategic and economic resource and might wish to be compensated appropriately.

A reasonable comparison might be made between the resource value of 1 kg of MOX fuel containing 4.5 w/o total plutonium and 1 kg of UO_2 fuel at 4.5 w/o U-235 enrichment. At the projected world market prices for uranium ore and enrichment (range 60 to 80 $/kg ore, 100 $/SWU) applicable between 2001 and 2011, this would range from 1500 to 1700 $/kgHM. Not all this is realisable, however, because the fuel fabrication cost for MOX fuel tends to be higher than that for UO_2. For the purposes of this paper, it will be conservative if we take a low range estimate for this differential. Following the OECD Fuel Cycle Economics Report[1], the minimum differential that might reasonably be expected is around 600 $/kgHM. Therefore the realisable value of 1 kg of MOX fuel could be no higher than 1100 $/kgHM. At 4.5 w/o total plutonium content, this corresponds to a maximum realisable value of 24000 $/kgPu.

[1] The Economics of the Nuclear Fuel Cycle, OECD Publications, 1995

The actual price to be paid for the plutonium might vary from zero up to the 24000 \$/kg figure in the worst case scenario. For an annual plutonium consumption of 10 t, the upper limit would be \$240 m/year, or \$2.4b over the full 10 years. It is appropriate to assume that the sponsoring body would pay this cost, and not the utilities.

3.2 Plutonium Metal/Oxide Conversion

The first requirement is to convert the plutonium metal pits into PuO_2 powder suitable for feeding into MOX fabrication plants. Assuming pessimistically a wet process would be needed, a plant with a 10 t/y throughput of plutonium is estimated to cost in the region of \$75m/year to run, including capital recovery over 10 years, operating and decommissioning costs. This estimate is based on experience gained with existing fuel cycle back-end facilities with similar processes and comparable throughputs. The corresponding unit cost is \$7500/kg Pu. Assuming that the MOX fuel would need around 4.5 w/o plutonium feed, this corresponds to a rounded \$340/kgHM of finished MOX fuel.

For the full 100 tPu programme, the metal conversion cost works out at \$750m. Again, the sponsoring body is assumed to pay this cost.

3.3 MOX Fabrication Price

The cost to an utility of procuring MOX fuel assemblies is dominated by the fabrication price; other costs including that of procuring the depleted uranium carrier and transport are small relative to fabrication and are only of minor significance for the purposes of this paper. The major capital expenditure in a programme of weapons grade plutonium disposition would be the construction of a MOX fabrication plant (or plants) with a combined capacity of up to 250 t/year. With the obvious commercial considerations that apply to existing and planned MOX fabrication plants, it is not possible to quote construction or operating costs. In any case, these will depend on where the plants were built and their specifications. An alternative approach is to use MOX fabrication prices in terms of \$/kgHM. These are set by the MOX fabricators at a level which will recover their capital and operating costs and also provide a return on investment.

As a guide to the range of MOX fabrication prices that should be assumed here, the OECD Fuel Cycle Economics Report provides the most authoritative guide. What is relevant here is the price that may apply over the period 2001 to 2011. The OECD report presents a sensitivity analysis with MOX fabrication prices ranging from 800 \$/kgHM to 1400 \$/kgHM, with the lower end of the range perhaps applying when MOX is better established commercially. For the purposes of this paper, it is preferable to choose the upper limit figure of 1400 \$/kgHM in order to be conservative.

This fabrication price is based on existing plants for the fabrication of PWR MOX assemblies from reactor grade plutonium. It may not be strictly valid for a plant processing weapons grade material, since meeting criticality constraints and other considerations may result in an additional cost for the operators. On the other hand, radiological protection would be less of an issue, which might mitigate the costs somewhat. However, there may also be other considerations which might affect the prices quoted, such as the fact that there would be a limited lifetime for the plants (since there is only a finite supply of weapons plutonium available and a definite target date for its use), and also there might be a risk of political factors changing that might result in the plant being unused or only partly used. For these reasons, potential operators of such a plant might demand an additional premium on fabrication. This is why the 1400 $/kgHM figure is taken here as the starting point, and for the various reasons just discussed, it is considered prudent to increase it, arbitrarily, to 2000 $/kgHM.

For a 1 GWy(e) utility, the annual MOX fabrication requirement of roughly 8 tHM (ie. 1/3-core MOX loading with an annual fuelling requirement of 25 tHM) would cost $16m. For the full 100 tPu programme, the total fabrication cost would be $4.4b, to be borne by the utilities.

This does not represent the net cost to the utility, however, because there are other factors to consider, the most important of which is the avoided cost of procuring the UO_2 assemblies which the MOX assemblies displace :

3.4 UO_2 Avoided Costs

The MOX fuel displaces UO_2 fuel that would otherwise need to be paid for. The total cost of a UO_2 assembly depends on the enrichment, the price of uranium ore, conversion and enrichment, with fabrication as a minor component. For the discharge burnups expected in the period 2001 to 2016, U-235 enrichments of at least 4.5 w/o will be required. With the projected ore and enrichment prices, the UO_2 costs range from around 1700 to 2000 $/kgU, or between $3.7b and $4.5b for the complete 100 tPu programme.

The UO_2 avoided costs offset most or all of the MOX fabrication costs. There are credible scenarios where the MOX options would give a utility a direct economic benefit. But there are also other factors which need to be taken into account :

3.5 Licensing & Operational Costs

The cost of establishing a licensing case for the use of MOX fuel in a particular utility's reactor is a significant item in absolute terms, possibly ranging from $1m to a few $m. This is a one-off cost for each utility, however, which may well be shared between several reactors for a multi-unit utility. In relation to the MOX procurement costs over the full 10 year period, such costs are negligible and can safely be disregarded here.

There are operational areas which may be affected by the use of MOX, such as fuel handling, fresh and spent fuel storage, fresh and spent fuel transport, security etc. Generally such costs are similarly small in relation to fuel procurement and can again be neglected.

3.6 Utility Inducements/Risk Compensation

It would most certainly be the case that utilities would need to be given a positive inducement to accept ex-weapons plutonium MOX into their plants. The use of MOX fuel potentially does impose an additional administrative burden on the management to deal with such issues as licensing, safeguards, radiological protection and public relations. An utility might also regard MOX from weapons plutonium as a potential financial risk because at any time political intervention might affect the availability of the plutonium and the fuel and this might force an utility to take strategic measures to guard against any loss of energy output.

For all these reasons, utilities may be expected to demand a degree of subsidy from the international/governmental bodies responsible for plutonium disposition. What would be a reasonable subsidy that would ensure utilities' cooperation ? Arbitrarily, a figure of 20% of the MOX fabrication cost is suggested here as a level which would be attractive to utilities.

Table 1 summarises the cost ranges obtained here, and identifies who would pay for each item. The bottom line is the net cost to the sponsoring bodies. At the low end of the range, the plutonium is assumed to be available at no cost. The sponsoring body pays the whole of the metal conversion cost and a subsidy to the utilities of 20% of the MOX fabrication price. At the high end of the range, the sponsoring body is assumed to pay the full market value of the plutonium in addition.

In terms of order of magnitude, these estimates are comparable with estimates recently published by the USDOE[2]. This gives a range of \$1.78b to \$2.09b as the cost of irradiating 50 t of ex-weapons plutonium in LWRs, with an avoided uranium benefit of \$1.39b to \$2.01b. This adds confidence that the estimates made here are a reasonable to use as a baseline for advanced systems.

4. DISCUSSION

Weapons plutonium disposition as MOX fuel in thermal reactors would be an entirely feasible strategy which accords with the objectives of all the interested parties. MOX technology is already established commercially, though not specifically for weapons plutonium. There are a number of detailed technical issues that would need to be addressed, such as the metal conversion technology, modified criticality requirements in the MOX fabrication plants and the impact of using high quality plutonium on the neutron physics characteristics of LWR cores. All these issues are resolvable, given the will to proceed.

The bottom line in Table 1 provides a cost base against which advanced systems for plutonium disposition should be compared. Even if the high range estimate of \$4b applies, it is difficult to see how any advanced system could go through the development, prototype and production implementation phases at a lower cost than the MOX strategy. Moreover, with any advanced systems it would be impossible to

---

[2] Technical Summary Report for Surplus Weapons-Useable Plutonium Disposition, USDOE, July 1996

achieve a commercial scale scenario in 15 years. Any advanced systems must therefore demonstrate clear and substantial technical and/or economic advantages.

The limitation of the MOX-based approach is that only the odd-numbered plutonium isotopes are fissile in a thermal flux. It may be difficult to justify an advanced reactor system that improves on this performance but fails to offer corresponding economic advantages over the LWR MOX option.

It is hoped that this paper will help to focus attention on the need to develop a robust case for advanced systems.

Table 1

Estimated Cost Ranges for a 100 tPu Disposition Programme

Item	Total Cost ($m)	Who Pays
Pu Acquisition	0 to 2400	Sponsoring Body
Pu Metal Conversion	750	Sponsoring Body
MOX Fabrication	4400	Utilities
UO_2 Avoided Costs	-3700 to -4500	Utilities saves
Licensing/Operational Incremental Costs	~ 0	Utilities
Net Cost of MOX Programme to Utility	**-100 to 700**	Utilities
Subsidy/Inducement to Utilities (20% of MOX fabrication)	880	Sponsoring Body
Net Cost to Sponsoring Bodies	**1630 to 4030**	Sponsoring Body

A REALISTIC PLUTONIUM ELIMINATION SCHEME WITH FAST ENERGY AMPLIFIERS AND THORIUM-PLUTONIUM FUEL

C. Rubbia, S. Buono[1], E Gonzalez, Y. Kadi and J.A. Rubio

Abstract

In a previous report [1] we have presented the conceptual design of a sub-critical device designed for *energy amplification* (production). The present note further explores the possibilities of the Energy Amplifier (EA) in the field of the *incineration* of unwanted actinide "waste" from Nuclear Power Reactors (PWR) and from the disassembly of Military Weapons.

The key idea which is developed is the one of using a Thorium-Plutonium mixture which is much more effective in eliminating Plutonium at acceptable concentrations (≤ 20%) than the conventional mixture of Uranium-Plutonium. The device operates as an effective Plutonium to ^{233}U converter. The latter can be later mixed with ordinary or depleted Uranium and it constitutes an excellent fuel for the PWRs. The EA sub-critical mode is preferred over the conventional Fast Breeder Reactor, because of the much smaller risks associated to the narrow criticality window (approximately ± 0.15 % in Δk) of a Fast Reactor, the negative void coefficient related to the presence of Plutonium and the inevitable presence of large amounts of highly toxic Plutonium.

It is shown that a cluster of EAs operated in conjunction with existing PWRs is a very effective and realistic solution to the ultimately complete elimination of the accumulated Plutonium and Minor Actinide stockpiles and it greatly alleviates the problem of definitive geologic disposal. Preliminary economical considerations show that Plutonium incineration when compared to direct geological disposal is not only environmentally more acceptable but also an economically profitable alternative.

Geneva, 12th December, 1995

---

[1]Sincrotrone Trieste, Trieste, Italy

E. R. Merz and C. E. Walter (eds.), Advanced Nuclear Systems Consuming Excess Plutonium, 89–134.
© 1997 *Kluwer Academic Publishers.*

1.- General considerations

The linearly growing accumulation of Plutonium and higher Actinides (Am, Cm, Cf, etc.) due to civil Nuclear Power and the stockpile of bomb-grade Plutonium from Military Arsenals are a cause for serious concern [2]. The projected situation in the year 2010 for the present civil nuclear power capacity is given in Table 1.1, taken from Schapira [3]. The surplus from destroyed nuclear warheads is somewhat uncertain. We will base our considerations on Ref. [3] and assume the amount of military grade Plutonium which needs to be eliminated is about 100 tons out of a total of about 300 tons, of which about 180 tons still in warheads.

Many schemes have been proposed in order to dissipate such a major liability to mankind. Underground geologic disposal is the only realistic solution so far retained for implementation and it begins to be deployed[1] in some countries. But it is not without concern and is a subject of strong public opposition, mainly in view of the very long period of perfect retention required, generally in excess of millions of years. Other schemes, like for instance sending the offending materials in space have been proposed but they do not seem practical in view of the large amounts of material involved and the risks of a crash of the vectors[2].

Studies on alternative methods based on nuclear transformation of the reprocessed fuel have gained a considerable momentum [3]. The majority of these

Table 1.1. *- Projected amount of Radioactive waste from ≈ 400 GWe Nuclear Reactors (world-wide) in year 2010. [3]*

Total spent fuel	300,000	tons
Plutonium isotopes	3000	tons
Neptunium isotopes(^{237}Np)	140	tons
Americium and higher Actinides	120	tons
Long lived Fission fragments		
^{99}Tc	250	tons
^{135}Cs	90	tons
^{129}I	60	tons

---

1 In some countries, deep underground disposal is forbidden by law. In others, only "domestically produced" waste can be disposed. For more details see [2].

2 The reliability of a normal launch now is estimated to be about 95%. The manned mission reliability is about 98%.

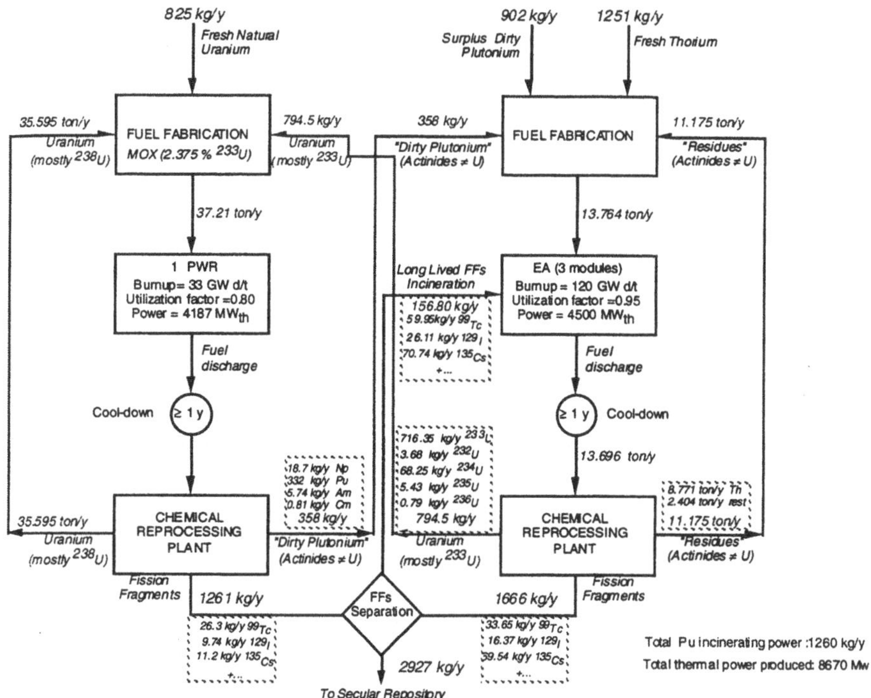

Figure 1.1 *General scheme of coupled EA-PWR. The mass flows refer to one year of operation. The power of the PWR is adapted to the figures for a single EA complex, as described in Ref. [1].*

schemes are based on the idea of "incineration" in which neutrons or other particles induce nuclear reactions which transform the bulk of the unwanted materials into more acceptable nuclear species. In the case of Actinides the leading processes are fissions of fissionable elements and transformation of poorly fissionable nuclides into fissionable ones by neutron capture(s). In the case of long-lived non-fissionable radio-nuclides, like for instance Fission Fragments (^{99}Tc, ^{129}I, etc.) in general, methods proposed transform long-lived into short-lived or stable species with a neutron capture (e.g. $^{99}Tc[2.1\times10^5 y]+n \rightarrow \gamma + ^{100}Tc[15.8\,\mathrm{sec}] \rightarrow \beta^- + ^{100}Ru[stable]$). These captures give particularly spectacular effects with Fission Fragments since nuclei which are the outcome of fission are already "neutron rich" and an extra neutron has a strong de-stabilising action.

Many schemes based on critical Reactors and on spallation target driven sub-critical devices have been discussed over the last several years [4]. The present proposal is directly derived from the Energy Amplifier (EA) of which we have recently released a conceptual design [1] in which a first description of these ideas has been already given. It is based on a coupled operation of an (existing) PWR and of a specially operated EA. We propose a fuel load for the EA specifically designed to achieve an effective incineration and made of a mixture (MOX) of the unwanted Plutonium plus *all* other associated higher Actinides[3], *imbedded in a Thorium matrix.* The presence of the Thorium, instead of the usual Uranium in the mixture eliminates the ^{239}Pu producing chain $^{238}U + n \rightarrow ^{239}U \rightarrow ^{239}Np \rightarrow ^{239}Pu$, replacing it with the ^{233}U producing $^{232}Th + n \rightarrow ^{233}Th \rightarrow ^{233}Pa \rightarrow ^{233}U$. While the former chain is producing additional ^{239}Pu, thus reducing the efficiency of incineration, the latter produces ^{233}U which can be reused and fully burnt in a Light Water Reactor once mixed with ^{238}U[4].

In short (Figure 1.1) the EA acts as a Pu $\rightarrow$ ^{233}U transformer and the PWR burns the ^{233}U, producing a smaller amount of Pu which is re-injected in the EA. The system PWR-EA operates indefinitely in a closed fuel cycle with an intermediate Fuel Reprocessing and Fuel Manufacturing phase. The cycle is fed externally by additional Pu, either from civil or military applications and smaller amounts of fresh Uranium and Thorium to compensate the amounts which are burnt and it discharges the Fission Fragments (FFs), which have a much shorter activity lifetime than the Actinides. Some specific long lived FFs (^{99}Tc, ^{129}I, etc.) can also be incinerated re-injecting them in the EA[5]. The net results [1] are (1) no more Actinides to be disposed in a Geologic repository and (2) the (ingestive) radio-toxicity of the FFs is after some 600 years in a "Secular Repository" about 10^4 times smaller than in the case of an open PWR cycle for the same produced energy. These discharged radio-elements have a negligible amount of α-activity and have a residual radio-toxicity which matches the best expectations of Fusion [1].

In order to illustrate the advantages of the Thorium-Plutonium mixture over the conventional methods based on Uranium-Plutonium we show in Figure 1.2 the Pu elimination efficiency as a function of the relative concentration of Pu in the Uranium-Plutonium fuel of a CAPRA scheme [5]. At a relative concentration of the order of 15%, the newly produced and the incinerated Pu exactly balance with no net improvement of the Pu stockpile. While appreciably below such a critical concentration the Reactor is a "Breeder", for higher concentrations it becomes an

---

3 We underline that in our method all Actinides, with the exception of the Uranium, are incinerated together. In the majority of the studies in the literature most often specific elements (e.g. Am, Np, Pu etc.) are separately targeted.

4 As pointed out later on, this is one of the many possibilities, which are application dependent. For instance the produced ^{233}U could be used instead as (initial) fuel for another, new EA.

5 Evidently by mass conservation the amount of fresh Th, U and Pu to be injected in the cycle is equals to the mass of the produced FFs, plus a significant mass due to the mc^2 term in the mass to energy transformation.

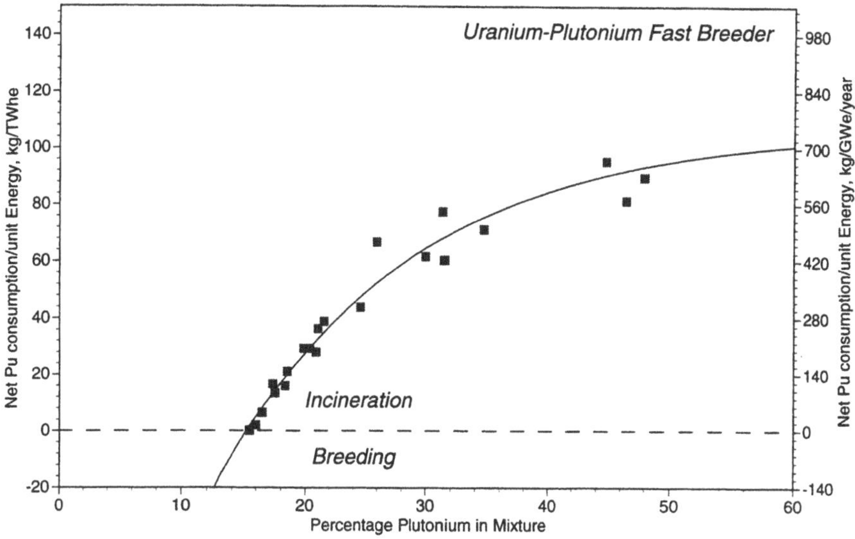

Figure 1.2: Plutonium elimination efficiency as a function of the relative concentration of Pu in the Uranium-Plutonium fuel for a fast Breeder. (Source: CEA, unpublished)

"Incinerator". The incineration efficiency becomes dominant only at Pu concentrations which are very high: evidently the best performance is achieved with a pure Pu Reactor.

Such very high concentration of Pu fuel has considerable problems since (1) its bare critical mass is very low (2) its reactivity is choked by the emergence of Fission Fragments after a small relative change of the Pu concentration[6] and (3) the Fast Breeder is harder to operate than with its standard fuel because of the increased negative void reactivity coefficient[7] (due to the high Plutonium concentration) and the reduced Doppler effect. Dissolving the unwanted Pu instead in a fertile Thorium matrix greatly alleviates such problems and a major relative fraction of the inserted Pu even at acceptable concentrations can be burnt without interventions or manipulations of the fuel, since the bred ^{233}U is an effective substitute to Pu in order to maintain a viable and constant criticality [1].

---

6 In a pure Plutonium fuel, fissioning about 10% of the mass produces fission fragments concentration sufficient to reduce the multiplication coefficient k by about 7%. Hence one is led to very frequent refills and manipulations. At the limit one is led to the "chemistry on-line" proposed by the Los Alamos group [6].

7 Expressed as $\Delta k/(\Delta\rho/\rho)$, i.e. a decrease in coolant density arising from the formation of voids in the core of a Fast Reactor produces an increase in reactivity.

Many variations besides the scheme of Figure 1.1 are of course possible. More generally, the ^{233}U is easily recovered chemically[8] at the end of the Thorium-Plutonium cycle in the EA and it constitutes a valuable product, since for instance

(1) It can be mixed with depleted or ordinary Uranium[9] to produce additional enriched fuel for PWRs without isotopic separation (Figure 1.1) or

(2) It can be re-injected in the EA in order to increase significantly the neutron inventory, as required in order to incinerate the long lived Fission Fragments (FFs), like for instance ^{99}Tc, ^{129}I, etc., produced both by the PWR and the EA. It is well known that the main problem with incinerating FFs is the availability of the appropriate fraction of the produced neutrons which must be sacrificed to this task.

(3) It can be used as "seeds" to develop a line of EAs purely operated on Thorium, as described in Ref. [1].

We believe that the fast EA coupled with the Thorium-Plutonium mixture is the winning formula and this for very specific reasons:

(1) It is well known that *fast neutrons* are more effective than thermal neutrons in incinerating Actinides from the discharge of a thermal PWR since the fission process is more abundant at high neutron energies and generalised to the majority of the elements involved in the incineration

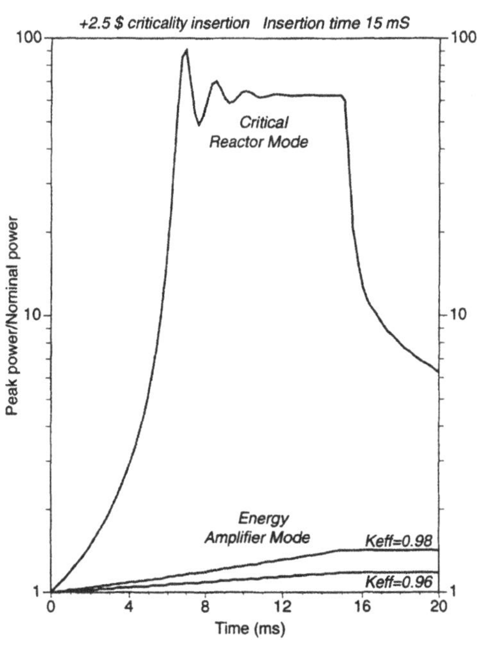

***Figure 1.3: Effect** of a fast (15 ms) 2.5 $ reactivity insertion in the device of Ref. [1], operated as a Fast Breeder or as a sub-critical device, with two possible multiplication coefficient values, namely k = 0.98 and k =0.96. The magnitude of the reactivity insertion corresponds roughly to the accidental extraction of all the control bars.*

8 ^{233}U is the dominant Uranium isotope, the rest being made of small amounts of ^{232}U, ^{234}U and ^{235}U. The effects of the activity of the ^{232}U descendants (^{208}Tl) are discussed further on.

9 Mixing with ^{238}U of course "denatures" the mixture and inhibits military diversions of the newly produced fuel.

chain. Therefore a fast neutron stage in the incineration cycle appears to us as almost a necessity.

(2) A *sub-critical operation,* namely an Energy Amplifier, offers considerable safety and operational advantages over a critical Fast Reactor. We show in Figure 1.3 taken from Ref. [1] the effect of an accidental fast reactivity insertion for a critical and a sub-critical operation. A significant fast accidental reactivity insertion ($\Delta k/k \approx 7 \times 10^{-3}$), of the order of magnitude of the full extraction of the control bars will cause a major, catastrophic power excursion, but in the case of the EA the effect is acceptable with mild consequences. The curves of Figure 1.3 have been calculated for the geometry of the EA of Ref. [1]. Rief and Takahashi [7] have reached the same conclusion in the case of a Sodium cooled Fast Breeder. The indicated reactivity insertion will produce the instant vaporisation of ≈ 85 % of the Sodium coolant in the critical case and no appreciable consequence for the sub-critical device.

(3) As already pointed out, *Thorium* matrix has the advantage of producing far less Plutonium than the Uranium. In the present scheme we have introduced Thorium only in the fast neutron part of the (closed) fuel cycle, since we would like to make a minimum number of changes in the thermal PWR operation. However the introduction of Thorium also in the thermal neutron segment is beneficial since it reduces the production of Plutonium in the burning of the ^{233}U. This point is further discussed in paragraph 5.

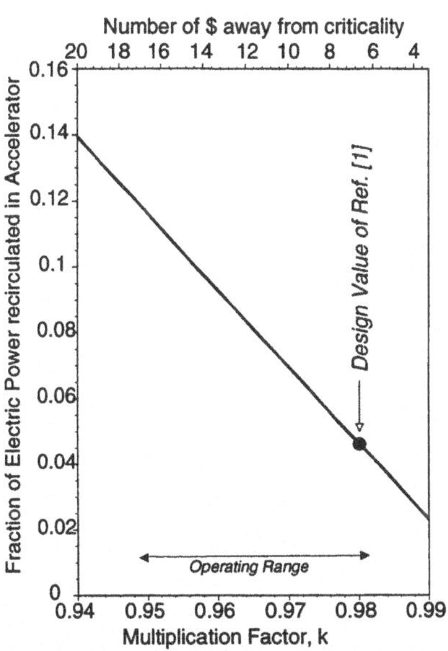

Figure 1.4 *Fraction of the total produced electric power recirculated in the Accelerator, as a function of the value of the multiplication coefficient* k. *A reasonable range of choices for the operating range of* k *is also indicated.*

It is evident that the added safety of operation offered by the EA is of particular value in this application in which one has necessarily to deal with large amounts of highly hazardous Plutonium. As an added safety it may be advisable to operate the Thorium -Plutonium mixture at a lower value of k than k = 0.98 which was chosen for the pure EA in Ref. [1]. For instance a value[10] of $k = 0.95 \div 0.96$ will ensure that during the very long, unperturbed incineration cycle (120÷200 GW × day/t) the EA behaves as safely as a "beam dump". The main drawback of a lower k is an increase of the power to be recirculated through the accelerator, which however remains within acceptable limits, as shown in Figure 1.4.

Plutonium or other Actinides complete elimination through fissions produces a very large amount of energy, namely 940 MW × day for 1 kilogram. In practice the energy yield is higher since for instance fissions occur in the Thorium matrix (few per cent) and in the ^{233}U bred in the Thorium. As shown later on this implies some 30% additional energy production, and therefore the energy produced by the incineration of 1 kg of unwanted Actinides is about 1200 MW × day. Typically a nominal 1.0 GW-electric PWR produces about 900 GW × day of thermal energy yearly and a total "Dirty Plutonium" waste (N, Pu, Am, Cm, Cf etc.) of 271 kg. Hence "incineration" of such a waste will inevitably lead to the production of some 271 × 1.200 = 325.2 GW × day, namely 325.2/900= 36.13 % of the power produced by the initial PWR, close to the "theoretical" limit in which only the Plutonium is burnt, namely 271 × 0.940/900 =28.30 % of the PWR power. The energy accumulated in the Actinide "waste" and which can be recovered with our method is therefore very large, since it amounts to about 1/3 of all the nuclear power produced so far. In particular an EA-incinerator will eliminate the Actinide waste at the rate produced by a PWR with only about 1/3 of the installed power.

Since only a fraction of the neutron interactions produce fissions in each reaction, in order to reach complete elimination, the fuel must undergo a long chain of nuclear transformations each with a specific, variable degree of fission probability. These reactions are sustained by the neutrons coming from the fission reactions. It is a fortunate circumstance that in our scheme the number of such neutrons is sufficient to complete the job throughout the evolution, essentially at a constant criticality [1]. As shown in Ref. [1] after a sufficiently long time an equilibrium situation is reached between the constant inflow of material to be incinerated and the reaction chain with the fission elimination channel.

A number of simplifying assumptions permits to calculate analytically the essential features of the evolution of the concentration vector $c_{(A,Z)}(\phi)$, where ϕ is the (average) neutron flux. We ignore the discontinuity of the refills and assume a constant inflow of the father element and neglect the (n,2n) and other channels which may introduce "loops" in the (A,Z) evolution plane. We assume that in the presence

---

[10] We recall that in the case of a Geologic Repository (for very long periods) the value $k \leq 0.95$ has been deemed acceptable.

of the neutron flux ϕ, for all elements there is only one transformation channel (either with neutron capture averaged cross section $\sigma_{capt}^{(i)}$ or radioactive decay with decay rate $\lambda^{(i)}$, whichever is dominant) and a dissipative, fission channel with spectrum averaged cross section $\sigma_{fiss}^{(i)}$. For very high values of A spontaneous fission and other forms of nuclear instability will contribute to such dissipative terms. The rate of transformation in a neutron flux ϕ is $\phi\sigma$ and the total rate $\mu^{(i)} = \phi(\sigma_{capt}^{(i)} + \sigma_{fiss}^{(i)})$ or $\mu^{(i)} = \phi\sigma_{fiss}^{(i)} + \lambda^{(i)}$ if the transformation is decay dominated. The survival, chaining coefficient, which represents the probability of continuation to the next step of the evolution chain is defined as $\alpha^{(i)} = \sigma_{capt}^{(i)} / (\sigma_{capt}^{(i)} + \sigma_{fiss}^{(i)})$ or $\alpha^{(i)} = \lambda^{(i)} / (\lambda^{(i)} + \sigma_{fiss}^{(i)} \times \phi)$ respectively. The procedure is schematically shown below:

Chain	$P \rightarrow N_1 \rightarrow$ $\downarrow$	$N_2 \rightarrow$ $\downarrow$	$N_3 \rightarrow$ $\downarrow$	$N_i \rightarrow$ $\downarrow$
Initial amount	$N_1(0)$	0	0	0
Removal rate	$\phi\sigma_{fiss}^{(1)}$	$\phi\sigma_{fiss}^{(2)}$	$\phi\sigma_{fiss}^{(3)}$	$\phi\sigma_{fiss}^{(i)}$
Transfer rate	$\phi\sigma_{capt}^{(1)}, [\lambda^{(1)}]$	$\phi\sigma_{capt}^{(2)}, [\lambda^{(2)}]$	$\phi\sigma_{capt}^{(3)}, [\lambda^{(3)}]$	$\phi\sigma_{capt}^{(i)}, [\lambda^{(i)}]$
Survival coeff. $\alpha^{(i)}$	$\frac{\sigma_{capt}^{(1)}}{\sigma_{capt}^{(1)} + \sigma_{fiss}^{(1)}}$	$\frac{\sigma_{capt}^{(2)}}{\sigma_{capt}^{(2)} + \sigma_{fiss}^{(2)}}$	$\frac{\sigma_{capt}^{(3)}}{\sigma_{capt}^{(3)} + \sigma_{fiss}^{(3)}}$	$\frac{\sigma_{capt}^{(i)}}{\sigma_{capt}^{(i)} + \sigma_{fiss}^{(i)}}$
Total rate $\mu^{(i)}$	$\phi\sigma_{fiss}^{(1)} + \lambda^{(1)}$	$\phi\sigma_{fiss}^{(2)} + \lambda^{(2)}$	$\phi\sigma_{fiss}^{(3)} + \lambda^{(3)}$	$\phi\sigma_{fiss}^{(i)} + \lambda^{(i)}$

Assuming first no refill ($P = 0$) and an initial number of nuclei $N_1(0)$, the time evolution is given according to the Bateman equation (i >1):

$$N^{(i)}(t) = N^{(0)}(t)\left(\prod_{j=1}^{j=i-1} \alpha^{(j)}\right) \times \left[\left(\prod_{j=1}^{j=i-1} \mu^{(j)}\right) \times \sum_{j=1}^{j=i} \frac{\exp(-\mu^{(j)} t)}{\prod_{\substack{k=1 \\ k\neq j}}^{k=i} (\mu^{(k)} - \mu^{(j)})}\right]$$

If alternatively, there is refill at the constant rate P per unit time and no initial nuclear sample, i.e. $N_1(0) = 0$, the formula becomes (i > 1)

$$N^{(i)}(t) = P\left(\prod_{j=1}^{j=i-1} \alpha^{(j)}\right) \times \left[\left(\prod_{j=1}^{j=i-1} \mu^{(j)}\right) \times \sum_{j=1}^{j=i} \frac{1 - \exp(-\mu^{(j)} t)}{\mu^{(j)} \prod_{\substack{k=1 \\ k\neq j}}^{k=i} (\mu^{(k)} - \mu^{(j)})}\right]$$

In practice, both refilling and initial nuclei are present and the actual number of nuclei will be simply the sum of the two above terms. Note that for a power density

$\rho \approx 100$ W/g, $\phi \approx 5 \times 10^{15}$ cm^{-2} s^{-1} and that the sum of cross sections is of the order of magnitude of $\approx 2 \times 10^{-24}$ cm^2, leading to an evolution time constant $1/\mu^{(i)}$ of the order of $\approx$ 3 years.

The asymptotic distribution is reached at the limit $t \to \infty$. At this stage the process is dominated by the refill term P and one can easily calculate the equilibrium amounts:

$$N^{(i)}(t \to \infty) = P \frac{\left(\prod_{j=1}^{j=i-1} \alpha^{(j)} \right)}{\mu^{(i)}} = N^{(1)}(t \to \infty) \frac{\mu^{(1)}}{\mu^{(i)}} \left(\prod_{j=1}^{j=i-1} \alpha^{(j)} \right)$$

The time required by $N^{(i)}$ to grow to $N^{(i)}(t \to \infty)(1 - 1/e)$ is approximately given by $\sum 1/\mu^{(j)}$ where the sum is extended up to i. Since the order of magnitude of the time constant is typically 3 years, equilibrium is reached after $\approx$ 3 (i-1) years where we have used i-1 to take into account that the step through the ^{233}Pa is fast. The fast decrease of $N^{(i)}(t \to \infty)$ with the rank in the chain is due to the product of the $\alpha << 1$ terms. To a fast decreasing degree of concentrations, the whole table of elements is eventually involved. As already pointed out, in practice the chain is not open-ended since spontaneous fissions and other instabilities ensure very small α–values toward the end. Fast neutrons are highly preferable over thermal neutrons, since the fission yield is much larger and opened to more elements.

Specific scenarios can be visualised. Evidently the EA incineration facility will be added to existing PWRs only many years after the beginning of operation. Therefore both the already accumulated Plutonium stockpile and the continuous production will have to be eliminated. Marchetti [8] has shown that the implantation of Nuclear Power in the western world follows an "epidemic equation" of the type $dN = aN(\overline{N} - N)dt$, namely the number of new adopters dN during time dt is proportional to the number of actual adopters N times the number of potential adopters $(\overline{N} - N)$, where $\overline{N}$ is the final number of adopters. Integrating the equation and replacing $\overline{N}(F = N / \overline{N})$ we get the solution $\log(F/1-F) = at + b$, where the time constant $\Delta T = 4.39/a$ is the time to go from F $\approx$ 0.1 to F $\approx$ 0.9 and the central date is $T_o = -b/a$. For the Western World T_o = 1981, ΔT = 19 years and a saturation value of 315 GWe. Parameters of the Marchetti's fit are listed in Table 1.2. By now all entries have reached in practice the asymptotic value, i.e. the market "niche " is saturated. The useful lifetime of the PWR is assumed to be 40 years. The nuclear energy supply from existing Reactors will then come to an end as shown in Figure 1.5. For instance in order to maintain the total nuclear power at its constant value, new power sources are required with T_o^{EA} = 1981+ 40 = 2021. The amount of accumulated Plutonium waste is shown in Figure 1.6.

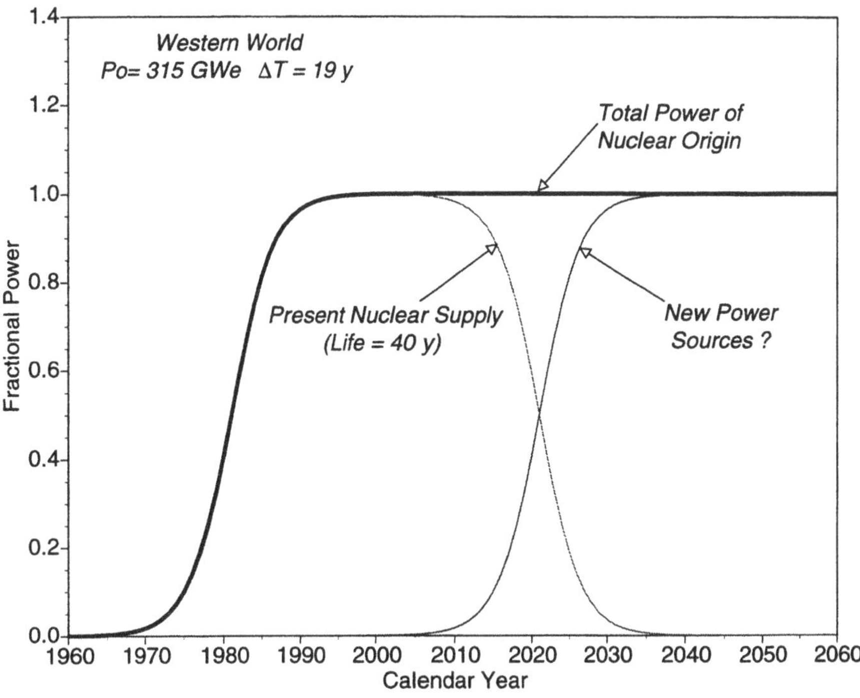

Figure 1.5: *Available power from existing nuclear reactors in the Western World and future evolution under the assumption of a 40 years lifetime for the power plants. The required new power sources in order to maintain the total power constant is also shown.*

Table 1.2.- *Parameters of the penetration of Nuclear Power in World nations according to Marchetti [8]. Values refer to the "Epidemic Equation".*

Country	$\bar{N}$ (GWe)	ΔT (years)	T_o (50%)
US(1)[11]	56	8	1974
US(2)	52	6	1985
USSR	70	20	1986
Sweden	12	14	1980
France	70	12	1983
Japan	20	15	1980
FRG	27	18	1983
Canada	16	20	1982
Western World	315	19	1981

[11] US exhibits two cycles.

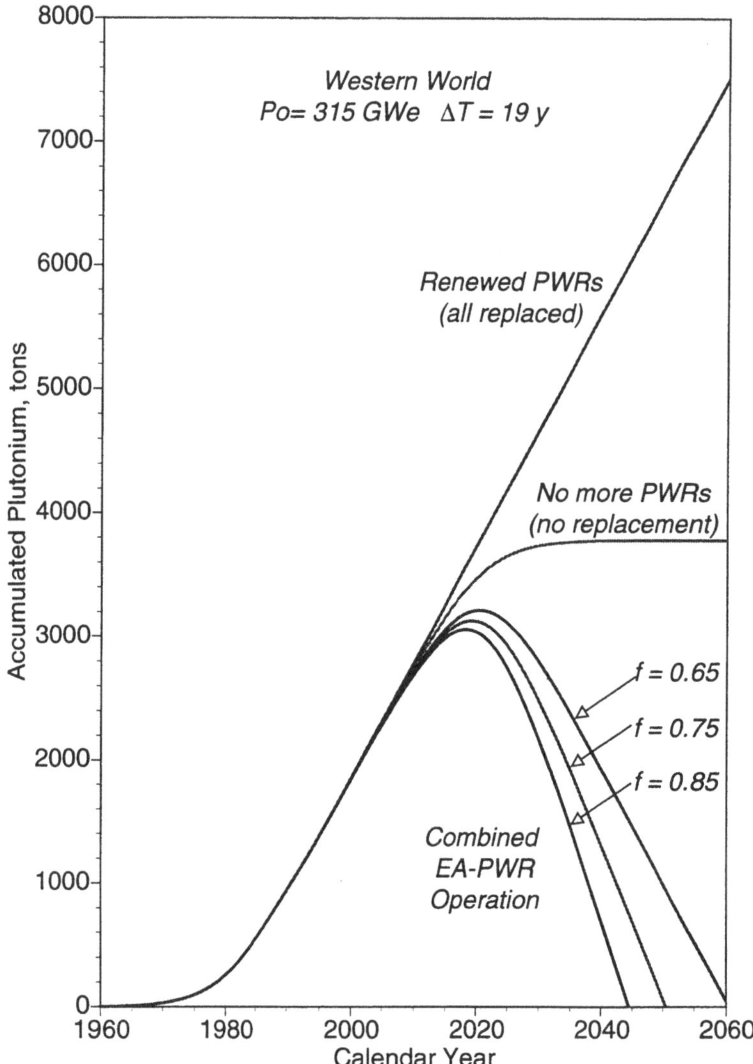

Figure 1.6 *Accumulated "Dirty Plutonium" waste in the Western World as a function of time for three possible scenarios, namely (1) All power plants are rebuilt in order to maintain the total power at the present level (2) No more nuclear plants are built and new power sources are of conventional, non nuclear type and (3) Nuclear power is renewed with EA-PWRs scheme with a fraction* f *of the total power produced by EAs.*

We can distinguish three limiting scenarios:

(1) If the replacement plants are ordinary PWRs and no incineration technique is applied, the waste accumulation will continue linearly with the spectacular amount of the order of 7000 tons[12] of Pu and minor Actinides by circa 2050.

(2) If the Nuclear option will come to a halt, namely no new Reactor is built in the future, the residual Pu waste will be still of the order of 3800 tons.

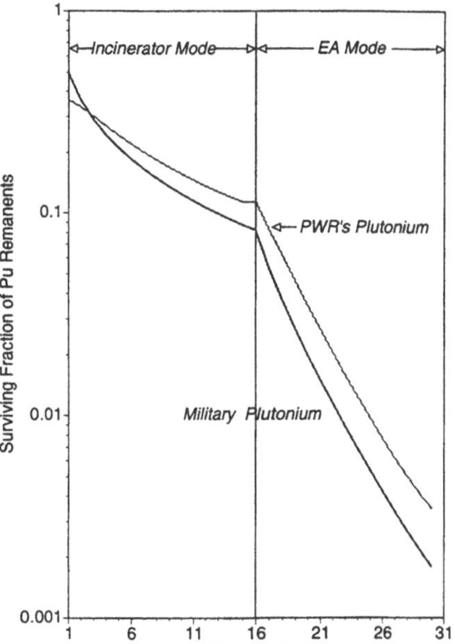

***Figure 1.7** : Residual stockpile of "Dirty Plutonium" and its descendants as a function of cycle in the EA. After a first phase of incineration in which the Pu is injected at each cycle, the EAs are operated as Amplifiers, with fuel recycling but no external Pu injection. (a) the injected Pu is "Dirty Plutonium" from PWR discharge; (b) the Pu is of military origin (bomb grade).*

(3) Introducing the EAs has spectacular effects. We assume that a fraction f of the total nuclear power is produced by EAs initially fed with Plutonium[13] and that the electric energy supply from nuclear sources is kept constant, the Western World's stockpile of "Dirty Plutonium" will be reset to zero by circa 2045 (T_o^{EA} + 24) for $f = 0.85$ and 2060 (T_o^{EA} + 39) for $f = 0.65$ (Figure 1.6).

We expect that the lifetime of an EA plant be longer than the one of a PWR, provided the necessary maintenance and replacement of components are carried out[14]. Assuming an EA useful lifetime of ≈ 80 years, the park of EAs will come to the end of the useful life by circa $T_o^{EA} + 80 = 2101$. Hence, after the existing Plutonium has been eliminated one may wish to continue with several EA cycles as a pure Thorium

---

12 For the open cycle and no reprocessing, the total amount of spent fuel "waste" is about 100 times larger, namely 0.7 million tons.

13 The EAs are operated with Pu-Th mixture in such a way that all the Pu (pre-existing and produced by the new 1-f fraction of PWRs) is eliminated. Extra power is produced with EAs operated in the energy Amplification mode, namely with Th and recycled ^{233}U. The required, initial ^{233}U stockpile is produced by Pu → ^{233}U transformation (EA).

14 In our view a well designed EA, after some part replacements could have a lifetime well in excess of 50 years.

burner as described in Ref. [1], namely with no Pu injection. This latter mode will be vastly cleaner than the PWRs [1] and in addition it will help to incinerate more completely the most neutron resilient radio-toxic residues accumulated in the EA fuel due to Pu burning initial cycles. As shown in Figure 1.7, the amount of residual "waste" drops by about one order of magnitude during the incineration phase and to about few units of 10^{-3} after the second phase. Since the burning of the unwanted Actinides is progressively growing parasitic with the reduction of the volume of the surviving waste, additional incineration may be pursued beyond the indicated 10^{-3} level of waste, if still considered excessive.

Assume finally a minimal programme in which one wants to eliminate *only* the accumulated waste, with no concern about maintaining the total nuclear power constant. If the useful lifetimes of the EA and of the PWR are respectively Γ_o^{EA} and Γ_o^{PWR}, the EA must be able to incinerate Pu at a rate $\eta = \Gamma_o^{PWR}/\Gamma_o^{EA}$ the PWR production rate. At the end of the operation one may wish to run n EA cycles as a pure Thorium burner . If $N = \Gamma_o^{EA}/\Gamma_{cycle}^{EA}$ represents the total number of cycles each of duration Γ_{cycle}^{EA} in the lifetime of the EA, the Pu incineration rate, again in units of the PWR production rate is $\eta = \left(\Gamma_o^{PWR}/\Gamma_o^{EA}\right)\left[N/(N-n)\right]$. Note that for a cycle time of 3 (5) years and a useful minimal plant lifetime of ≈ 90 years[15], $N \approx$ *30 (18)* namely only a limited number of refills and major manipulations are necessary over the lifetime of the EA. Assuming that the lifetime of the PWR to be 1/2 of the one of the EA, we could set $n = N/2 \approx 15(9)$ and therefore $\eta = 1$, namely EA and PWR burn and produce waste at the same rate with a power production in the EA which is about f = 0.33 of the one of the PWRs[16] and a second, subsequent phase of pure EA burning is carried out in order to complete the elimination of the unwanted waste.

The combined EA-PWR cycle is very effective also in eliminating Military Surplus Plutonium. The elimination of the totality of the estimated surplus of 100 tons of ^{239}Pu will produce a total of 400 GW × year in the EAs. *Two clusters of 3 EA units each for a thermal power of 4500 MW and an electric power of 1870 MW could eliminate it in about 45 years.* As mentioned above, an additional period of similar duration with the EAs operated as "amplifiers", in which the ultimate burn-up is performed "parasitically" will ensure the final reduction of the residue to the level of few 10^{-3} of the initial mass (≤ 200 kg) (Figure 1.7). The 86.18 tons of ^{233}U produced in

---

15 We have slightly stretched the useful lifetime of the PWRs and of the EAs, since we assume that if the technology will have to be terminated, one would try to get the maximum benefit from the investment.

16 We have taken into account that the thermal efficiency of an EA is higher than the one of a PWR because of the higher operating temperature [1]. If one assumes equal efficiencies, then f = 0.40.

the transformation can be used to produce as much 86.18/0.02375 = 3629 tons of useful fuel for PWRs (2.375% enrichment). The produced fresh fuel is sufficient to run a standard 1000 MW (electric) PWR for about 100 years.

When compared to the relatively established MOX-Fuel technology applied to Thermal PWRs [9], the present scheme has the advantage that while the former method allows essentially a single pass through the Reactor and then it becomes "waste", with a reduction of about a factor 2 in the unwanted Actinides, the present scheme allows many passages without appreciable loss in reactivity, eventually completed by a final phase as energy generating EA (without fresh Plutonium injections) and therefore (as explained in detail later on) the surviving Plutonium and higher Actinides are several orders of magnitude smaller than the total injected stockpile.

Finally Plutonium incineration when compared to deep geological storage is not only environmentally more acceptable but also economically profitable, since the cost of the energy produced by the EAs is competitive in price to the one of PWRs [10]. This point will be discussed more completely in a forthcoming paper [14].

2.- The Energy Amplifier

A conceptual design of a fast neutron operated EA is described in detail in Ref. [1] to which we refer for all details. In short the EA is made of two main coupled elements, (1) a specially designed proton accelerator of high energetic efficiency and high reliability, capable of delivering a beam typically of 10 ÷ 20 mA at a kinetic energy of 1 GeV (Figure 2.1) and (2) a Lead cooled beam dump in which the beam power is greatly increased by the secondary interactions of thermonuclear cascade in a sub-critical medium and extracted by conventional heat exchangers (Figure 2.2).

In order to enhance the "passive" nature of the "beam dump" we have chosen for that design a pure convective cooling[17]. In spite of the large power involved, we have reached a sensible set of parameters, mainly because of the unique properties of the molten Lead as coolant. A general layout is shown in Figure 2.3 and the list of main parameters is given in Table 2.1 both taken from Ref. [1].

---

17 Evidently pumps, especially electromagnetic pumps, could be used instead, if further analysis proves them more convenient.

***Table 2.1** : Main parameters of the Energy Amplifier (Thorium operated)*

Gross Thermal Power/unit	1500	MW
Primary Electric Power	625	MW
Type of plant	Pool	
Coolant	Molten Lead	
Sub-criticality factor k, (nominal)	0.98	
Doppler Reactivity Coefficient, $(\Delta k/\Delta T)$	-1.37×10^{-5}	
Void coefficient (coolant) $\Delta k/(\Delta \rho/\rho)$	+ 0.127	
Nominal energetic Gain	120	
Accelerator re-circulated Power	30	MW
Fraction Electric Power recirculated in Accel.	0.0465	
Control Bars	none	
Scram systems(3)	CB_4 rods	
Seismic Platform	yes	
Main Vessel		
Gross height	30	m
Diameter	6 m	m
Material	HT-9	
Walls thickness	70	mm
Weight (excluding cover plug)	2000	ton
Double Liner	yes	
Proton Beam and Spallation Target		
Accelerator type	Cyclotron	
Number of beams	1	
Accelerator overall efficiency[18]	43%	
Kinetic energy	1.0	GeV
Nominal current	12.5	mA
Nominal beam Power	12.5	MW
Maximum current	20	mA
Spallation Target material	Molten Lead	
Beam radius at spallation target	7.5	cm
Beam window	Tungsten, 3.0 (1.5)	mm
Max. power density in window	113	W/cm^2
Max. Temp. increase in window	137	°C
Window expected lifetime	≥ 1	year

[18] Beam power/Mains Load.

Fuel Core				
Initial fuel mixture		$ThO_2 + 0.1^{233}UO_2$		
Initial fuel mass		28.41		ton
Cladding material		low act. HT-9		
Specific power		52.8		W/g
Power density		523.		W/cm^3
Average Fuel Temperature		908		°C
Maximum Clad Temperature		707		°C
Dwelling time (eq. @ full power)		5.0		years
Average Burn-up		100.0		GWd/t
Breeder Core				
Initial fuel mixture		ThO_2		
Initial fuel mass		5.6		ton
Cladding material		low act. HT-9		
U^{233} stockpile at discharge		242.7		kg
Power density at end cycle		3.0		W/g
Primary cooling system				
Approximate weight of the coolant		10,000		ton
Pumping method		Nat. Convection		
Height convection column		25		m
Convection generated primary pressure		0.637		bar
Heat exchangers		4×375		MW
Decay heat removal		RVACS		
Inlet temperature, Core		400		°C
Outlet temperature, Core		600		°C
Coolant Flow in Core		53.6		ton/s
Coolant speed in Core, average		1.5		m/s
Decay Heat Passive Cooling (RVACS)				
Riser channel gap width		18		cm
Downcomer channel gap width		57		cm
Trigger Temperature	500	600	700	°C
EA Coolant max Temperature rise	110	83.5	64.5	°C
Time to max.Temperature rise	17.5	11.2	9.5	hours
Outlet air Temperature (@ max. temp.)	273	302	334.3	°C
Outlet air Speed (@ max. temp.)	13.4	14.2	15.2	m/s
Air flow Rate (@ max. temp.)	52.8	56.1	60	m^3/s
Extracted Heat (@ max. temp.)	8.57	9.65	10.84	MW

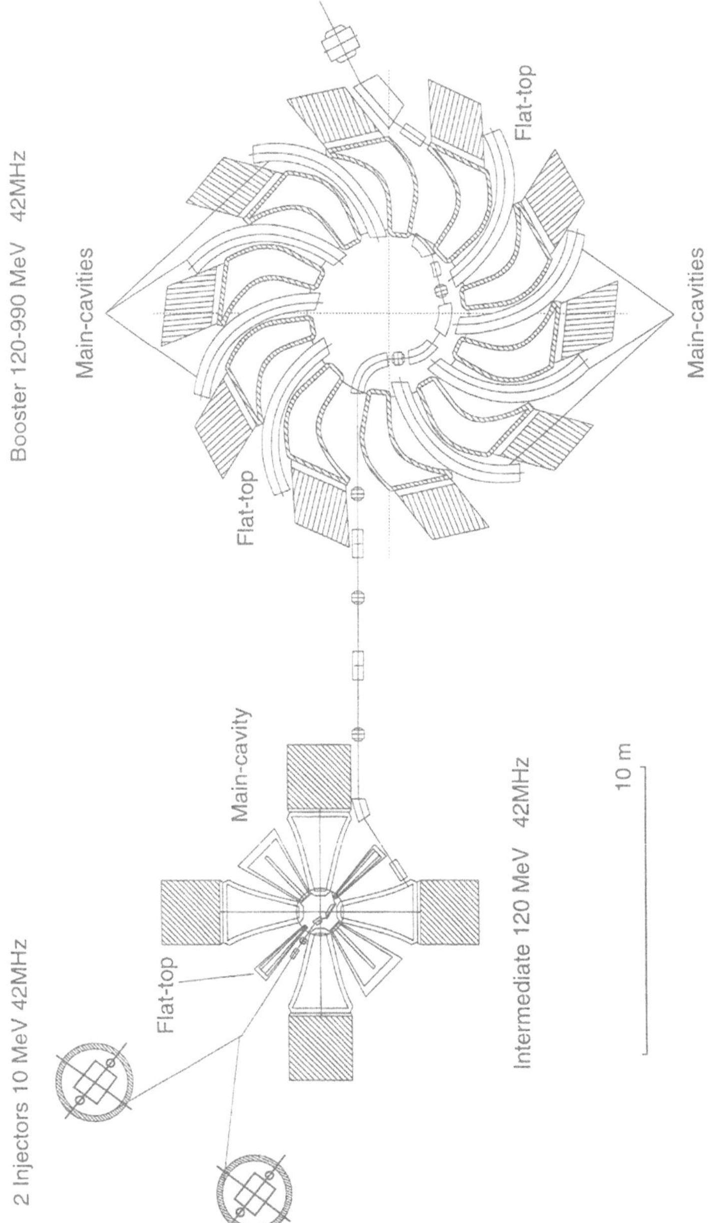

Figure 2.1: *General scheme of the particle Accelerator. Note the modest scale of the components. For more details we refer to Ref. [1].*

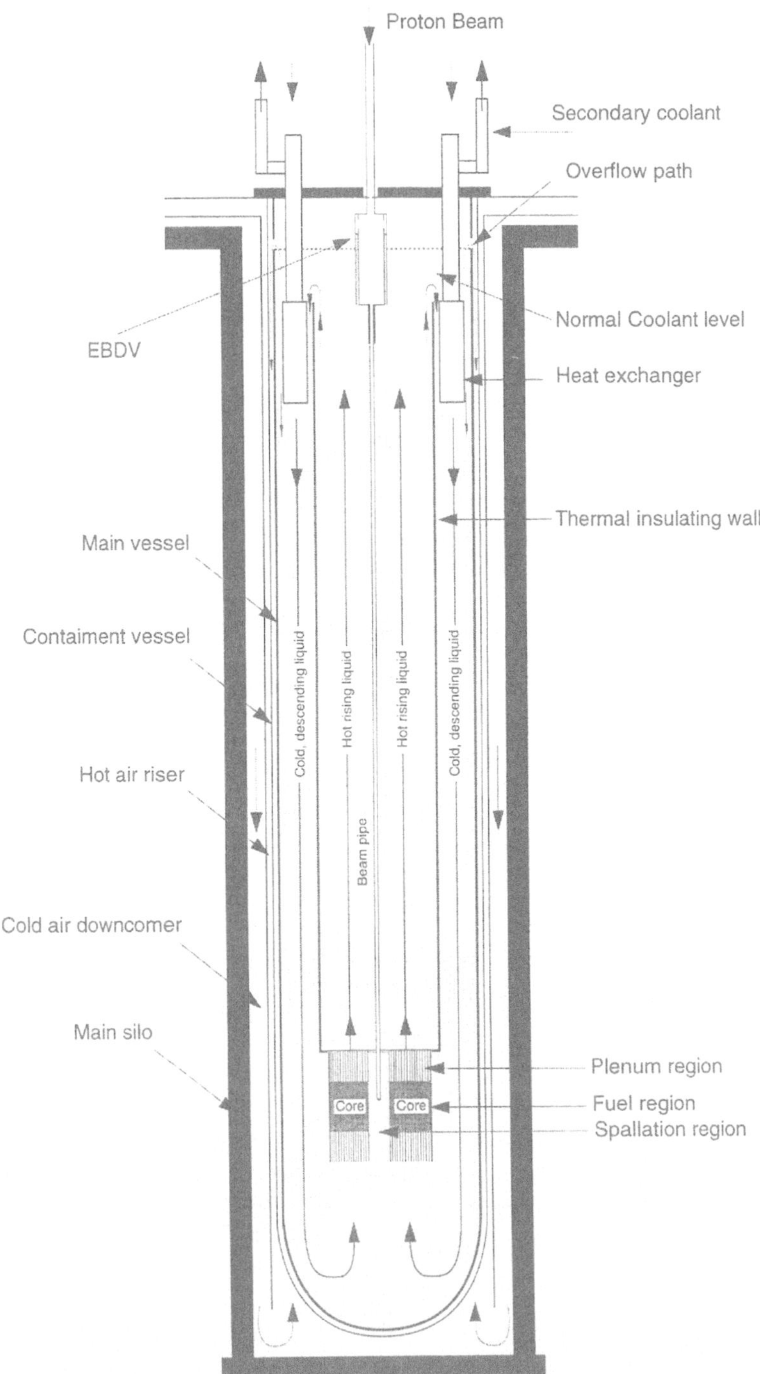

Figure 2.2: General layout of the Energy Amplifier/Beam Dump [1].

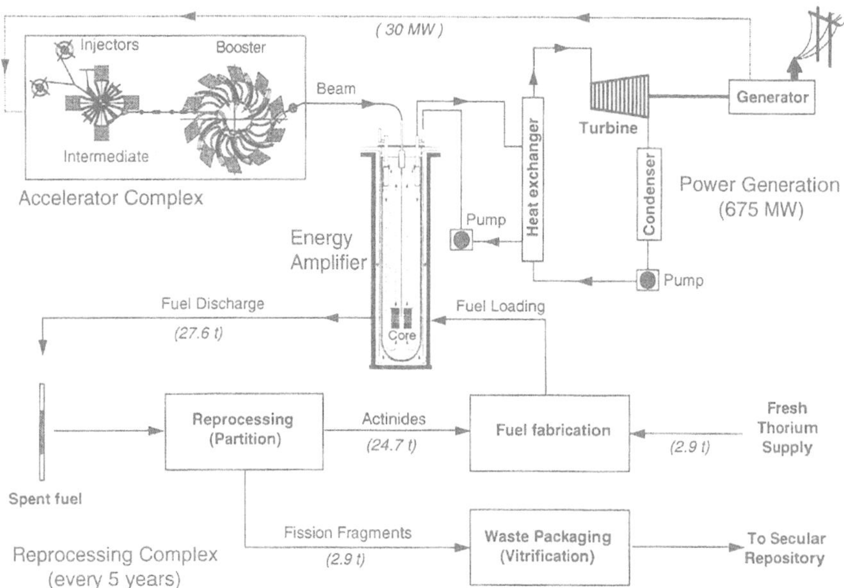

Figure 2.3: *Principle diagram of the Energy Amplifier complex [1].*

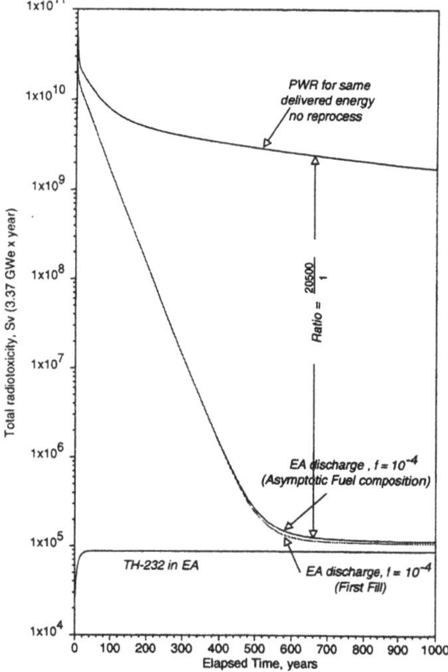

Figure 2.4: *Ingestive radiotoxicity in the "waste" of the Secular Repository as a function of time. The corresponding values for a standard open cycle PWR is also shown.*

For simplicity we have used in this paper the same parameters of Ref. [1] namely a nominal unit power of 1500 MW_{th}. The system is modular and several units can be clustered if needed. The fuel is mixed oxide (MOX) cladded with low activity steel[19], essentially identical to the one used in Fast Breeders. The operating temperature is of the order of 650 °C.

The fuel (see Figure 2.3) is kept sealed inside the EA for an extended burn-up, typically in excess of 100 GW × day/t, and then "regenerated" removing the fission products (FFs), and topping it with additional fresh primary fuel which in the simplest mode is natural Thorium. The power produced is controlled by the proton beam current and there are no control bars. The device is operated at all times well below criticality and it is intrinsically safe against melt-down or other thermal run off due to misfunctioning of the accelerator or other unplanned accidents.

All produced Actinides are re-injected in the EA in an indefinite, closed cycle. In a more advanced version, FFs are also reprocessed in order to extract some of the very long lived radio-nuclides (^{99}Tc, ^{129}I, etc.) which are also "incinerated" in the EA, thus further reducing the activity of the FFs after 500÷1000 years. Even without this precaution, the (ingestive) radio-toxicity after a first-cool down period of about 600 years is reduced to about 5×10^{-5} of the one of the waste from an ordinary PWR for the same produced energy (Figure 2.4).

3.- Plutonium incinerating mode of the EA

The EA can operate with a variety of fuels. In the present application we start with a mixture of "Dirty Plutonium", which is the standard discharge from a PWR, topped with natural Thorium. The assumed composition of the "Dirty Plutonium" is given in Table 3.1. However the performance of the EA is largely independent of the exact mixture used [1]. The neutron inventories (Table 3.2) are only slightly different from the ones with pure Thorium operation [1]. The neutron spectra are essentially the same as for standard EA operation (Figure 3.1) and change insignificantly during the evolution of the fuel composition, from the initial load to the asymptotic mixture. The corresponding effective cross sections which are used as input to the evolution calculations [1] are listed in Table 3.4.

---

19 The exact choice of the cladding depends on the effects of Lead corrosion, which needs further investigations (see Ref. [1]).

Table 3.1. *Assumed composition of "Dirty Plutonium". From Ref. [11]*

Radio-Nuclide	Relative Concentration	Radio-Nuclide	Relative Concentration
^{237}Np	0.075066	^{241}Am	0.48572E-02
^{239}Np	0.75434E-08	^{242}Am	0.43789E-04
^{236}Pu	0.92361E-06	^{243}Am	0.91257E-02
^{238}Pu	0.02204	^{242}Cm	0.48940E-03
^{239}Pu	0.52988	^{243}Cm	0.72123E-05
^{240}Pu	0.21747	^{244}Cm	0.33522E-02
^{241}Pu	0.10193	^{245}Cm	0.20386E-03
^{242}Pu	0.03550	^{246}Cm	0.22925E-04

The relevant safety parameters are listed in Table 3.3. The most notable difference is that the void coefficient $\Delta k/(\Delta\rho/\rho)$ notoriously positive for standard EA operation is now negative, and it amounts to $\Delta k/(\Delta\rho/\rho) = -0.035$ at the beginning of the fill, when no ^{233}U is present[20]. Such a void coefficient[21] is far less of a concern than in a conventional Sodium cooled Fast Breeder since (1) the boiling point of Lead is very high (1743 °C at normal pressure) and (2) the sub-criticality margin of the EA is vastly larger than the maximum change due to extreme void effect. However it cannot be excluded that an accidental loss of the molten Lead might occur[22].

In such a hypothetical event, part or the totality of the core and the surrounding region will be made void of Lead coolant and moderator. We have simulated the emptying of the beam dump in a variety of conditions and looked at the changes in the reactivity coefficient *k*. (Figure 3.2). We find that during the emptying process the resulting void effect is always negative. We also remark that, as expected, at the end of the cycle (EOC), because of the beneficial presence of the ^{233}U the reduction is more pronounced than at the beginning of the cycle (BOC).

---

[20] Evidently during operation while the initial Plutonium load is slowly incinerated, the newly bred ^{233}U moves the operating point slowly towards the standard EA operating conditions, therefore improving the void coefficient.

[21] Here the void coefficient is defined as the variation of k due to a relative change in coolant density ρ. A positive value implies that k decreases with decreasing density.

[22] We underline the extreme unlikeness of such a happening. A small crack in the vessel (doubly lined) will produce a leak which will stop as soon as the molten Lead flow comes into contact with cold environment, since the solid Lead will act as a plug. Only a massive destruction of the container and of the concrete walls of the pit will permit the loss of a major fraction of the coolant. Note the large excess of molten Lead above the core, which has to be lost before the core becomes exposed. The whole installation is seismically protected. [1].

***Table 3.2** : Neutron capture inventory of the F-EA for the Thorium-"Dirty Plutonium" cycle. (first fill ; k_0 = 0.9896 ; 16.5% MOX).*

Zone-Wise			Fraction
Core			0.8970
Blanket			0.0425
Plenum			0.0257
Diffuser			0.0279
Beam Tube + Window			0.0005
Main Vessel			0.0064
Leakage			0.0018
Material-Wise			**Fraction**
Fuel (Th + U)			0.8577
Breeder (Th)			0.0395
Lead of which	percentage	Abs. Fraction	
Diffuser	(46.81 %)	0.0279	
Plenum	(12.08 %)	0.0072	
Core	(39.09 %)	0.0233	
Blanket	(02.01 %)	0.0012	
Lead Total			0.0596
Structures of which	percentage	Abs. Fraction	
Cladding	(84.06 %)	0.0364	
Window	(01.15 %)	0.0005	
Main Vessel	(14.78 %)	0.0064	
Structures Total			0.0433
Leakage			0.0018

***Table 3.3:** - Safety-related physics parameters of the F-EA.*

	Th-U Asymptotic	Th-Dirty Pu Asymptotic		Th-Weapon Pu Asymptotic
	EOC	BOC	EOC	BOC
Doppler Coefficient, $\Delta k/\Delta T$	- 2.27E-5	- 1.12E-5	- 1.38E-5	- 1.66E-5
Change in $\%\Delta k/k$, emptying				
Top Core	- 0.1029	- 0.0552	- 0.0714	- 0.0680
Half Core	- 4.0513	- 2.3235	- 3.1485	- 2.9560
Full Core	- 6.4844	- 0.3468	- 2.9342	- 2.3909
All Regions	- 11.923	- 4.1138	- 7.4858	- 6.7542

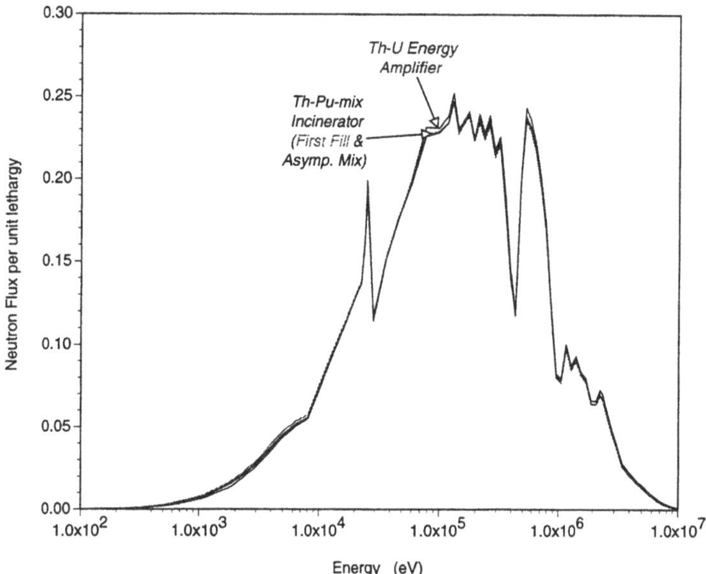

***Figure 3.1:** Neutron spectra in Fuel Core for different fuels in the EA. Spectra are normalized to equal integrated flux.*

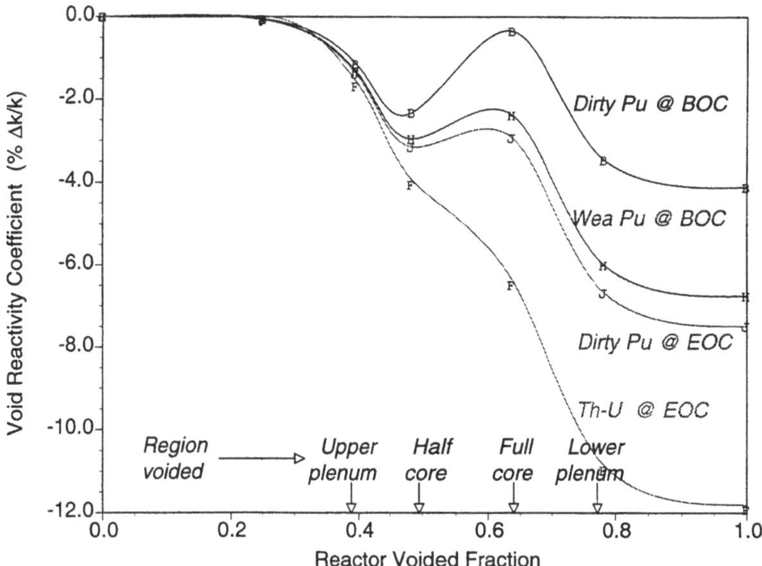

***Figure 3.2:** Relative variation of the multiplication coefficient* k *as a function of the fraction of the vessel voiding in a variety of conditions.* (EOC), (BOC) *are respectively the end and the beginning of the cycle. There is a small difference between the fillings with PWRs Plutonium waste* (Dirty Pu) *and the one with surplus Weapon Plutonium* (Wea Pu). *The case of the standard Thorium-Uranium operated EA* (Th-U) *is also shown as a reference.*

Table 3.4: *- Averaged cross sections (barn) of Actinides relevant to the fast EA.*

Element	Capture	Fission	Elastic	(n->2n)	(n->n')	Total
^{230}Th	0.167803	0.025697	13.218194	0.001447	1.123411	14.536536
^{232}Th	0.308828	0.008570	10.087153	0.001402	0.819348	11.073583
^{231}Pa	2.957017	0.225269	8.817699	0.000978	1.213704	13.214634
^{233}Pa	0.976008	0.055470	7.792018	0.000407	1.833241	10.657163
^{232}U	0.652538	2.068427	8.993785	0.000750	0.487607	12.203108
^{233}U	0.257299	2.684459	7.637661	0.000746	0.419488	10.999662
^{234}U	0.559472	0.310451	9.705321	0.000150	0.791574	11.366984
^{235}U	0.507261	1.813811	8.607527	0.000965	0.688819	11.618405
^{236}U	0.410061	0.088795	10.565082	0.000727	0.970194	12.034825
^{237}U	0.417513	0.618110	8.794017	0.001956	0.584155	10.415781
^{238}U	0.308613	0.036678	10.165911	0.001064	0.913203	11.425482
^{237}Np	1.470774	0.296474	8.992845	0.000231	0.699920	11.459522
^{238}Np	0.400021	2.091224	8.931683	0.000723	0.285945	11.709574
^{239}Np	1.827500	0.430915	8.857689	0.000319	0.925176	12.041578
^{236}Pu	0.356359	1.424113	9.138505	0.000296	0.346369	11.265685
^{238}Pu	0.513813	1.073784	9.522254	0.000077	0.484591	11.593491
^{239}Pu	0.461561	1.751134	8.810845	0.000290	0.636930	11.640947
^{240}Pu	0.468515	0.353373	9.840660	0.000225	0.646495	11.248711
^{241}Pu	0.521293	2.426248	8.400412	0.001785	0.432803	11.780483
^{242}Pu	0.425562	0.239415	10.411948	0.000573	0.740759	11.809869
^{243}Pu	0.356255	0.779540	8.941603	0.004175	0.738341	10.819971
^{244}Pu	0.185471	0.200118	10.316672	0.001769	0.904391	11.608458
^{241}Am	1.753717	0.246073	9.221643	0.000016	0.628412	11.849841
^{242}Am	0.568606	3.072716	8.130913	0.000364	0.335152	12.107742
^{241m}Am	1.401599	0.188868	9.556294	0.000071	1.038566	12.185221
^{241}Cm	0.190791	3.025927	8.138980	0.000032	0.261699	11.617401
^{242}Cm	0.291108	0.142550	9.883826	0.000017	0.805147	11.122590
^{243}Cm	0.352351	2.896037	8.587559	0.001300	0.268673	12.105916
^{244}Cm	0.819713	0.392233	10.052383	0.000347	0.581884	11.845643
^{245}Cm	0.301665	2.315751	8.479863	0.001682	0.934156	12.033182
^{246}Cm	0.214099	0.237992	10.388494	0.000473	0.848259	11.689314
^{247}Cm	0.309619	1.923196	8.803992	0.002529	0.439674	11.479051
^{248}Cm	0.238365	0.274690	10.860856	0.000578	0.884598	12.259142
^{249}Bk	1.247552	0.150108	9.798133	0.000135	1.282295	12.478257
^{249}Cf	0.615083	2.538316	8.764503	0.000441	0.489086	12.407440
^{250}Cf	0.381355	1.031658	8.567427	0.000943	0.544598	10.526023
^{251}Cf	0.303871	2.307346	8.484874	0.002975	0.503585	11.602709
^{252}Cf	0.309327	1.337872	10.545552	0.000414	0.581779	12.774944

There is a small difference between the fillings with PWRs Plutonium waste (Dirty Pu) and the one with surplus Weapon Plutonium (Wea Pu). We conclude that even in extreme conditions the behaviour of the EA is safely sub-critical.

The maximum duration of the burn-up is determined by many practical considerations like the radiation damage effects on the fuel and the emergence of FFs captures. The duration of each fuel cycle T_{cycle} is related to the length of the burn-up B_0 by the average power density ρ in the fuel, $T_{cycle} = B_o / \rho$. The mass of the fuel M_{fuel} is determined by the total power P to be produced, $M_{fuel} = P/\rho$. In a first approximation the incineration rate $\dot{M}_{Inci} = dM_{Inci}/dt$ is only dependent on P, namely independent of ρ and M_{fuel} for a given B_0. A large variety of choices in these parameters is therefore possible. In the energy production version of Ref. [1], a relatively low fuel power density has been chosen $\rho \approx 55$ watt/gr (oxide), in order to ensure a maintenance free condition for a long period, $T_{cycle} = 5$ years and $M_{fuel} = 25$ ton. In our specific application shorter cycles may be advisable. In this case one might adopt the value $\rho \approx 100$ watt/gr (oxide) which is ordinarily used in the Fast Breeders (Superphenix, MONJOU, EFR, ALMR etc.) and therefore $T_{cycle} = 3$ years and $M_{fuel} = 15$ ton). As already pointed out, these choices do not affect the incineration rates, as long as the nominal power is set.

The evolution programme used to make predictions of this work takes into account the neutron losses elsewhere than in the fuel according to Table 3.2 and the emergence of the Fission Captures, treated as a neutron sink. A fraction of the neutron inventory can be preassigned for incineration of long lived FFs. At this stage this feature is also treated as a neutron sink. For more details we refer to Ref. [1], where a more sophisticated method based on full Montecarlo simulation has been also used. It has been shown that the presently used evolution method is in good agreement with the full Montecarlo method and that its precision is adequate for the present analysis. In the programme

i) We specify first the initial value of the multiplication coefficient k_0. The initial concentration of "Dirty Plutonium" topped up with ^{232}Th is set accordingly.

ii) The burning is then followed for a specified period of time at a constant power density or constant neutron flux or at constant accelerator current, as preferred.

iii) At the end of the cycle the fuel is "discharged", cooled for a specified time period (with only decays active) and a new refill is prepared with (1) the full Actinide residue from the preceding run minus the Uranium isotopes which are separately recovered, (2) a fresh amount of "Dirty Plutonium" chosen in order to start with the initially chosen multiplication coefficient k_0 ; (3) fresh ^{232}Th to reach correct total initial fuel mass.

We repeat then for an arbitrary number of cycles points ii) and iii) in succession, thus following the incineration procedure. Results of the calculations are now discussed. They show that the short and long term evolutions are extremely well behaved.

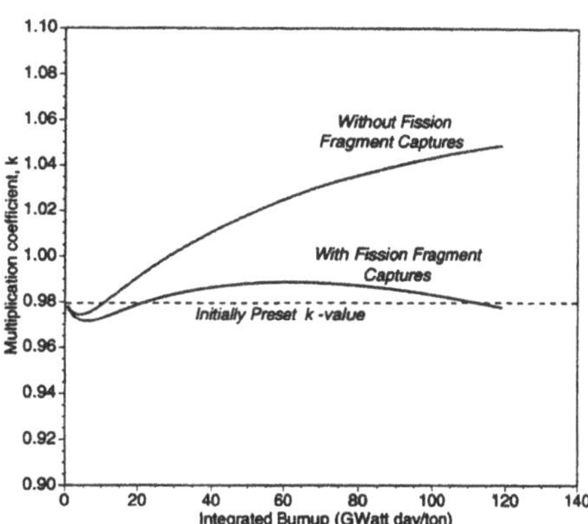

Figure 3.3: *Variation of the multiplication coefficient* k *with and without FFs captures. The stabilizing effect of the captures is evident. The choice of the initial value of* k *is arbitrary.*

i) The multiplication coefficient k remains remarkably constant over a long burn-up, because of the compensation between the rising amount of ^{233}U, the loss of mass and fissionable mass of the higher Actinides and the growing capture rate of the FFs. This is visualized in Figure 3.3. The duration of the cycle is therefore determined primarily by the emergence of radiation damage mainly in the fuel cladding and by the fact that the Pu to ^{233}U conversion becomes less efficient since the fraction of ^{233}U which is burnt in the EA grows with time. We have chosen as a guideline 120 GW × day/t, but larger values may be attainable.

ii) The continuous recycling with periodic addition of fresh "Dirty Plutonium" can be repeated indefinitely without loss of criticality coefficient k (Figure 3.4) and with a steady amount of fresh additions (Figure 3.5). The first load is evidently larger than the others, since significant reactivity is maintained in the Actinide residue which is transferred from the previous cycles.

iii) Asymptotic mixture of the fuel is reached after a few cycles and maintained indefinitely. The content of Actinides (Figure 3.6) after an initial growth,

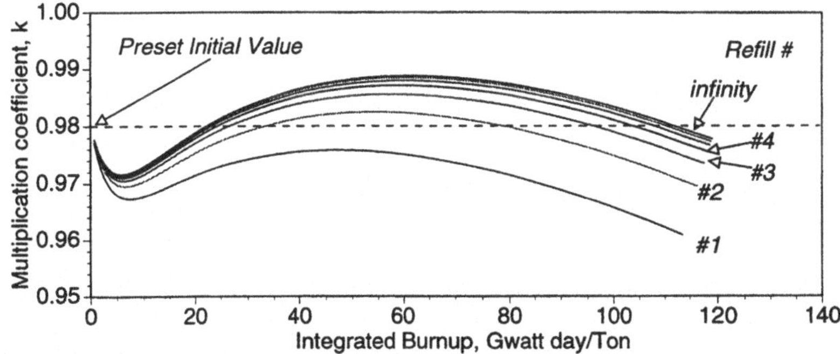

Figure 3.4: *Variation of the multiplication coefficient* k *as a function of burn-up for different, successive fillings and a specified, arbitrary initial value.*

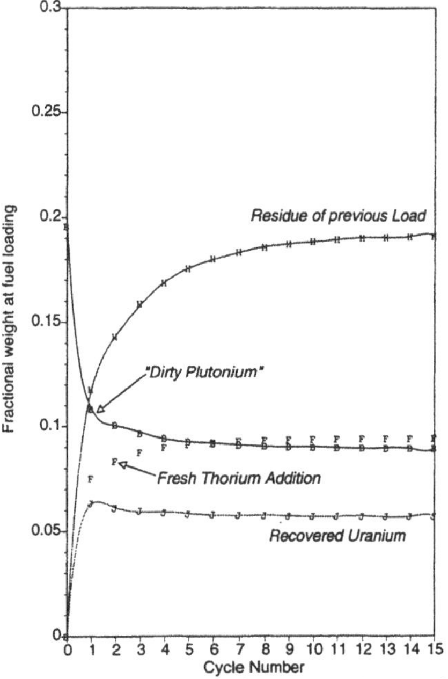

Figure 3.5: Fractional composition at the beginning of the cycle, as a function of the cycle number. The amount of (1) new "Dirty Plutonium", (2) the residue from the previous cycle, excluding Thorium, (3) the fresh Thorium addition are shown.

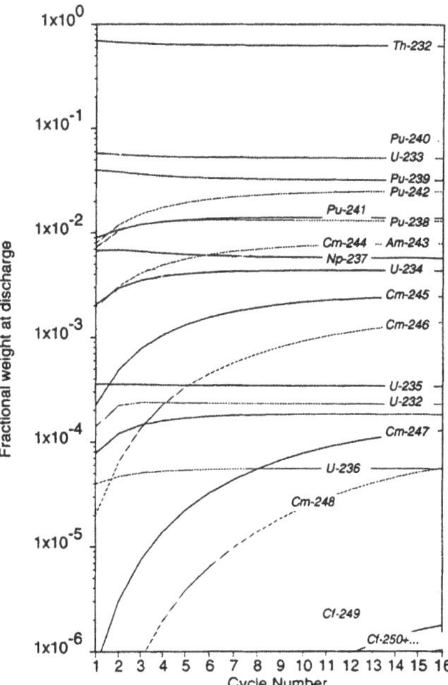

Figure 3.6: Fractional weight in the fuel of Actinides as a function of the cycle number. Asymptotic concentrations are evident. Note that notwithstanding the changes in composition, the multiplication coefficient remains remarkably constant (Figure 3.4).

stabilizes in the residue because of equilibrium between incineration due to fissions and production due to captures in the father element(s).

iv) The operation of the EA is normally controlled by the intensity of the beam current from the Accelerator. The beam current variation for the constant power of 1500 MW_{th} is shown in Figure 3.7. With the exception of the first fill which requires a somewhat higher current and could eventually be run at somewhat reduced power, an accelerator current with a nominal value of some 12 mA and a maximum value of about 18 mA is required. This is well within the design parameters of the accelerator of Ref. [1].

v) The power produced is primarily due to fissions. In Figures 3.8a and 3.8b we give the fractional power contribution for each of the elements in the mixture as a function of the burn-up. For definiteness we show the case of the asymptotic mixture, but the behaviour is not very different for the early fillings. At the beginning of the cycle, burning is concentrated on the Plutonium (fresh and remnants) and on the higher Actinides. During the cycle, while the reactivity of these elements decreases, freshly produced ^{233}U takes over. At the end of the

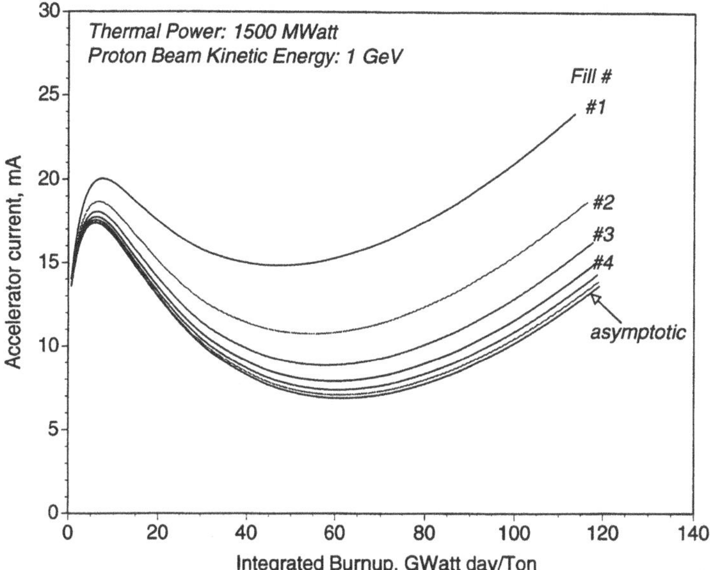

Figure 3.7: *Accelerator current in mA for protons of 1 GeV kinetic energy in order to maintain the delivered power to the constant value of 1.5 GW. Note that the first fill requires a larger current: it may be preferable to operate at somewhat lower delivered power.*

cycle, some 45% of the produced power is due to ^{233}U. The integrated fractional power produced by the burning of the Thorium/Uranium component amounts to 0.30 of all fissions[23] after 120 GW × day/t, corresponding to about 70% of the theoretical limit[24] of pure "Dirty Plutonium" burning (Figure 3.9).

The scheme allows continuous supply and incineration of "Dirty Plutonium" with an essentially constant residue which is continuously recycled. The accumulated amount of incinerated material and of the fresh ^{233}U as a function of the number of cycles performed is shown in Figure 3.10. We remark the almost linear elimination rate of the "Dirty Plutonium", amounting to 9.20% of the initial metal fuel mass for each cycle (120 GW × day/t) and the almost linear production rate of ^{233}U amounting to 5.801 % of the fuel mass. Hence the Pu → U conversion efficiency is 63%. The isotopic composition of the produced Uranium is primarily ^{233}U with noticeable contaminations of ^{232}U (4.63×10^{-3}), ^{234}U (0.0859), ^{235}U (6.83×10^{-3}) and ^{236}U (1.00×10^{-3}). Once the produced Uranium is diluted with ordinary Uranium to constitute new fuel for the PWRs, the relative concentration of the γ-emitting ^{232}U drops to about 100 ppm.

---

23 The fraction of the power produced by Thorium fissions alone is ≈ 2%.

24 We recall that the complete fission of 1 ton of "Dirty Plutonium" will produce 940 GWatt × day.

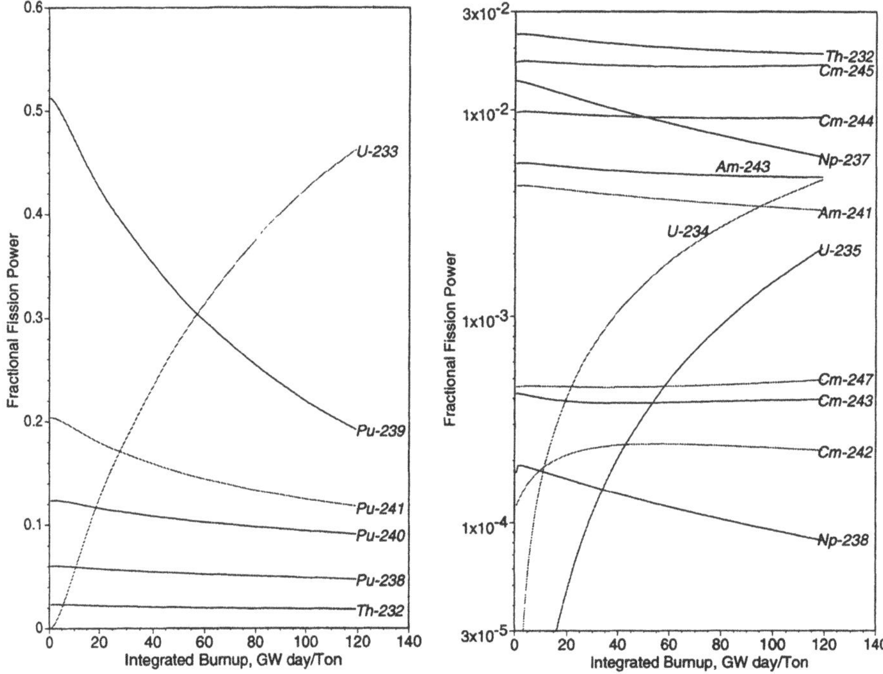

***Figure 3.8a,b** :Fractional power contributions (fissions) as a function of the burn-up for an asymptotic fuel composition. (a) large concentrations; (b) small concentrations.*

At a constant energy production rate of 1500 MW_{th} the EA will accumulate a yearly energy production of 547.5 GW × day, corresponding to the elimination of 547.5/120 × 0.0920 = 0.420 tons of "Dirty Plutonium" independently of the actual amount of fuel mass. The correspondingly produced Uranium is 547.5/120 × 0.05801= 0.265 tons which can be used to manufacture about 13 tons of PWR fuel with 1.87 % of ^{233}U. We remark that for a nominal 1000 MW_e PWR[25] [10], corresponding to an integrated thermal energy of 900 GW × day, the burnt fuel Uranium metal mass is 27.271 ton/year and the total "Dirty Plutonium" (namely Pu, Np, Am and Cm) accumulated amounts to 0.271 Ton/year [11].

The main conclusions of these calculations are that a single (1500 MW_{th}) EA module operated as an incinerator:

i) It can match the total Pu, Np, Am and Cm production rates of 0.420/0.271 = 1.55 nominal (1000 MW_e) PWRs and in addition

---

[25] Basis for the calculations are a thermal efficiency of 0.325, a dwell time of 1100 days, a burn-up of 33 GWatt × day/t, and a capacity factor 0.80. The fuel contains initially 3.3% of ^{235}U. These exemplificative figures come from Ref. [11].

ii) It produces through the Pu $\rightarrow$ U conversion enough fresh fuel for 13/27.271 = 0.48 PWRs.

iii) Incineration of the "Dirty Plutonium" of PWRs will finally produce an incremental energy in the EAs equal to 547.5/(900 $\times$ 1.55) = 0.392 of the one already produced in the PWRs. We believe that this extra energy is produced to very competitive costs.

iv) Ample allowance (10% of the full neutron stockpile) has been set aside for FFs incineration. Although the method is not completely perfected [12] we expect that it should be sufficient to eliminate the main long lived FFs produced namely ^{99}Tc , ^{129}I and although with previous mass separation ^{135}Cs.

For more details on possible combined strategies we refer to paragraph 5.

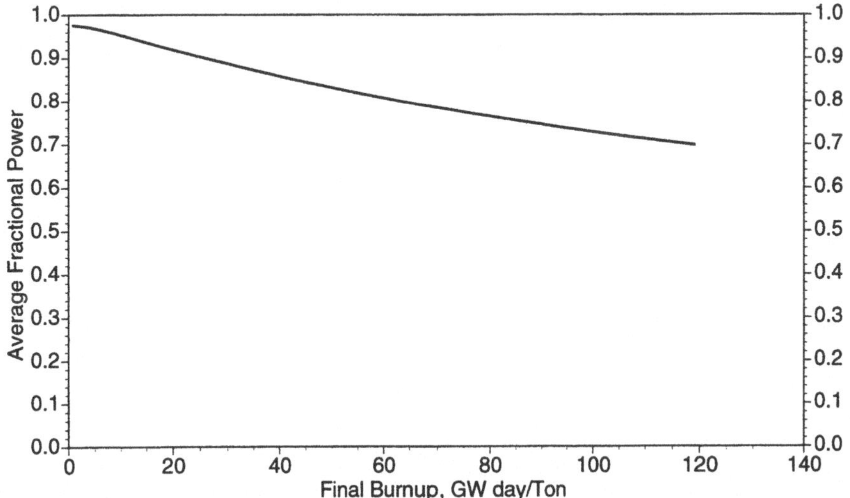

***Figure 3.9**: Fractional power generated by the "Dirty Plutonium" and its descendants, averaged over the fuel cycle, for different final burn-ups. For small burn-ups almost all the power comes from Pu burning, the small difference ($\approx$ 2.5%) being due to fissions in Thorium. At larger final burn-ups, burning in the freshly produced Uranium contributes significantly.*

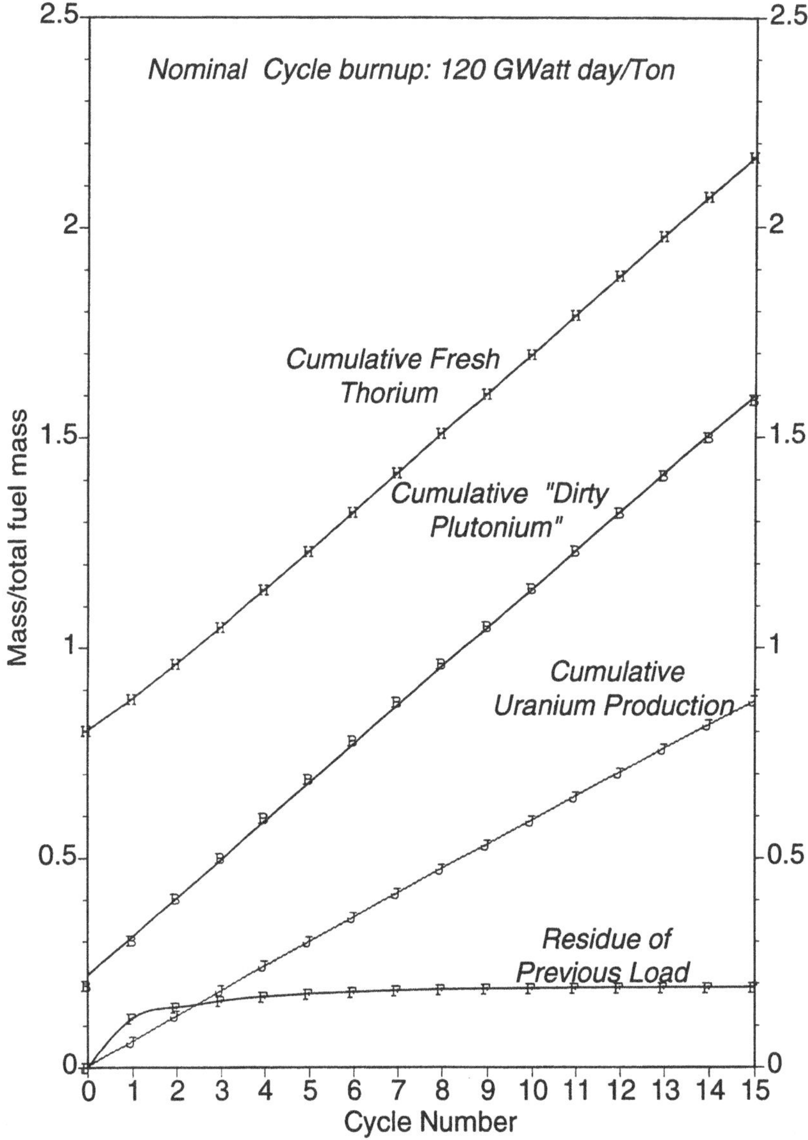

***Figure 3.10**: Cumulative amount of incinerated material and of the fresh ^{233}U as a function of the number of cycles.*

4- Fuel utilization in an ordinary PWR

The stockpile of the produced ^{233}U accumulated in the discharge of the EA constitutes a valuable asset. At any rate it represents a substantial toxicity and it must be burnt. We shall consider here the possibility of using it as a fuel in an ordinary PWR. In short the PWR is used to incinerate the ^{233}U "waste" from the EA, while the EA will incinerate the "Dirty Plutonium" from the PWR and eventually some of the long lived FFs produced by both units.

Since the purpose of the exercise is the one of having no Actinide in the repository, following Figure 1.1, the idea we pursue is the one of recycling the spent fuel from the PWR in the same way as for the EA, namely dividing it in three streams:

(1) all Actinides other than Uranium and eventually some of the most offending long lived radio-nuclides are recycled in the EA-fuel,
(2) the FFs are disposed in the secular repository and
(3) the Uranium is re-used for new PWR-fuel fabrication.

The PWR is evidently an existing machine. It is important that the minimum changes from ordinary operation are introduced. Recycling the bulk of the Uranium at each new fuel fabrication is reasonable since the ^{233}U enriched fuel has already many of the features of the customary MOX, including the presence of a substantial γ-activity, primarily due to the already mentioned presence of ^{232}U. The programme ORIGEN [13] has been used to compare the performance of this new and of the ordinary PWR fuels. As reference for ordinary fuel we have taken 3.3% ^{235}U-enriched Uranium in the form of Uranium Oxide. The nominal burn-up has been set to 33 GW $\times$ day/t and the power is kept constant at all times. The dwell time of the fuel in the Reactor is three years, with three separate refill phases, performed once a year[26]. The utilization factor is 0.80. We shall consider the first fill and the evolution of the fuel composition due to recycling in the PWR.

(1) For the first fill we have adjusted the concentration of the Uranium produced by the EA (^{233}U-rich mix) in a natural Uranium fuel in order to obtain the same average reactivity $< k_\infty >= 1.165$ as the reference fuel over the specified burn-up period.

---

[26] At each refill the oldest third of the fuel is extracted and substituted with new fuel.

(2) For the subsequent fills, the bulk of Uranium comes from the previous discharge. The amount of ^{233}U-rich mix is kept constant and the fuel mass is preserved by injecting an appropriate amount of fresh Uranium[27]. As already pointed out, after several refills, element concentrations tend to equilibrium values.

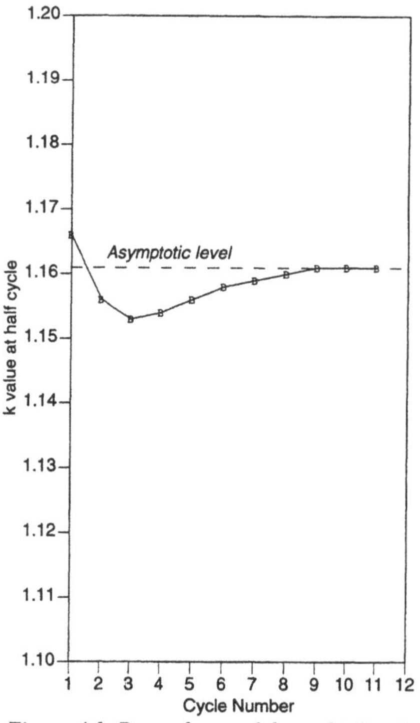

Figure 4.1: Dependence of the multiplication coefficient k_∞ half way through the fuel cycle. The asymptotic level is also shown.

It is a very fortunate circumstance that the behaviour of this "asymptotic mixture" in a PWR is very similar to the one of the initial fill. It is this feature that makes our scheme possible and attractive when compared with other procedures which suffer from a rapid "decay" of the properties of the recycled fuel. The relevant yearly quantities, normalized to 1 GW nominal thermal power are given in Table 4.1 for successive fuel cycles. The multiplication factor is remarkably independent of the fuel cycle[28], as shown in Figure 4.1.

The differences of the reactivity k_∞ as a function of the burn-up within the cycle is shown in Figure 4.2, for the first and the 11-th combined cycle (representative of an asymptotic mixture)[29] and compared with the ^{235}U reference case. The difference in behaviour between the two fuels and the standard reference case are extremely small and they become even smaller with the asymptotic mixture. The slightly higher value of k_∞ of ^{233}U and its slightly faster fall-off are further compensated when one considers fuel shuffling exercises which are needed to uniformize the reactivity of the

---

[27] Since the composition of the bulk of the recycled Uranium is ^{238}U, the exact amount of fresh Uranium is not very critical and it is mainly added for mass conservation. In the calculations we have used Natural Uranium but depleted Uranium could be used instead.

[28] It could be eventually further trimmed changing slightly the ^{233}U-Mix, although it does not seems required, since the control bars could easily correct this small difference.

[29] Note that each fuel cycle will last at least 3 years of dwell time in the PWR and 1 year for reprocessing. Hence the 10 cycles will span over 40 years which corresponds to the lifetime of a PWR.

fuel during burn-up. Variations of Uranium concentrations during burn-up are displayed in Figure 4.3. Note that at the 11-th cycle and beyond concentrations of ^{234}U, ^{235}U and ^{236}U remain essentially constant, as the resultant of burning and formation by neutron capture.

***Table 4.1:** Inventory of PWR operation at initial load and after 33 GW × day/t.*

	Cycle Number									
	0	1	2	3	4	5	6	7	8	10
General parameters										
k_{∞}@ 1/2 burn	1.166	1.156	1.153	1.154	1.156	1.158	1.159	1.16	1.161	1.161
Flux × (10^{14})	3.36	3.22	3.19	3.14	3.10	3.07	3.05	3.03	3.02	3.01
Inventory at initial load , Kg/ GW(t)										
Natural U		195.7	196.5	197.8	197.4	197.4	197.1	196.1	196.4	197.2
^{233}U-Mix	189.8	189.8	189.8	189.8	189.8	189.8	189.8	189.8	189.8	189.8
Inventory at discharge, Kg/ GW(t)										
Actinides ≠ U	84.75	85.33	85.20	84.91	84.70	84.74	84.81	84.97	85.11	85.43
He,Pb,Th	0.023	0.032	0.037	0.039	0.040	0.041	0.042	0.042	0.042	0.043
Np	1.583	2.191	2.496	2.775	3.064	3.351	3.621	3.874	4.098	4.473
Pu	81.29	81.32	80.89	80.32	80.00	79.73	79.56	79.46	79.39	79.31
Am	1.591	1.578	1.539	1.494	1.456	1.428	1.408	1.394	1.384	1.372
Cm	0.262	0.258	0.245	0.231	0.219	0.211	0.204	0.200	0.197	0.194
FF -Total	301.2	300.9	300.9	301.2	301.2	301.2	301.2	301.2	301.2	301.2
^{99}Tc	6.32	6.23	6.23	6.26	6.26	6.26	6.28	6.28	6.28	6.28
^{129}I	2.23	2.33	2.35	2.34	2.34	2.33	2.33	2.33	2.33	2.33
^{135}Cs	2.59	2.58	2.60	2.63	2.65	2.65	2.67	2.67	2.67	2.67

It is well known that a substantial amount of the produced energy is due to fissions in Plutonium, rather than Uranium. We display in Figure 4.4 the relative contributions of these two elements as a function of the burn-up. It appears that for the first fill the contribution due to Plutonium is slightly larger than the reference case. For the subsequent cycles, since the concentration mix of the fuel is getting richer with higher Uranium isotopes, the effects of Plutonium are slightly smaller. All these differences are small and do not represent problems. The Plutonium isotopic composition during burn-up is extremely similar for all cases, since it is driven by the ^{238}U bulk material. Neptunium production is about 50% the one of the standard fuel.

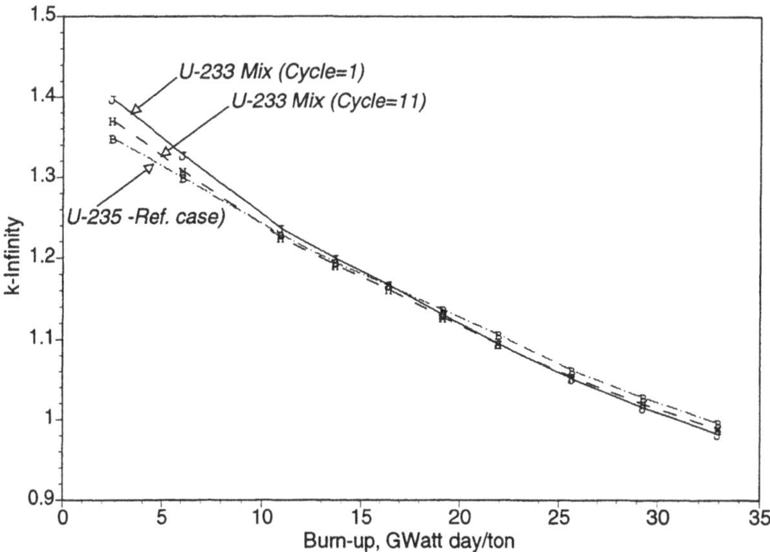

***Figure 4.2**: Reactivity k_∞ as a function of the burn-up within the cycle for the first, the 11-th cycle (representative of an asymptotic mixture) and for the ^{235}U reference case.*

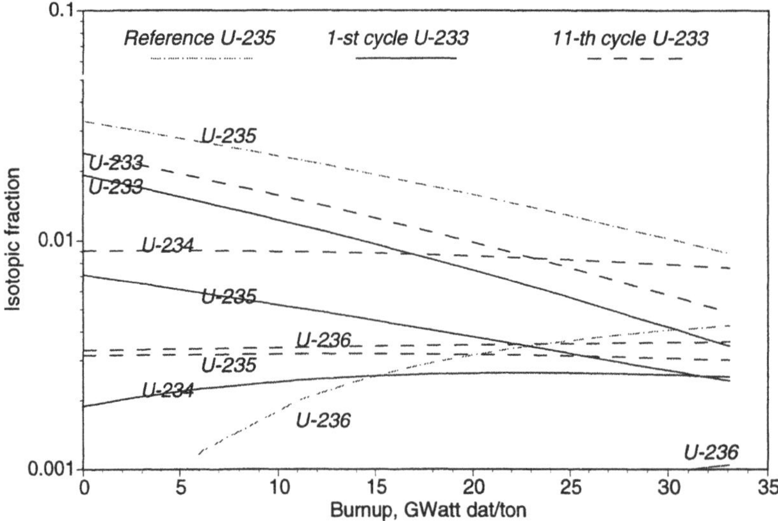

***Figure 4.3**: Uranium isotopic concentrations a function of burn-up for the first, the 11-th cycle (representative of an asymptotic mixture) and for the ^{235}U reference case.*

Looking in more detail at the evolution of the fuel composition as a function of the fuel cycle, we give in Figure 4.5 the fuel composition at the load of the n-th cycle, as a function of n. The ^{235}U concentration drops from the initial concentration of natural Uranium and settles to a level of ≈ 0.32 %, very close to the one of ^{236}U (0.34 %). The other noticeable isotopes, besides the bulk of ^{238}U are: ^{234}U (0.09 %) and the highly toxic and γ-emitting ^{232}U (2.0×10^{-4}).

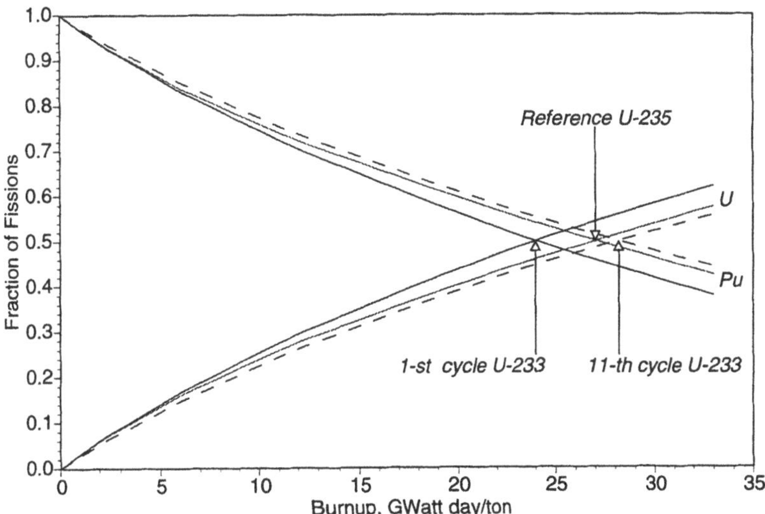

***Figure 4.4**: Relative contributions to produced energy of Uranium and Plutonium isotopes as a function of the burn-up for the first, the 11-th cycle (representative of an asymptotic mixture) and for the ^{235}U reference case.*

Finally since the aim of the exercise is the one of eliminating the Plutonium "waste", we briefly mention the possibility of replacing also in the PWR the Uranium fuel matrix with a Thorium matrix. In this way the Plutonium production will be minimized at production and the role of the EA limited to the fresh fuel production for the PWR through Pu $\rightarrow$ U conversion of some, already existing Plutonium. The initial PWR fuel for the first fill is made of Thorium Oxide mixed with ^{233}U-rich Oxide mix from the EA in a percentage of 3.651 %, chosen (as previously) such as to equalize the mean k_∞ to the one of the reference mixture. The dependence of the fuel reactivity is very close to the one of the reference case (Figure 4.6). At the end of the cycle, Fission Fragments are separated from Actinides with the help of reprocessing and destined to the Secular Repository. The fuel for the next cycle is made with the total Actinide residue from the previous cycle (with eventually some of the most toxic long lived FFs), topped with fresh Thorium in order to maintain the active mass of fuel constant. However only a smaller fraction of ^{233}U-rich Oxide mix must be added, since some of the ^{233}U is bred in the PWR during the burn-up. The concentration of ^{233}U depends on the balance between the amount burnt and the one bred through the process $^{232}Th + n \rightarrow {}^{233}Th \xrightarrow{(\tau_{1/2}=22.3m)} {}^{233}Pa \xrightarrow{(\tau_{1/2}=27d)} {}^{233}U$

As well known for thermal neutrons and at sufficiently small neutron flux the relative ^{233}U concentration tends to a breeding equilibrium of $C_\infty = 1.35$ % [1]. In the present configuration the externally added fraction is important ($C_0 = 3.29$ % of ^{233}U) and the concentration $C(t)$ tends as a function of time roughly exponentially to the breeding equilibrium, namely $C(t) = (C_0 - C_\infty)\exp(-\lambda t) + C_\infty$, as shown in Figure 4.7. The exponential constant (fitting of ORIGEN prediction) is $1/\lambda$ = 38.82 (GW × day/t)$^{-1}$. The net fraction of the ^{233}U which has been burnt and which has to be replaced by a fresh supply from the EA is therefore $\Delta C(T) = (C_0 - C_\infty)(1 - \exp(-\lambda T))$, where T is the design fuel burn-up. With our parameters $\Delta C(T)$ = 1.11%, namely only 33.8% of the initial load. *The ^{233}U-rich mix from the EA in the PWR has therefore an energy yield with a Thorium matrix which is 2.96 times better than with a Uranium matrix.* The trans-uranic Actinides produced are very small[30], and they can be safely neglected[31].

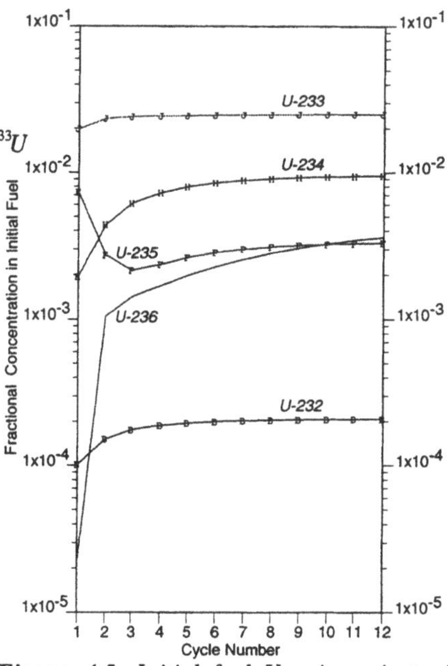

Figure 4.5: Initial fuel Uranium isotopic concentrations as a function of the cycle number.

However, operation with a Thorium based fuel of an ordinary PWR is significantly more complicated, since the long lifetime of the ^{233}Pa produces, as well known, a reactivity variation as a function of shut-down and of power variations which has a magnitude and features which resemble the well known Xe-effect. As already pointed out this effect is absent in Fast Neutron devices and in particular in the EA. The phenomenology is well described in the literature [11]. The maximum change in reactivity switching off the Reactor (which grows with the 27 days ^{233}Pa half-life), is approximately equal to the ratio of ^{233}Pa and ^{233}U concentrations, $(\Delta k/k)_{t=\infty} \approx N(^{233}Pa)/N(^{233}U)$ and it is shown in Figure 4.8 as a function of the burn-up.

---

30 At the end of the cycle: Np (75.87 gr/GWatt(t)) and Pu (21.29 gr/GWatt(t)). Other Actinides (Am, Cm, etc.) are truly negligible.

31 They could eventually be separated and incinerated in the EA with negligible consequence to the inventory stockpile.

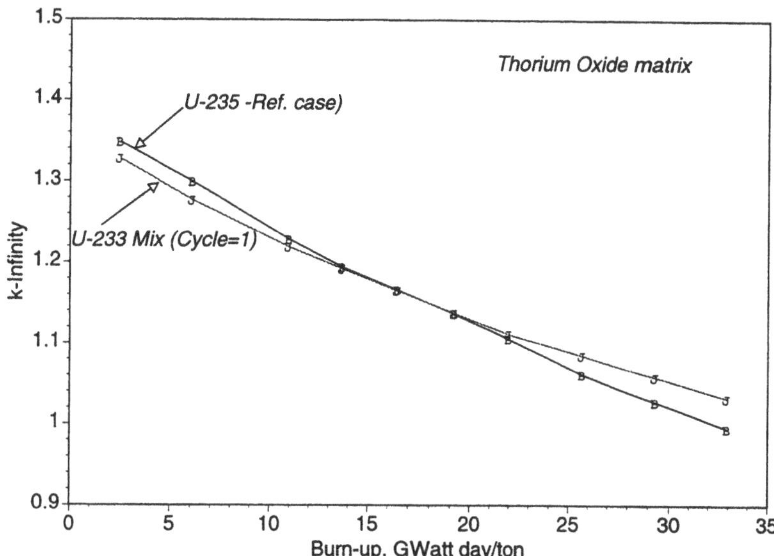

***Figure 4.6**: Reactivity k_∞ as a function of the burn-up within the cycle for the first cycle with Thorium Oxide Matrix and for the ^{235}U reference case. The ^{233}U-rich mix from the EA is mixed at the percentage of 3.651 %,*

In the most unfavourable conditions it may amount to a $\Delta k/k = +0.08$ change with a time constant of $27 \times 1.406 = 38$ days. Finally at the neutron flux conditions[32] of the present example, there is a significant probability that the ^{233}Pa may be captured before it decays $^{233}Pa + n \rightarrow {}^{234}Pa \xrightarrow{\tau_{1/2}=6.7h} {}^{234}U$, thus missing the highly fissionable ^{233}U step and causing a reduction in $\Delta k/k$. For a more quantitative analysis we refer to Ref. [1].

Finally the presence of ^{233}U in a Thorium bulk material could open the way to military diversions of the highly fissile ^{233}U by simpler chemical separation of the fuel. This question has been amply discussed in Ref. [1], to which we refer for more details. In short, the presence of a strong γ-emitter in the ^{232}U, decay chain will make the realization of such a bomb very hard, at any rate harder than just extracting the "Dirty Plutonium" from the reactor fuel during standard operation. Hence the use of the EA-derived fuel should not increase the diversive risks with respect to the same PWR already operating with ordinary fuel. Should it not be considered sufficient, it is possible to "denaturate" the Uranium in the fuel [11], by adding a significant amount of ^{238}U in such a way that the resultant critical mass of the (chemically) extracted Uranium is considerably increased. In this way the production of Plutonium waste,

---

32 The neutron flux is $\phi = 2.7 \times 10^{14}$ cm^{-2} s^{-1} at the beginning of burn-up and grows approximately linearly to $\phi = 3.8 \times 10^{14}$ cm^{-2} s^{-1} at the end of the cycle

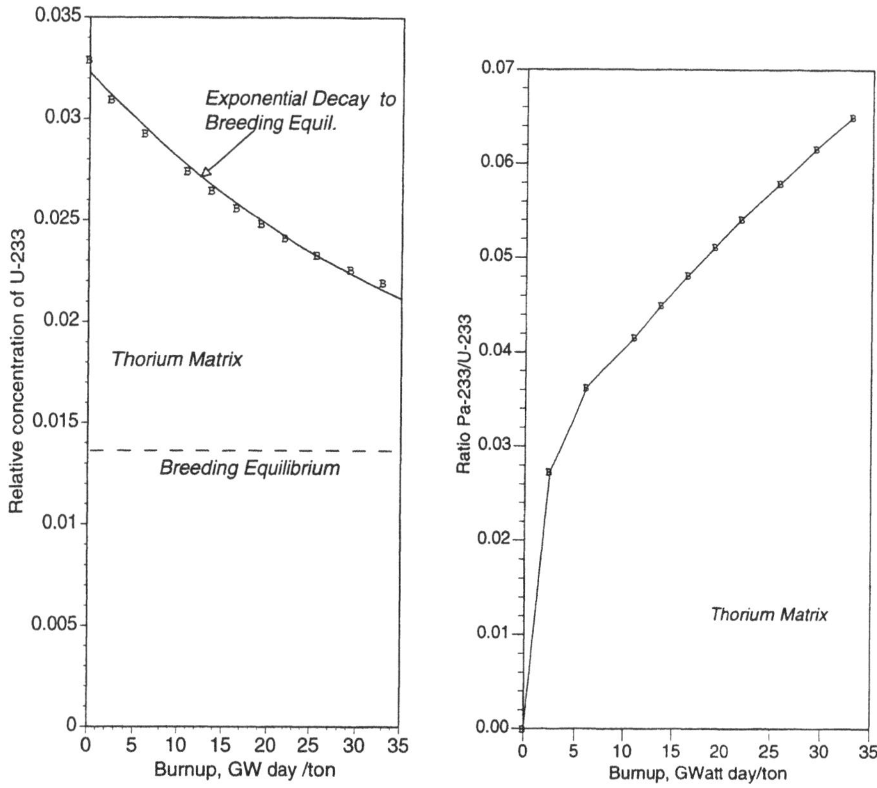

Figure 4.7: *^{233}U relative concentration in the fuel as a function of the burn-up within the cycle for the first cycle with Thorium Oxide Matrix. The fit formula is given in the text.*

Figure 4.8: *$^{233}Pa/^{233}U$ ratio of atomic concentrations in the fuel as a function of the burn-up within the cycle for the first cycle with Thorium Oxide Matrix.*

although larger than without "denaturation" will be much less than the one with Uranium based fuel.

5- Mass flow in the combined EA-PWR incineration cycle

The general mass flow is shown in Figure 1.1. Values shown are for asymptotic conditions. We have chosen a cluster of 3 EA units for a total thermal power of 4500 MW and an electric power of 1870 MWe. The fuel burn-up is set to 120 GW × day/t.

The EA units are fed with fresh metal Thorium at the rate of 1.251 tons/y and "Dirty Plutonium" at the rate of 1.260 tons/y. They produce 794.5 kg of fresh metal Uranium[33] and 1.66 Tons of Fission Fragments (the slight mass defect is due to the energy produced). The fuel fabrication is based on known MOX-technologies and the chemical reprocessing is only required to separate out of the spent fuel the three main components, namely (1) Uranium (794.5 kg/y) (2) Actinides other than Uranium (8.771 ton/y of Th and 2.404 ton/y of Pu and other Actinides) and (3) the fission fragments (1.66 tons/y), fuel claddings plus miscellanea.

The PWR size has been chosen in order to match the rate of ^{233}U-rich Mix from the EAs[34]. The corresponding total thermal power is 4187 MW(t) for a utilization factor of 0.80 and a burn-up of 33 GW $\times$ day/t. The total amount of fuel of each early batch is 37.213 tons. The dwell time of each fuel batch in the PWR is three years.

As already mentioned, at each cycle fresh Uranium (approximately 825 kg) and new ^{233}U-mix from the EA must be added in the fuel fabrication. We have kept constant the amount of ^{233}U-mix and slightly adjusted the amount of added fresh Uranium in order to maintain the fuel mass constant. At the end of the cycle, by reprocessing, all other Actinides other than Uranium and the FFs are collected. The spent fuel (with a slight loss due to the mass to energy conversion) is reprocessed in the same way as the one of the EA and partitioned[35] in (1) Uranium (35.59 tons/y) (2) Actinides other than Uranium, called "Dirty Plutonium" (358 kg/y of Pu and other Actinides) and (3) the fission fragments (1.261 tons/y), fuel claddings plus miscellanea. The Actinides are actually subdivided as follows: Np (18.7 kg/y), Pu (332.3 kg/y), Am (5.74 kg/y) and Cm (0.81 kg/y).

The "Dirty Plutonium" locally generated is incinerated together with the surplus of external origin which has to be incinerated. Hence the "surplus" incineration capacity of the complex drops to 0.902 ton/year. The PWR nominal power associated to such a surplus Pu yield is of the order of 4187/358 x 902= 10.55 GW(t), namely about the "waste" of 3 additional 3.51 GW(t) power stations in addition to the one already fed with ^{233}U. In total a single 4.5 GW(t) EA can match the "waste" rate related to a total of 10.55 + 4.187 = 14.737 GW(t) produced in PWRs.

---

33 The actual isotopic composition is as follows: 716.35 kg/y of ^{233}U; 3.68 kg/y of ^{232}U; 68.25 kg/y of ^{234}U; 5.43 kg/y of ^{235}U; 0.79 kg/y of ^{236}U. See Ref. [1] for more details.

34 We do not mean that we must have 1 PWR which has exactly this power. This is the equivalent power which has to be generated with the manufactured fuel in order to burn all the produced fuel. In practice it can be achieved in a number of ways and with different mixtures of new and standard fuel.

35 Actual numbers refer to the 11-th cycle. For more details we refer to paragraph 4.

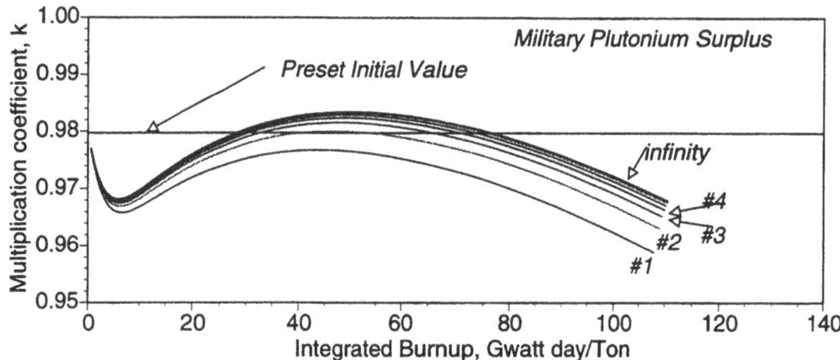

***Figure 6.1**: Variation of the multiplication coefficient* k *as a function of burn-up for different, successive fillings and a specified, arbitrary initial value.*

These are idealized conditions and are for illustration purposes. For more realistic schemes we refer to paragraph 1.

Other very interesting schemes are possible if, as already mentioned, the PWR is operated with a Thorium based Fuel. We recall that the ^{233}U-rich mix from the EA in the PWR has therefore an energy yield with a Thorium matrix which is 2.96 times better than with a Uranium matrix. Hence the same 14.737 GW(t) of power could be produced in PWRs using the ^{233}U-mix from the only one single EA module of roughly 1.50 GW(t)[36]. The "Dirty Plutonium" incineration rate required to keep the system going will obviously be only 1260/3 = 420 kg/y!

6- Elimination of Military Plutonium

The possibility of extending our scheme to the elimination of military Plutonium is briefly considered. The Plutonium is now essentially pure ^{239}Pu, which implies a much higher fission cross-section. During incineration higher Pu isotopes and minor Actinides are formed and progressively incinerated. But since they are not supplied with the initial incinerable material, their concentration is significantly lower.

---

36 Following Ref. [1] we have defined a EA power station a cluster of three modules.

The multiplication coefficient with periodic Pu fillings with initial concentration set to produce a specified initial value of the multiplication coefficient k_0 is very constant over an extended burn-up and many cycles, reaching an asymptotic concentration of acceptable properties (Figure 6.1). In this respect the two fillings of "Dirty Pu" and pure ^{239}Pu have rather similar behaviour. Required concentrations of the mix as a function of the cycle number is shown in Figure 6.2 . The accumulated amount of incinerated material and of the fresh ^{233}U as a function of the number of cycles performed is shown in Figure 6.3 .

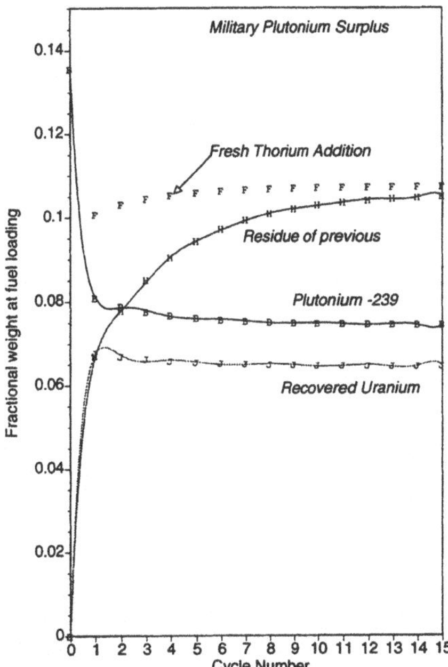

Figure 6.2: Fractional composition at the beginning of the cycle, as a function of the cycle number.

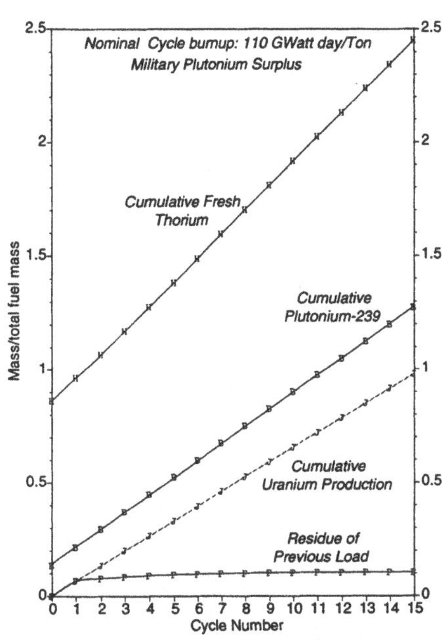

Figure 6.3: Cumulative amount of incinerated material and of the fresh ^{233}U as a function of the number of cycles.

We remark the almost linear elimination rate, amounting to 7.56% (9.20%) of the initial metal fuel mass for each cycle (110 GW × day/t) and the almost linear production rate of ^{233}U amounting to 6.52 % (it was 5.801 % in the "Dirty Pu" case) of the fuel mass. Hence the Pu → U conversion efficiency is quite large, 86.18 % (it was 63%). The energetic efficiency in burning ^{239}Pu is of the order of 70% for a cycle lasting 110 GW × day/t. The contribution of various Actinides to the (fission) power is shown in Figures 6.4a and 6.4b. It is not very different than the corresponding figures for "Dirty Plutonium".

At a constant energy production rate of 1500 MW_{th} a single EA unit will accumulate a yearly energy production of 547.5 GW × day, corresponding to the elimination of 547.5/110 × 0.0756 = 0.376 ton/y of ^{239}Pu independently of the actual amount of fuel mass and cycle duration. The correspondingly produced Uranium is 547.5/110 × 0.0652= 0.324 tons which can be used to manufacture about 12.5 tons of PWR fuel with 2.3 % of ^{233}U. This means that the elimination of the totality of the 100 tons of ^{239}Pu will produce a total of 400 GW × year in the EAs. Two clusters of 3 EA units each for a total thermal power of 4500 MW and an electric power of 1870 MWe could *eliminate in about 45 years the 100 tons of unwanted ^{239}Pu.* An additional period of similar duration with the EAs operated as "amplifiers", in which the ultimate burn-up is performed "parasitically" will ensure the final reduction of the residue to the level of few 10^{-3} of the initial mass (≤ 200 kg) (Figure 1.7). The 86.18 tons of ^{233}U produced in the transformation can be used to produce as much 86.18/0.02375 = 3629 tons of useful fuel for PWRs. The produced fresh fuel is sufficient to run a standard 1000 MW (electric) PWR for about 100 years.

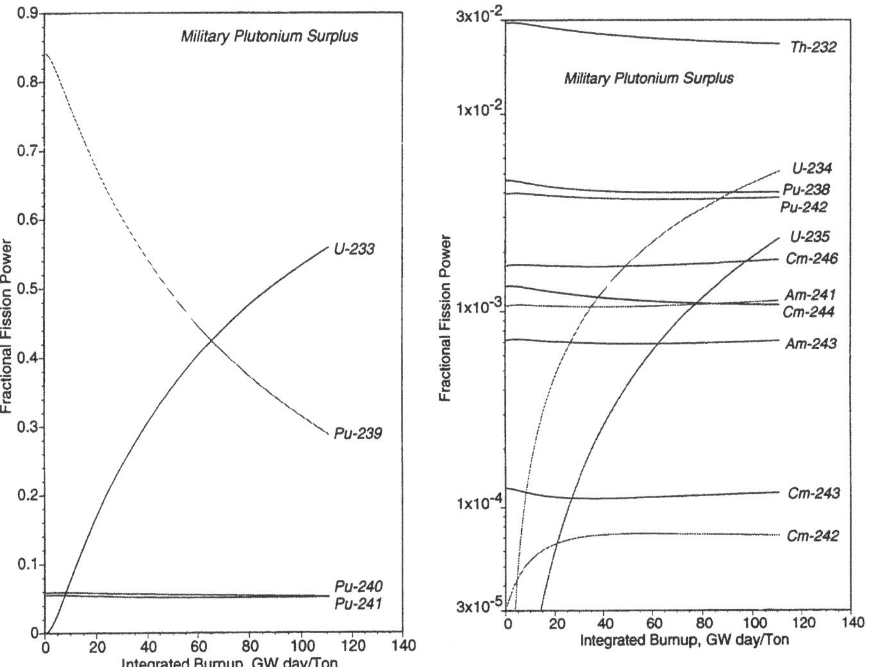

Figure 6.4a,b *Fractional power contributions (fissions) as a function of the burn-up for an asymptotic fuel composition. (a) large concentrations; (b) small concentrations.*

Acknowledgements.

We would like to acknowledge the dedicated work of **Susan Maio** and **F. Saldaña**. We would like to thank many of our CERN colleagues for frequent discussions of various aspects of the project and for their continuing and enthusiastic support and in particular: **F. Carminati, N. Fiétier, J. Galvez, C. Gelès, R. Klapisch, P. Mandrillon, J.P. Revol** and **Ch. Roche**.

We would like to acknowledge the continuing support of the CERN Management (in particular of **H. Wenninger**), of the Sincrotrone Trieste (in particular **G. Viani**), and the DGXII of the European Commission, in particular its Director General, **P. Fasella**, as well as **H.J. Allgeier, G. C. Caratti** and **W. Baltz**.

References.

[1] C. Rubbia et al, "Conceptual Design of a Fast Neutron Operated High Power Energy Amplifier", CERN/AT/95-44 (ET), 29th September 1995. See also C. Rubbia, "A High Gain Energy Amplifier Operated with Fast Neutrons", AIP Conference Proceedings 346, International Conference on Accelerator-Driven Transmutation Technologies and Applications, Las Vegas, July 1994.

[2] Report from the Committee on International Security and Arms Control, "Management and Disposition of Excess Weapons Plutonium", National Academy of Science Press, Washington D.C., 1994.
French Law from 30th December 1991 (Curien Law).

[3] J.P. Schapira, Report IPNODRE-94-04 (1994)

[4] T. Takizuka et al., Specialist Meeting on Accelerator-driven Transmutation for Radwaste and other Applications (Stockholm, 24-28 June 1991);
G. Van Tuyle et al., BNL Reprt 52279 (1991);
H. Takahashi, Fusion Technology 20 (1991) 657.
V. D. Kazaritsky et al., in Accelerator Applied to Nuclear Waste, 8th Journées Saturne, Report LNS/Ph/94-12
See also Ref. [6]. Many peapers have been written on the subject. We refer to Ref. [1] for a more comprensive list.

[5] J. Rouault et al., "Physics of Pu Burning in Fast Reactors: Impact on Burner Core Design", Proc. ANS Topical Meeting on Advances in Reactor Physics, Knoxville, 11-15 April 1994.

[6] C. Bowman et al., Nucl. Instr. and Methods, A330, 336 (1992). More recent developments are to be found in the AIP Conference Proceedings 346, International Conference on Accelerator-Driven Transmutation Technologies and Applications, Las Vegas, July 1994

[7] H. Rief and H. Takahashi, "Control of Accelerator Driven Sub-Critical Systems Fuelled by Actinides", Proceedings of a Technical Committee Meeting on Safety and Environmental Aspects of Partitioning and Transmutation of Actinides and Fission Products, held in Vienna, 29 November - 2 December 1993, IAEA-TECDOC-783, January 1995.

[8] C. Marchetti "The secular Dynamics of the Energy Systems", 2nd International Conference on Energy Environment and Technological Innovation, Rome, October 1992

[9] Global '95, Int. Conf. on Evaluation of Emerging Nuclear Fuel Cycle Systems, Versailles, 11-14 September 1995.

[10] C. Roche and C. Rubbia, "Some Preliminary Considerations on the Economical Issues of the Energy Amplifier", CERN/AT/95-45 (ET), 29th October 1995.

[11] M. Benedict, Th. H. Pigford and H.W. Levi, "Nuclear Chemical Engineering", McGraw-Hill, 1981.

[12] S. Andriamonje et al., "Experimental Study of the Phenomenology of Spallation Neutrons in a Large Lead Block", Proposal to the SPSLC, SPSLC/P291, 5th May 1995.

[13] M.J. Bell, "ORIGEN, -The Oak Ridge Isotope Generation and Depletion Code ", ORNL-4628, May 1973.

[14] "A realistic estimate of the costs of an Energy Amplifier Scheme", paper in preparation, same authors as present paper.

WEAPONS GRADE PLUTONIUM DESTRUCTION IN THE GAS TURBINE MODULAR HELIUM REACTOR (GT-MHR)

by
David Alberstein
Director, Plutonium Consumption Program
General Atomics
San Diego, California, USA

Abstract

The Gas Turbine Modular Helium Reactor (GT-MHR), when fueled with surplus weapons grade plutonium, has the unique capability to destroy 90% of the initially charged plutonium-239 and 65% of the intitially charged total plutonium in a once through reactor cycle while generating electricity at plant efficiencies of nearly 50%. The plutonium content and quality in the spent fuel is so low that there is little or no military or commercial incentive for reprocessing and recycle. The spent fuel is well suited for disposal as whole elements in a geologic repository. The unique inherent passive safety characteristics of the GT-MHR result in a design that is meltdown proof and insensitive to operator errors. The high efficiency of the GT-MHR results in minimal environmental impact and substantial advantages in plant economics.

Since the summer of 1994, the Russian Federation Ministry for Atomic Energy (MINATOM), General Atomics (GA) in the United States, and more recently Framatome in France have been participating in a cooperative program to develop the GT-MHR for disposition of surplus weapons grade plutonium in Russia. The near term objective of this program is to construct a GT-MHR plant at Seversk (Tomsk-7) that would burn weapons grade plutonium and would replace the power provided by the plutonium production reactors at that site. Fuel development and conceptual design activities are underway at several Russian institutes and are scheduled to be completed and fully documented in October 1997. International support for this effort has been increasing in recent months. Framatome joined the conceptual design effort in January 1996, and organizations in several other major nations with nuclear power capability have expressed interest in participating.

Russia is interested in the GT-MHR for plutonium destruction because it offers a uniquely high level of reactor safety, it is the only nuclear reactor capable of using the direct cycle gas turbine for production of electricity at plant efficiencies of nearly 50%, the uranium fueled version of the reactor has high commercial potential as an export commodity that will address balance of payments issues, it achieves a higher level of plutonium destruction (and a higher amount of electrical energy per unit mass of plutonium consumed) without recycle than any other reactor technology, and its fuel cycle offers superior diversion and proliferation resistance.

This paper provides a description of the GT-MHR and its plutonium disposition characteristics, including the diversion and proliferation resistance characteristics of the fuel cycle. Information

E. R. Merz and C. E. Walter (eds.), Advanced Nuclear Systems Consuming Excess Plutonium, 135–146.
© 1997 *Kluwer Academic Publishers.*

is also presented on deployment cost and schedule, on the gas-cooled reactor experience base, and on the status of the current GT-MHR development program.

Plutonium-Fueled GT-MHR Description

The GT-MHR is a passively safe, helium cooled, graphite moderated, advanced reactor system that is based on existing technology. Thermal energy is converted to electric power by use of a direct Brayton cycle helium gas turbine power conversion system.

The GT-MHR reference plant consists of four 600 MWt (286 MWe) modules capable of providing a total net electrical generating capacity of 1144 MWe. As shown in Figure 1, which shows the arrangement of one module, the reactor core is contained in an uninsulated steel reactor pressure vessel that is connected by a cross vessel to a vessel that contains the power conversion system. The reactor and power conversion vessels are, respectively, 8.4 m and 8.5 m in diameter. The modules are located below grade in a 39 m deep, high pressure, low leakage containment with characteristics typical of those of commercial light water reactors.

Refractory coated particle fuel, shown in Figure 2, is used in the plutonium-fueled GT-MHR. The fuel is in the form of tiny (200 μm diameter) plutonium oxide fuel kernels coated first with a porous graphite buffer layer followed by layers of silicon carbide and pyrolytic carbon. This system of particle coatings is referred to as a TRISO coating. The coated fuel particle total diameter is about 635 μm. The particles are mixed with graphitic material and formed into cylindrical fuel rod compacts 12.45 mm in diameter and 49.3 mm long. The fuel rod compacts are inserted into hexagonal prismatic graphite fuel element blocks, as shown in Figure 2. The completed fuel element is 0.79 m high and 0.35 m wide across the flats, and it weighs about 115 kg. The fuel element configuration is identical to that successfully demonstrated in the Fort St. Vrain reactor in the United States. A standard fuel element contains about 22 million coated fuel particles.

During the past 30 years, coated fuel particles have been successfully fabricated and demonstrated in high temperature reactors and in numerous successful irradiation tests in Russia, the U.S., Germany, and Japan. Successful irradiation of fuel with high fissile material content to conditions which exceed the requirements of the GT-MHR for plutonium disposition has been demonstrated. Highly enriched uranium coated particle fuel designed for irradiation at more than 1250°C, peak fuel burnup up to 750,000 MWd/t, and fast neutron fluence up to 8 x 10^{25} n/m^2 (E > 0.18 Mev) was reviewed and approved by the U.S. Nuclear Regulatory Commission for use in Fort St. Vrain. Extensive operation and test data on coated fuel particles confirm their performance up to and beyond the maximum temperatures that are expected to occur in design basis accidents.

Six irradiation tests have been conducted using near weapons grade (up to 88% plutonium-239) plutonium coated particle fuel. Five of these tests were performed in the DRAGON reactor in the late 1960s, and the sixth was performed by General Atomics in the Peach Bottom gas cooled reactor in the early 1970s. For each of these tests, the fuel consisted of TRISO coated plutonium fuel kernels. The fuel for the DRAGON tests was fabricated by Belgonucleaire in Europe. For the Peach Bottom test, coated particles and fuel rod compacts were fabricated at Oak Ridge National Laboratory. Conditions in these six tests included irradiation at temperature up to 1450°C, peak fuel burnup up to 747,000 MWd/t, and fast neutron fluence up to 2.2 x 10^{25} n/m^2 (E > 0.18 Mev). These temperature and burnup conditions are expected to envelope those experienced by the coated fuel particles in a plutonium-fueled GT-MHR. For all of these tests, the plutonium fuel performed well, establishing the feasibility of high burnup, TRISO-coated, plutonium fuel for the GT-MHR and confirming the high levels of plutonium destruction that the GT-MHR can achieve.

The reactor core consists of 1020 hexagonal prismatic fuel elements stacked in a ten element high annular array of 102 columns, as shown in Figure 3. One third of the fuel elements is replaced annually. Because the plutonium-fueled GT-MHR uses no fertile fuel material, excess reactivity control and negative temperature feedback are provided through use of erbium oxide poison rods located in selected fuel holes in the graphite fuel elements. The core has a strong negative temperature coefficient of reactivity.

The direct Brayton cycle power conversion system results in considerable simplification of balance of plant design relative to conventional Rankine steam cycle plant designs. The entire power conversion system is located in the power conversion vessel. The turbomachine consists of a generator, gas turbine, and two compressor sections submerged in helium and vertically mounted on a single shaft supported by magnetic bearings. The power conversion system includes three compact heat exchangers: a highly effective recuperator and water-cooled precooler and intercooler. Extensive technology data bases exist for these individual components at their required service conditions, and integrated testing of the first complete power conversion system using a fossil-fired heat source is planned prior to installation in the first GT-MHR.

The GT-MHR is unique in that it is the only reactor capable of using the Brayton cycle and achieving plant efficiency near 50%. The GT-MHR process flow is shown in Figure 4. Helium coolant exits the reactor core at 850°C and 7.01 MPa, flows through the center hot duct within the cross vessel, and is expanded through the turbine in the power conversion vessel. The turbine directly drives the electric generator and the high and low pressure compressors. Helium exits the turbine at 510°C and 2.64 MPa and flows through the highly effective recuperator to return as much energy as possible to the cycle, and then through the precooler to reject heat to the ultimate heat sink. Relatively cold helium at 26°C enters the first of the two compressor sections, exits to an intercooler where additional heat is rejected to the ultimate heat sink, and then enters the second compressor section, from which it exits at 7.24 MPa and 112°C and passes through the recuperator. Helium at 490°C and 7.07 MPa flows from the recuperator exit, through the outer annulus within the cross vessel, and back to the core inlet and downward through the core to complete the loop.

The GT-MHR retains virtually all fission products within the fuel particle coatings under all accident conditions. This retention is accomplished through a combination of inherent safety characteristics and selection of passive design features. These include: (1) helium coolant, which is single phase, inert, and has no reactivity effects; (2) graphite core, which provides high heat capacity and slow thermal response, and structural stability at very high temperatures; (3) refractory coated plutonium oxide particle fuel, which allows extremely high burnup and retains fission products at temperatures much higher than normal operation and peak accident conditions; (4) negative temperature coefficient of reactivity, which inherently shuts down the core above normal operating temperatures; and (5) an annular, low power density core (about 6.3 MW/m^3) in an uninsulated steel reactor vessel surrounded by a reactor cavity cooling system, which enable passive heat transfer to the ultimate heat sink while maintaining fuel temperatures below damage limits. A high pressure containment is also provided for additional margin and defense in depth. These safety design features result in a reactor that can withstand loss of coolant circulation or even loss of coolant inventory and maintain fuel temperatures below damage limits without operator intervention or use of active safety systems. No credible severe accident scenario involving a large release of radionuclides from the core has been identified for the GT-MHR.

Plutonium Disposition Capability

Evaluations of the plutonium disposition capability of the GT-MHR show that it is capable of destroying more than 90% of the initially loaded plutonium-239 and more than 65% of the initially loaded total plutonium in a single pass through the reactor. As shown in Figure 5, this is the highest level of plutonium destruction without recycle attained by any reactor alternative for plutonium disposition. This high level of plutonium destruction is achievable because of the high burnup capability of the coated particle plutonium fuel and because the plutonium fueled GT-MHR uses no fertile fuel, so no new plutonium is bred. As a result, the GT-MHR is the most effective option for maximizing the energy content recovery from the material and for destroying weapons grade plutonium without reprocessing. One 4-module GT-MHR plant is capable of dispositioning about 1 metric tonne of weapons grade plutonium per year to this high level of destruction. This capability, along with the high efficiency of the plant, maximizes the amount of electrical energy recovered per unit mass of plutonium processed.

During the recent joint U.S./Russian study of plutonium disposition options, Russia examined two alternatives for disposition of 50 metric tonnes of weapons grade plutonium using the GT-MHR. In the first variant, three 4-module plants would be built, one plant at Seversk (Tomsk-7), one at Krasnoyarsk, and one at Mayak. The first reactor module would be started up about 8.5 years following the decision to proceed, and the twelfth module would be started up about six years later. These plants would be capable of converting 50 metric tonnes of plutonium to highly depleted spent fuel within 25-30 years following the decision to proceed with the project.

In the second variant, one plant containing four reactor modules would be built at Seversk. The first reactor would be started up about 8.5 years following the decision to proceed, and the fourth module would be started up about two years later. This plant would operate with weapons grade plutonium fuel during its entire design lifetime and would be capable of converting 50 metric tonnes of weapons grade plutonium to highly depleted spent fuel over about 56 years following the decision to proceed.

The same capacity fuel fabrication facility would be used for both variants. This would enable the fabrication of 50 metric tonnes of weapons grade plutonium into coated particle fuel compacts to begin within 5 to 6 years and to be completed, at a rate of 3 tonnes per year, within about 25 years of the decision to proceed. The fresh fuel, in the case of the second variant, would be stored until needed to fuel the GT-MHR reactors. MHR fresh fuel is sufficiently resistant to diversion and reprocessing to meet the requirements for plutonium disposition. The concentration of plutonium in fresh fuel elements for the GT-MHR is about a factor of 30 less than that in light water reactor mixed oxide (MOX) fresh fuel, and the processes for separation of plutonium from fresh fuel are relatively complex and not well developed. Each fresh fuel compact contains only 0.2 grams of weapons grade plutonium in a volume of 6×10^{-6} m^3. The fuel in each compact is contained in about 6,000 coated fuel particles.

Retrieval of plutonium from GT-MHR spent fuel would be difficult for all the reasons that apply to other forms of spent fuel (high radioactivity, large mass, extensive safeguards and security, etc.) but also because of additional reasons that are uniquely characteristic of this fuel and add to the difficulty of plutonium retrieval. The plutonium dilution in each spent fuel element is very high (200 to 250 grams in an element that weighs 115 kg); the technology for separating plutonium from MHR spent fuel has not been fully developed; and the isotopic composition of the discharged plutonium (about 28% plutonium-239, 30% plutonium-240, 33% plutonium-241, and 9% plutonium-242) is, due to the high level of burnup obtained, particularly unattractive for use either in explosives or in fuel cycle applications. Hence, there is little or no incentive for recycling of the spent fuel of the GT-MHR for either military or commercial purposes.

The refractory spent fuel elements are an excellent waste package that is suitable for direct disposal, without further modification, in whole element form inserted into a canister, in a geologic repository. The TRISO coated particle fuel has the benefit of long term containment of radionuclides without relying on the performance of the waste package and surrounding geologic media. Quantitative assessments show that the TRISO coating should maintain its integrity for hundreds of thousands of years in a repository environment. Leaching tests have demonstrated that coated particle fuel retains radionuclides better than borosilicate glass. Although the spent fuel volume of the plutonium fueled GT-MHR is larger than that of other reactor options for plutonium disposition (resulting in excellent diversion and proliferation resistance), the decay heat generation rate from the spent fuel is lower. Because the loading density of spent fuel in a geologic repository is governed by decay heat generation rate and not by spent fuel volume, the total repository volume required to accommodate GT-MHR spent fuel is similar to that for other reactor spent fuels.

Deployment Cost

During the joint U.S./Russian study of plutonium disposition options, Russia prepared an estimate of the cost to deploy the GT-MHR for plutonium disposition in Russia. Table 1 presents a summary of overall capital expenditures for each of the two alternatives evaluated during the joint study. Costs are presented in 1991 dollars. As indicated in the table, the capital cost for one 4-module GT-MHR plant, including engineering development and the fuel fabrication facility, is about $1.34 billion. The cost for three 4-module plants is about $3.12 billion.

For comparison purposes, these costs are comparable to costs of other options for disposition of weapons grade plutonium in Russia, and they are about 1/3 of the cost required to develop and deploy the GT-MHR for plutonium disposition in the U.S.. It should be noted that the low cost of the fuel fabrication facility for the GT-MHR results from the fact that the GT-MHR uses pure plutonium fuel with no fertile material. This involves significantly lower amounts of heavy metal processing for fuel production and less scope of fuel fabrication facility construction as compared to other reactors with mixed oxide fuel.

Gas-Cooled Reactor Experience Base

There is extensive world-wide experience with gas-cooled reactors. More than 50 gas-cooled reactors have been built and operated since 1956. These include five high temperature, helium-cooled reactors: the DRAGON test reactor in Great Britain, the Peach Bottom Unit 1 reactor in the U.S., the AVR reactor in Germany, the Fort St. Vrain reactor in the U.S., and the THTR in Germany. All five of these reactors have used highly enriched uranium (93% U-235) fuel in the form of coated particles. Many of the design features of the GT-MHR are based on features of these earlier high temperature reactors that were successfully demonstrated. In addition, construction of the HTTR reactor has been completed in Japan, and construction of the HTR-10 reactor is underway in the People's Republic of China. HTTR component testing is underway, and startup testing is scheduled to begin in late 1997.

The design of the GT-MHR is also based on more than 30 years Russian experience in designing gas cooled reactors, including the development of detailed designs of several plants both for electricity generation and for process heat applications. Russia has built and operated a number of experimental, test, and process development facilities for reactor fuel, core design, and plant system design, and has developed analytical procedures and codes. The Experimental Design Bureau of Machine Building (OKBM) in Nizhny Novgorod, the Russian Research Center-Kurchatov Institute (RCC-KI), the A.A. Bochvar All Russian Scientific Research Institute of Inorganic Materials (ARSRIIM) in Moscow, and the Scientific and Industrial Association Research Institute (LUTCH) in Podolsk have been the principal participants. As result of these programs, a large cadre of engineers and scientists that are fully familiar with gas-cooled reactor technology exists at a number of sites within Russia.

Current International Development Program Status

Discussions in late 1992 between General Atomics (GA) and Russian officials of opportunities in the nuclear energy field identified strong common interests in the development of gas-cooled reactors. In April 1993, General Atomics signed a Memorandum of Understanding with the Russian Federation Ministry for Atomic Energy (MINATOM) to enter into a cooperative agreement

> *"to cooperate in the promotion, implementation, and execution of a joint U.S./Russian design and development program of a Modular Helium Reactor (MHR) with a direct cycle Gas Turbine (GT-MHR) and subsequently construct, test, and operate a prototype in Russia."*

The GT-MHR would be capable of using either uranium or weapons-grade plutonium as its fuel. The primary objective of this effort as originally conceived was to develop the GT-MHR as an export commodity for both the U.S. and Russia. Design and construction of the prototype would be done in Russia to U.S. and other international standards to ensure the suitability of the design for licensing in the U.S. and in other potential world markets. Plutonium disposition was also noted in the Memorandum of Understanding to be a mission for which the GT-MHR was well suited.

In November 1993 a detailed proposal for a joint U.S./Russian program for the development of the GT-MHR was formally submitted by MINATOM to the U.S. Government, requesting support of the work in Russia using Nunn-Lugar funding. Annual U.S. funding support to Russia of $25 million over a period of five years ($125 million in total) was requested. One of the program benefits specifically identified in the Russian proposal is that it:

> *"supports U.S. and Russian proliferation avoidance objectives by providing the capability for the substantial destruction of weapons grade plutonium while producing electric power without the need for fuel reprocessing. The proposed reactor could be used for the destruction and degradation of plutonium made available from the dismantling of nuclear weapons in Russia conditional to a similar program for plutonium disposition being implemented in the United States."*

The program was proposed to be conducted in two phases: 1) a design and development phase, and 2) a prototype construction, startup, and initial operation phase. The cost of the first phase was estimated to be about 1/3 of the cost to perform this work in the United States, with about 85% of the work to be performed in Russia. The cost of the second phase, prototype construction, was again estimated to be approximately 1/3 of the cost to perform the work in the United States. Russia proposed that most of the second phase cost be financed by international financial institutions, with repayment being made from the sale of electricity generated by the plant.

Subsequent to the March 1994 agreement between the U.S. and Russia to cease the production of weapons grade plutonium in Russia by the year 2000, MINATOM in June 1994 proposed that the emphasis of the proposed cooperative GT-MHR program be shifted to plutonium disposition, with development of the uranium fueled version of the reactor as an export commodity considered as a benefit that would still produce a return on funds invested in the development of the first plant. It was proposed that the first modular helium reactors under the cooperative program be built at Tomsk-7 (now known as Seversk), the site of some of Russia's plutonium production reactors, and that they be fueled with surplus Russian weapons grade plutonium.

This proposal would not only reduce the Russian stockpile of surplus weapons grade plutonium, but it would also provide alternative employment for former Russian nuclear weapons program technical staff and generate electricity for use in the surrounding region. The Seversk site is particularly well suited for the plutonium disposition mission because of the extensive experience of the staff with handling of plutonium. All facilities required for weapons grade plutonium disposition could be located at this site, thereby minimizing the need for transportation of plutonium in various forms.

Since the summer of 1994, the MINATOM, GA, and more recently Framatome in France have been participating in a cooperative, cost-shared program to develop the GT-MHR for disposition of surplus weapons grade plutonium in Russia. The near term objective of this program is to design and construct a GT-MHR plant at Seversk that would burn weapons grade plutonium and would replace the power provided by the plutonium production reactors at that site. Fuel development and conceptual design activities are underway in Russia at OKBM, RRC-KI, ARSRIIM, and LUTCH and are scheduled to be completed and fully documented in October 1997. As part of the recent joint U.S./ Russian evaluations of plutonium disposition alternatives, Russia has prepared a report on the use of the GT-MHR for disposition of weapons grade plutonium. Russia has proposed that its cooperative program with GA and Framatome be expanded to include other governments and industrial participants, and Russia has offered to match funding provided to it by other program participants for work to be done in Russia. Organizations in several other major nations with nuclear power capability have expressed interest in participating, and discussions are underway to determine conditions under which they might participate.

With adequate programmatic support, production of plutonium oxide coated fuel particles can begin at the Seversk Chemical Complex in Russia within five to six years. The first GT-MHR reactor module can be operational at Seversk, dispositioning weapons grade plutonium and producing 286 MWe, in 8.5 years, and additional reactor modules of the four module plant can be brought on line at six month intervals. This deployment schedule is based upon a relatively low level of funding in 1996 and 1997 and implementation of the international partnership program by the beginning of 1998, with sufficient funding being available to complete the detailed design in the four year period from the beginning of 1998 to the end of 2001. Construction of the first reactor module in the plant would be performed in the three and one-half year period from the beginning of 2002 to the middle of 2005.

Conclusions

The GT-MHR is an effective system to maximize the level of destruction of surplus weapons grade plutonium without recycle. The plant produces nuclear electricity at high levels of efficiency while providing a superior level of nuclear safety. The GT-MHR fuel cycle has high resistance to diversion and proliferation, making the system particularly well suited for international deployment for plutonium disposition.

Russia has made it clear that it intends to burn its surplus weapons grade plutonium in reactors and has expressed a strong desire to conduct a cooperative program with the U.S. to deploy the GT-MHR for disposition of surplus weapons grade plutonium. The high level of plutonium destruction without recycle and the high level of diversion and proliferation resistance offered by the plutonium fueled GT-MHR will ensure the safe and secure destruction of the stockpiles of surplus weapons grade plutonium while extracting a substantial portion of the useful energy content from this material.

To initiate this cooperative GT-MHR program, General Atomics and MINATOM have each set aside one million dollars of their own funding to support conceptual design activities in Russia. Recently, Framatome has joined the cooperative program, and organizations in other nations with nuclear power capability have expressed interest in supporting this effort. The conceptual design program is underway in Russia and is on schedule to be completed by October 1997.

At a time when worldwide concern regarding diversion and proliferation of surplus weapons-useable fissile materials is very high, this program offers a unique opportunity to reduce the world's inventory of weapons grade plutonium and thereby increase world security. Russia has demonstrated its commitment to use of the GT-MHR for plutonium destruction by committing its own funding and resources to initiate the design development program and by offering to the U.S. Government to cost share the first phase of the program. It has begun conceptual design activities on its own initiative, with cooperation by General Atomics. Other members of the world nuclear community have recognized the merits of this program and have joined the cooperative effort or are exploring how they can join it. No other such program is underway for other plutonium disposition options.

TABLE 1
CAPITAL EXPENDITURES FOR
PLUTONIUM-FUELED GT-MHR DEPLOYMENT IN RUSSIA

Description	Alternative 1 Three 4-modules US M$ (1991)	Alternative 2 One 4-module US M$ (1991)
Engineering Development and Design	275	178
Fuel Development	97	97
GT-MHR Plant Construction	2640	958
Fuel Fabrication Facility Construction	58	58
Spent Fuel Storage Facility Construction	51	51
TOTAL CAPITAL EXPENDITURES	3,121	1,342

FIGURE 1

GT-MHR MODULE ARRANGEMENT

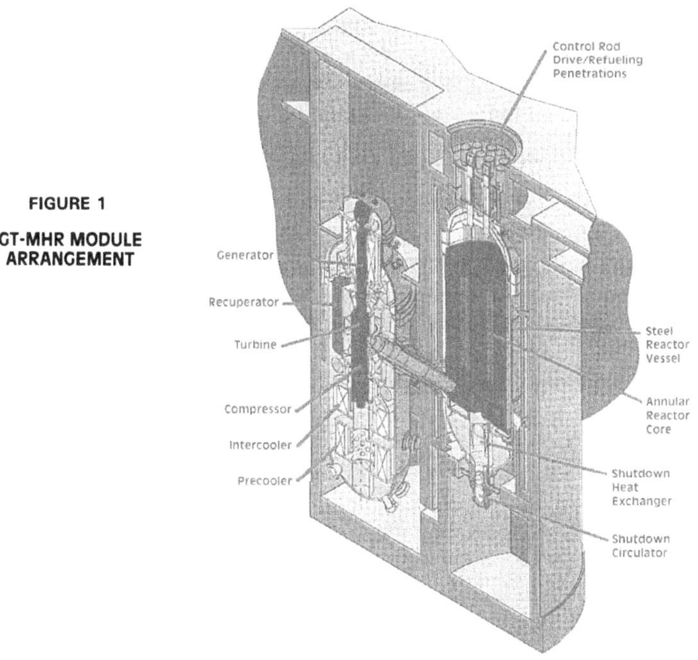

FIGURE 2

PLUTONIUM-FUELED GT-MHR FUEL ELEMENT COMPONENTS

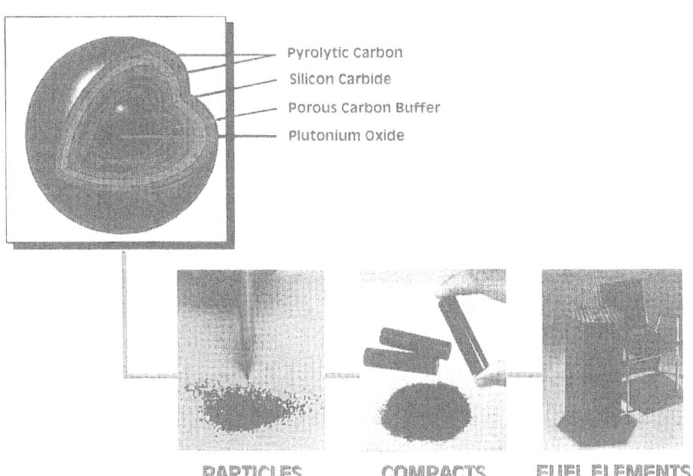

FIGURE 3

GT-MHR CORE LAYOUT

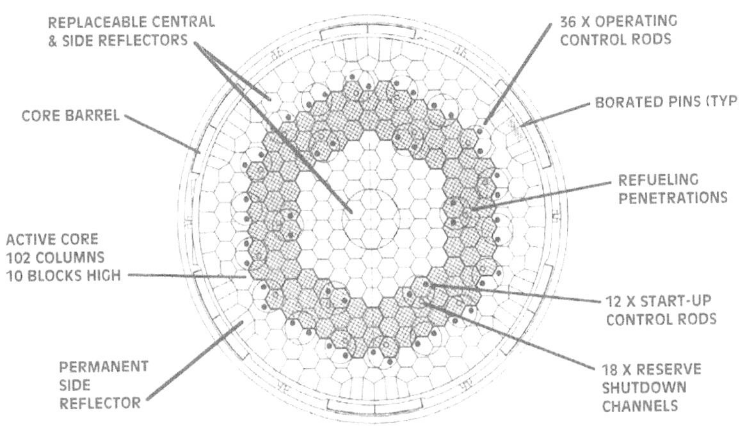

FIGURE 4

GT-MHR POWER CONVERSION PROCESS FLOW DIAGRAM

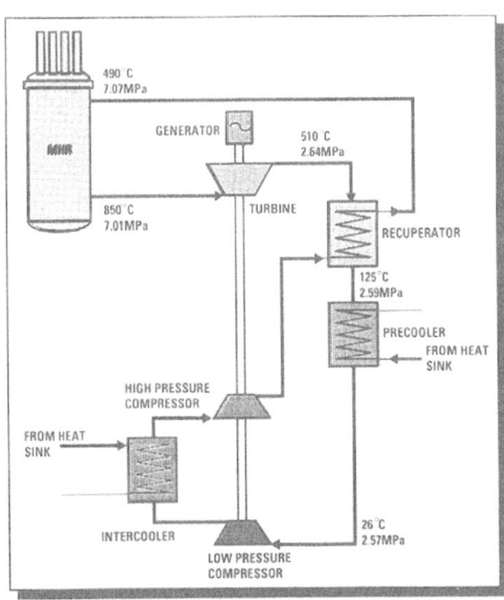

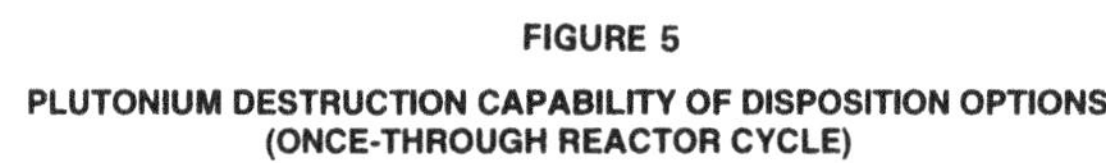

FIGURE 5

PLUTONIUM DESTRUCTION CAPABILITY OF DISPOSITION OPTIONS (ONCE-THROUGH REACTOR CYCLE)

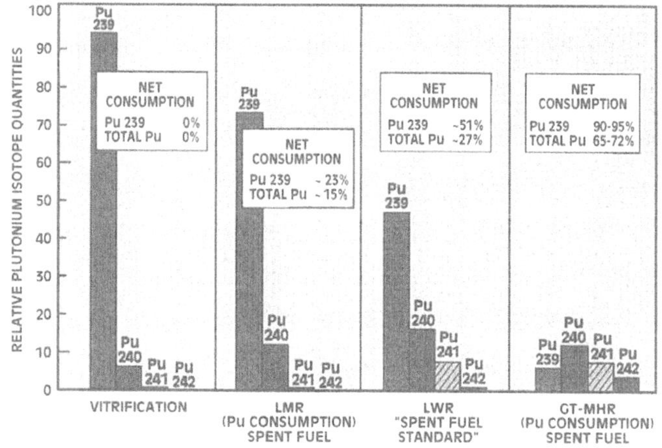

ACCELERATOR DRIVEN SYSTEMS - SOME SAFETY AND FUEL CYCLE CONSIDERATIONS

H. RIEF [1)], J. MAGILL [2)] and H. WIDER [3)]
Joint Research Centre of the European Community
1) Institute for Systems, Informatics and Safety (ISIS), I-21020 Ispra, Italy, retired;
2) Institute for Transuranium Elements, Postfach 2340, D-76125 Karlsruhe
3) Institute for Systems, Informatics and Safety (ISIS), I-21020 Ispra, Italy.

Abstract

A fast spectrum accelerator-driven system (ADS) may become a very versatile tool in the nuclear industry and allow a number of new applications. Such a system may serve as a means of transmuting large quantities of actinides and long lived fission products into short lived waste. Other possibilities, include the cross-progeny fuel cycle burning Pu in a thorium matrix and generating ^{233}U which can be used to refuel light water reactors. In addition, use of a thorium cycle could result in reduced actinide production and a much lower radio-toxicity of the waste.

In this paper we focus on the use of thorium hosted weapon and reactor grade plutonium in a fast ADS from the viewpoint of neutron economy and Pu annihilation rate. We consider a sub-critical system (k_{eff} = *0.98 / 0.99*) such that super-prompt critical excursions, which may result from coolant voiding, molten cladding and fuel movement, or positive temperature coefficients due to a large actinide inventory, etc. are not possible. Such a system would combine the intrinsic safety of an ADS with the advantages of a reactor with a large delayed neutron fraction (imitated by the spallation neutrons) and a rather flat power distribution for nearly uniform burn-up across the whole fuel region. In a system of this kind, a proton beam power of 10-20 MW would produce enough spallation neutrons to obtain a 1 GWe power plant. In this case, a much less expensive and a rather compact multistage cyclotron arrangement might be sufficient to produce the required proton beam power. As safety is the main concern, the JRC embarked in a series of safety studies of near critical systems. Initially we used a simplified kinetics code, which accounted only for negative Doppler feedback in an adiabatic system, to calculate the effect of fast transients. Later, after some modifications, the JRC's sophisticated European Accident Code, (EAC-2) has been used. This includes additional feedback due to axial fuel expansion, sodium voiding and fuel motion. The analysis showed that for fast, or medium to fast reactivity ramps, the ADS has a major advantage in coping with serious reactivity insertions when compared to conventional reactors. On the other hand, "slow" core melting may occur, if in a loss of heat sink or coolant flow accident the accelerator beam is not switched off. Therefore passive means for shutting-off the proton beam are of primary importance.

E. R. Merz and C. E. Walter (eds.), Advanced Nuclear Systems Consuming Excess Plutonium, 147–161.
© 1997 *Kluwer Academic Publishers.*

1. Introduction

Among the unconventional approaches for the proliferation-proof disposition of plutonium, both from strategic and commercial origin, arguments favouring Accelerator Driven Systems (ADSs) and the Cross Progeny fuel cycle combining ^{233}U - ^{238}U and ^{232}Th - Pu are discussed in this paper. In the recent past, the transmutation of plutonium in Accelerator Driven Systems has attracted much attention, but it was also subject of misunderstanding.

In 1985 the JRC Ispra launched a study investigating the possibility of transmuting minor actinides in an accelerator driven subcritical system and presented a quantitative analysis at the 1986 ICENES [2]. The idea was to propose and evaluate, an alternative to underground disposal, as well as the project of a fast critical Japanese actinide burner, which suffered from a small delayed neutron fraction. In this paper it was shown that the actinides generated by several uranium fuelled LWRs can be burned (transmuted) in a single transmutation plant using a 10 - 20 mA proton beam impinging on a spallation target. The obvious reason for choosing a subcritical system was to avoid fast transients leading to dangerous power bursts, as they might be caused by an accidental reactivity insertion such as positive void coefficients, fuel movement, etc.

A few years later, similar studies were started in Japan in the frame of the Omega project ("Options of Making Extra Gains from Actinides") and in the US at LASL and BNL, and in Europe at CERN [1]. Both, thermal and fast systems were analyzed with respect to actinide burning and fission product (F.P.) transmutation. From the very beginning it was obvious that a fast system would be more suitable for actinide burning, although F.P. transmutation requires in most cases neutron energies in the resonance region.

Comparing the ratio of fission to absorption for thermal and fast neutrons in Table 1 it can be seen that a fast neutron flux is essential for efficiently burning even number actinides. Also for the odd number actinides such as ^{237}Np and ^{241}Am a significant increase in fissioning can be expected with the use of a fast spectrum (Table 2).

In principle, a fast critical reactor could be used to burn actinides. But in order to operate it safely certain restrictions in the fuel composition and the geometrical lay-out have to be respected (in most designs a flat, pancake like fuel arrangement is required to avoid a positive void coefficient). In particular, the percentage of minor actinides and even plutonium has to be kept below a certain limit as these materials have an insignificant or vanishing Doppler coefficient and a small delayed neutron fraction. With their abundant presence they would therefor hamper the control of the reactor.

It has been shown that accelerator driven systems are not subject to these restrictions [14]. The subcriticality provides more freedom in the choice of the geometrical configuration and the neutron spectrum, which should be as hard as possible to enhance actinide fissioning. For these reasons it is also possible to fuel an ADS with actinides including large quantities of recycled plutonium. From a safety point of view it is possible to choose the most critical core configuration, so that any accidental rupture or a melt-down accident would not lead to re-criticality.

If actinides are fissioned in an ADS a surplus of neutrons is generated. In an actinide fission approximately 3 neutrons are generated of which one is required to make a fission in the next generation assuming that $k_{eff} < 1.0$. The remaining 2 neutrons are either lost by leakage and "parasitic" absorption or they can be directed into fertile material (^{238}U or Th) where they breed fissile material to be fed, for example, into LWRs.

TABLE 1. Even Number Actinides

Nuclide	Thermal			Fast		
	n, gamma	n, f	fiss / abs	n, gamma	n, f	fiss / abs
Pu_240	179	.614	0.34%	0.469	0.353	43%
Pu_242	31.3	.443	1.4%	0.426	0.239	36%
Np_238	13.2	129	91%	0.400	2.09	84%
Am_242	76.2	364	83%	0.568	3.07	84%
Cm_242	5.38	0.525	8.9%	0.291	0.143	33%
Cm_244	14.1	0.910	6.0%	0.820	0.392	32%
Cm_246	2.79	0.602	18%	0.214	0.238	53%

TABLE 2. Odd Number Actinides

Nuclide	Thermal			Fast		
	n, gamma	n, f	fiss / abs	n, gamma	n, f	fiss / abs
Pu_239	154.8	95.0	63%	0.462	0.353	43%
Pu_241	32.5	9.81	23%	0.521	2.43	82%
Np_237	28.2	0.538	1.9%	1.47	0.296	17%
Np_239	10.8	0.433	3.9%	1.83	0.431	19%
Am_241	77.6	1.05	1.3%	1.75	0.246	12%

2. Fuel Cycle Considerations for Various Scenarii of Pu Transmutation

2.1 LIGHT WATER REACTORS

2.1.1. *Pu Burning in MOX Fuelled LWRs*

With the operation of over 400 *U*-fuelled reactors, the world stockpile of *Pu* is steadily increasing and will reach, depending on burn-up criteria, in 50 years between 3500 - 5000 tonnes. At present only LWRs offer the possibility of transmuting substantial quantities of *Pu*. To this end UO_2 based MOX fuel containing some 3 - 6% *Pu* is fabricated by a well established technology in some countries. The *Pu* can be either *R-Pu* from the first cycle, which in the near future will reach a total of 1400 t of which almost 300 t are separated from waste, while the estimate for *W(eapons)-Pu* is around 270 t [6]. Despite the fact that considerable amounts of *Pu* are regenerated by neutron capture in U^{238}, this strategy, if rigorously applied, could after the year 2020 keep the *Pu* stock-pile more or less constant at a level around 2000 t. As a benefit, a gain of valuable energy and a saving of natural resources would result. In the case of MOX made from *W-Pu*, it would be transformed into *spent fuel*. This strategy of recycling is, however, limited to two cycles. Thereafter, the even *Pu* isotopes grow to such a percentage that they hamper the neutron economy and may lead to a positive temperature coefficient. Transmutation of this type of plutonium requires therefore other solutions.

2.1.2. *Pu Transmutation in an Inert Matrix*
It is almost self-evident that the possibility of incinerating *Pu* in an inert matrix was investigated with priority. The motivation here is to avoid the reproduction of *Pu* by a fertile matrix. The absence of a Doppler effect in this fuel implies, however, that only a fraction of the core can be in this form. While in the inert matrix more than 90% of the *Pu* can be burned, there remains a substantial amount of *Pu* in the rest of the fuel and again we face the problem that this *Pu* cannot be incinerated in a LWR after the second recycle.

2.2. FAST REACTORS

Twice recycled Plutonium could be transmuted in fast reactors, but for reasons mentioned above, only in relatively small portions and with the additional problem that new *Pu* would be generated in the (inevitable ?) ^{238}U fertile matrix.

An ADS dedicated to the burning of *Pu+M(inor)A(ctinides)* might offer a way out of this dilemma. If intrinsically sub-critical it can cope with poor grade fuel beyond the Spent Fuel Standard i.e. fuel with a large percentage of ^{242}Pu, *Am* and *Cm*. In this context the following scenarii might be considered:

2.2.1. *Pure Fuel burning*
Pure fuel consisting of *Pu+MA* has a k_∞ value of 3.1 (see Figure 1). It has therefore to be used either in small quantities (a few kilograms only) to allow for neutron loss by leakage and/or through the use of absorbers. These absorbers could be long lived fission products (like ^{99}Tc, ^{129}I, etc.) which by neutron capture are transmuted to short lived or stable isotopes.

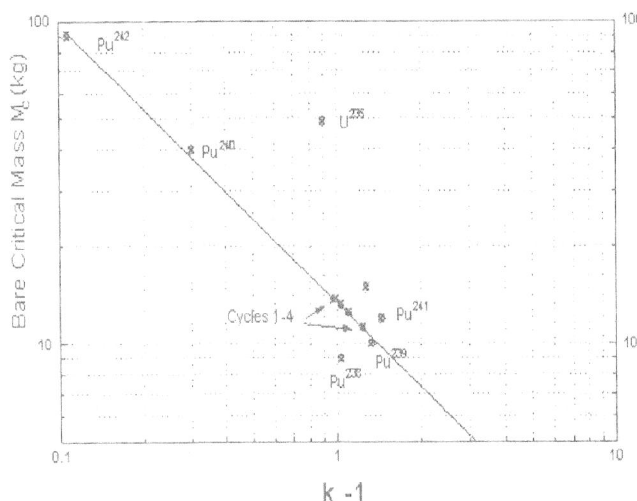

Figure 1. The bare critical mass of different fissionable materials.

2.2.2. *Pu Burning in a Fertile Matrix:*
Most ADS concepts described in the literature use the excess neutrons generated in the subcritical assembly to breed fissile material in a fertile matrix, usually ^{233}U in thorium. In this strategy the ^{233}U would be used to fuel LWRs.

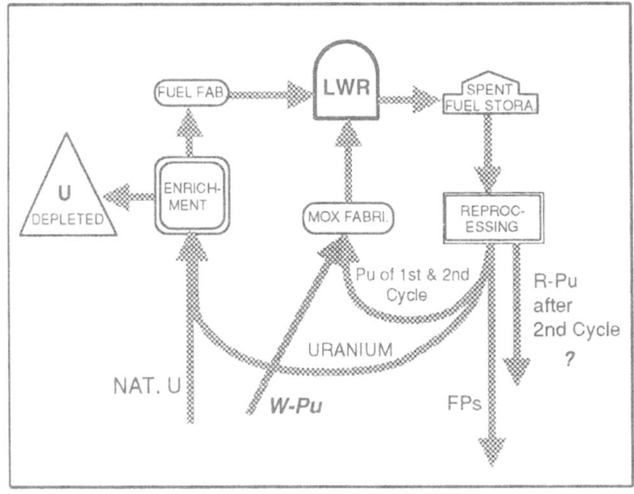

Figure 2. The classical fuelling scheme of LWRs

2.3. THE CROSS PROGENY FUEL CYCLE

Among the various ways of consuming "spent fuel *Pu*", we focus interest on the *cross progeny* [5,15] fuel cycle in which LWRs play an important role. It seems plausible to assume that over the next half century LWRs will dominate nuclear power generation. On the one hand, it is almost certain that the lifetime of the existing 350 LWRs will be extended beyond the originally predicted 30 years as in many cases their radiation damage turns out to be less then expected. On the other hand, most of the 50 reactors now under construction and in planning are of this type or they will be EA-PWRs. Even if from a scientific-technical point of view *fast advanced liquid metal reactors* were the preferred choice, it is difficult to imagine that there would be a large scale replacement of LWRs by ALMRs.

In Fig. 2 we show the fuelling scheme of a LWR with *Pu* recycling. It can be seen that the scheme has two drawbacks. One is the large demand for natural uranium. In the enrichment plant for each unit of fresh fuel, depending on the initial enrichment, 5 - 7 units of worthless depleted uranium are produced and have to be deposited safely. Already today the stockpile of depleted uranium is much larger then the low cost *reasonably assured U resources*[28]. Furthermore, it is well known, that in nuclear power generation the initial part of the fuel cycle (mining) constitutes the largest radiation burden for workers of the nuclear industry.

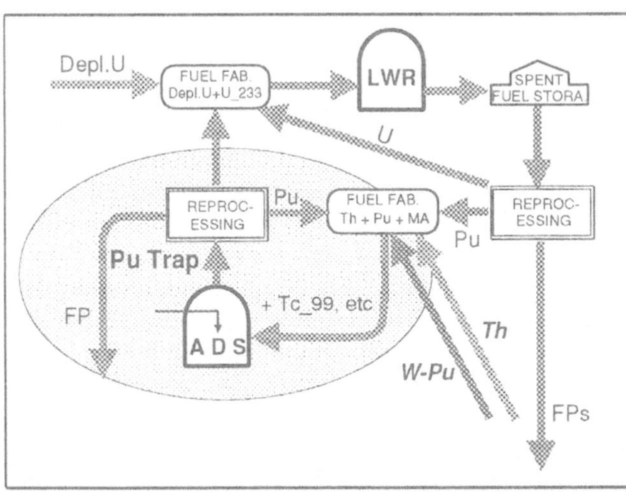

Figure 3. The ^{233}U - ^{238}U : ^{232}Th - Pu Cross Progeny fuel cycle

The second shortcoming is that despite recycling

Table 3: Actinide concentration for Pu recycling in a thorium matrix; cycle length is 5y and 1y cooling

Nuclide	Fresh Fuel	1st Cycle	BoC 2	EoC 2
Th	900 kg	778 kg	897 kg	772 kg
Pa	-	477 g	472 g	590 g
U*	-	67.6 kg	0	68 kg
Np	-	10 g	10 g	21 g
Pu	100 kg	48.7 kg	100 kg	52 kg
Am	-	2.48 kg	2.45 kg	3.8 kg
Cm	-	158 g	156 g	484 g
Total	1000 kg	897 kg	1000 k g	897 kg
k	1.036	(1.155) 1.169	0.9711	(1.151) 1.165

Nuclide	BoC 3	EoC 3	BoC 4	EoC 4
Th	892 kg	768 kg	887 kg	764 kg
Pa	584 g	613 g	607 g	617 g
U*	0	68 kg	0	67.5 kg
Np	20 g	27 g	27 g	31 g
Pu	103.5 kg	55.4 kg	107 kg	58 kg
Am	3.8 kg	4.7 kg	4.7 kg	5.4 kg
Cm	479 g	866 g	857 g	1.2 g
Total	1000 kg	897 kg	1000 kg	897 kg
k	0.9624	(1.153) 1.167	0.9685	(1.153) 1.170

U* consists mainly of ^{233}U + some minor fraction of ^{234}U and higher order U isotopes.

W-Pu and *R-Pu* at the maximum possible rate, there remains still an impressive amount of *Pu* (and minor actinides) in the spent fuel.

Both facts, the high consumption of natural uranium as well as the requirement to reduce the large quantities of plutonium in the spent fuel, call for new solutions which will briefly be discussed in what follows.

An interesting possibility of maintaining the *Pu* stockpile at low levels, is offered by the cross-progeny fuel cycle which has been recently analyzed in detail in the context of ADSs [17, 24]. In this scheme ^{233}U is produced from thorium by burning weapons- or reactor-plutonium in a hard spectrum ADS, which transmutes simultaneously MA and long-lived FPs. The ^{233}U produced in the ADS fuels LWRs in addition to ^{235}U and the *Pu* reprocessed from the LWR fuel returns to the ADS. In [24] it was shown that uranium fuel containing 2.7% ^{233}U extends the burn-up by a factor of two as compared to a 2.7% ^{235}U enrichment. In order to obtain the same burn-up with ^{235}U an initial enrichment of 4% would be required. The extension of the life-cycle of the fuel would reduce the requirement for *Pu* reprocessing and the *Pu* stockpile by a factor of two. Furthermore, this fuel cycle reduces the long-lived high-level radioactive waste inventory and has the advantage that it is rather proliferation resistant due to the high energy gamma-ray daughter nuclei in the thorium cycle.

From an economic point of view one has to compare the cost of generating ^{233}U by an ADS against the cost of enriching ^{235}U to be employed as fuel in LWRs. In an ADS as an extra bonus energy is produced while generating ^{233}U which after separation from the thorium could be mixed with depleted and therefore very inexpensive uranium. This scheme *might be more attractive than isotope separation* considering a long or even medium term time span.

Within this context, we have considered recycling *Pu-Th* in an ADS. The calculations were made using ORIGEN [4]. Starting with a 10% mixture of *R-Pu* (i.e. 1 tonne of fuel contains 900 kg ^{232}Th and 100 kg *R-Pu*). Table 3 shows the actinides present at the end of the first cycle (corresponding to 5 years irradiation followed by 1year decay). In this we have burned 122kg *Th* and 51 kg Pu. The question now arises - how do we continue into the second cycle ?

One possibility is to remove the entire 67.6 kg *U* (predominantly ^{233}U + ^{234}U) from the spent fuel thereby lowering the k_∞ of the fuel considerably. (This ^{233}U is then used to fuel conventional PWR's instead of ^{235}U). The remaining mass of 830 kg has to be replenished to 1000 kg. We do this adding 51 kg *Pu* (to recover the initial *Pu* content) and 122 kg *Th*. It can be seen that even after four cycles the production rate of U^{233} and the burning of *Pu* remains fairly constant so that it is reasonable to assume that this cycling can be continued. Since we are burning roughly 50 kg Pu/t in five years at a power level of 55 MW/t, the elimination rate is approximately 0.18 t of Pu/GW_{th} y, or about 300 kg/y for a 30t core.

In another scenario it is proposed to abandon U^{238} at all and to use U^{233} in a thoria matrix in thermal systems, where its content is stretched by conversion. The U^{233} missing at the beginning of each new cycle would be generated by a hard-spectrum ADS thorium breeder. This version suffers, however, from the fact that thermal U^{233} - thorium system can only be operated efficiently if the neutron flux is below $5 \cdot 10^{13}$. At higher flux "parasitic" capture in ^{233}Pa hampers the neutron economy.

3. Some Physics Considerations of how to Choose k_{eff} of the Subcritical System

3.1 THE DIRECT USE OF SPALLATION NEUTRONS *($k_{eff} = 0$)*

First we consider the case in which spallation neutrons are used as they are generated in a proton target without any further multiplication by fissionable isotopes. For best efficiency we assume that the fission products would be placed around the target and depending on the material to be transmuted, either the fast neutrons would be used as they are emitted from the target or they would be slowed down by a moderators to energy bands with higher transmutation cross sections as for example the resonance or the thermal region.

Assuming that it is possible to make all the spallation neutrons available for the transmutation process the following amount of energy E_{fp} is necessary to transmute the fraction q_{fp} of radio-nuclei per fission process in a nuclear energy system

$$E_{fp} = q_{fp} \frac{P_b}{n_{sp}} \frac{1}{\eta_b \eta_T} \; [MW] \tag{1}$$

where q_{fp} = fraction of fission products to be transmuted

P_b = proton energy
n_{sp} = number of neutrons generated by one proton
η_b = efficiency of converting electricity into proton beam energy (= 0.5)
η_T = efficiency of converting thermal energy into electricity (= 0.33)

In the case of a 1.5 GeV proton beam emitting 45 neutrons per spallation in a lead target the transmutation of ^{99}Tc, ^{129}I, ^{135}Cs, ^{90}Sr, ^{85}Kr and ^{93}Zr (constituting 28 % of all fission products) would require 57.6 MeV to transmute the fission product fraction of one fission process. This is 29 % of one fission, and thus, of the total power production of the nuclear energy system under consideration! Because of the very optimistic assumptions made in this estimate the real percentage of energy required would even be higher. Together with the cost for reprocessing it would make this type of accelerator transmutation prohibitively expensive, at least in a commercial nuclear energy system.

Accelerator Driven Subcritical Assemblies with $k_{eff} < 1.0$: To improve neutron economy the spallation neutrons can be multiplied in a subcritical assembly. In such a system the main part of neutrons is generated by fission in a reactor-like subcritical facility. Technically this is realized by surrounding a proton target by fissionable material. In most designs a circulating liquid lead is proposed to remove the heat released in the target and the subcritical system. It should be noted that the heat production per spallation neutron is considerably lower than in a fission process (30 MeV against 80 MeV).

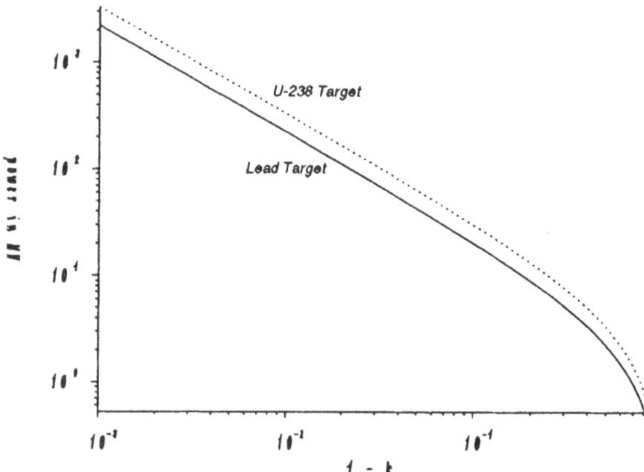

Figure 4. Power production of a 1 GeV, 1mA Accelerator-Driven Assembly as a function of the Subcriticality (1-k_{eff}).

The power production P_{fi} of a subcritical assembly fed by spallation neutrons can be quantified by:

$$P_{fi} = n_{sp} \frac{a \cdot k}{\nu(1-k)} \frac{i}{C} E_f \qquad (2)$$

where:

k = multiplication factor; a = importance of the target position and target neutron energy distribution (usually $a>1$ for a central target position, values around 2.5 are possible; ν = mean number of neutrons in a fission process; E_f = power release per fission (=3.1·10^{-10}W); n_{sp} = neutron yield from one proton; i = proton current; C= proton charge (= 1.6 10^{-19} A sec).

In Figure 4 the power production of an accelerator driven facility is shown as a function of sub-criticality *(1-k)*. It was assumed that a proton beam of 1 GeV and 1 mA impinges on a *Pb-Bi* target releasing 30 neutrons per spallation with an importance of

$a=1.6$. It leads to

$$P_{fi}(1\ mA) \cong 4\frac{k}{(1-k)}\ [MW] \tag{3}$$

It can be seen that near criticality a 1 mA current already generates a relatively high fission power. For $k = 0.97$ about 100 MW can be achieved.

The excess neutrons from the subcritical system as well as its fission power which can be transformed into electricity, are now exploited to run the transmutation process.

Expression 4 developed by Takahashi *et al.* [21] quantifies the energy required to transmute a fraction q_{fp} of fission products in such a system. A positive sign of E_{fp} means that there is even a surplus of energy, while a negative sign indicates the need to add energy to the system from outside.

$$E_{fp} = \frac{n_{sp}\frac{k}{\nu(1-k)}E_f - \frac{P_b}{\eta_b\eta_T}}{n_{sp}\left[\left(1-\frac{k}{\nu}\right)\eta_{fp} + \frac{k}{1-k}\left(\left(1-\frac{k}{\nu}\right)\eta_{fp} - \frac{q_{fp}}{\nu}\right)\right]}\ [MW] \tag{4}$$

where

$$\eta_{fp} = \frac{\Sigma_a(FP)}{\Sigma_a(FP + Fuel + StructMat.)}$$

The condition for break-even or a positive energy balance is given by

$$k \geq \frac{1}{1 + \frac{n_{sp}E_f\eta_b\eta_T}{P_b \cdot \nu}}$$

Note that this expression is independent of the proton current and to a large extent also of the type of system considered. For a lead target and a proton beam of 1 to 2 GeV *break-even* requires a k value near 0.75. But the amount of nuclei transmuted depends on the power of the system and therefore on the proton current as shown by Equation 2.

3.3. THE CHOICE OF THE AMOUNT OF SUBCRITICALITY

As Accelerator Driven Systems can be operated at different levels of subcriticality, there arises the question of what is a good choice. Some authors distinguish between systems which have a multiplication factor above and below the fuel storage criterion ($k_{eff} = 0.95$). This distinction is rather arbitrary and only justified by legal and psychological reasons.

It is assumed that the public would accept more easily a system with a subcriticality below the fuel storage criterion in which the various accident scenarios would remain far below criticality. The price one pays for this is a low self-multiplication which decreases by $k_{eff}/(1 - k_{eff})$ [27]. Operation of such a system requires therefore, a relatively expensive high current proton accelerator. This in turn will lead to considerable radiation damage in the beam window and the target and its surrounding materials [25], so that these elements have to be replaced frequently and lead to a reduced plant availability. Furthermore, the shielding requirements along a high current LINAC and also behind the target would be substantial. If the beam is not split, the power distribution will become highly peaked near the spallation source leading to strongly varying coolant requirements inside the system.

On the other hand, if a sub-criticality near k_{eff} = 0.98 / 0.99 can be chosen such that all fast transients as they may result from coolant voiding, fuel movement, positive temperature coefficients due to a large actinide inventory, etc. can be avoided, then such a system would combine the intrinsic safety of an ADS with the advantage of a reactor with a large delayed neutron fraction (imitated by the spallation neutrons) and a rather flat power distribution, so that a more equal burn-up across the whole fuel region can be obtained. In a system of this kind a proton beam power of 10 - 20 MW would produce enough spallation neutrons to obtain a 1 GWe power plant. In this case, a much less expensive and a rather compact cyclotron might be sufficient [21] to produce the required proton beam power.

3.4. CONSIDERATIONS CONCERNING THE POWER DISTRIBUTION IN SUBCRITICAL SYSTEMS

As an example of the power distribution in a subcritical system we analyzed an ADS experiment carried out at CERN by Rubbia and coworkers [18]. In this experiment a subcritical water moderated assembly containing a uranium spallation target is coupled to a proton beam of variable energy (1 to 3 GeV).

In our calculations we employed the Monte Carlo program *LAHET* [13] coupled to the code *MCNP4* [3]. The subcritical assembly was mocked up respecting all important geometrical details, such as the water tank containing 270 fuel rods consisting of natural uranium, surrounded by an aluminum cladding, as well as the natural uranium target of dimension 67 × 5 × 5 cm^3 embedded in the assembly. In order to determine a 3-D map of the the power distribution in the system (see Figure 5), the 270 fuel rods were subdivided into 5 pieces each, along the z-axis.

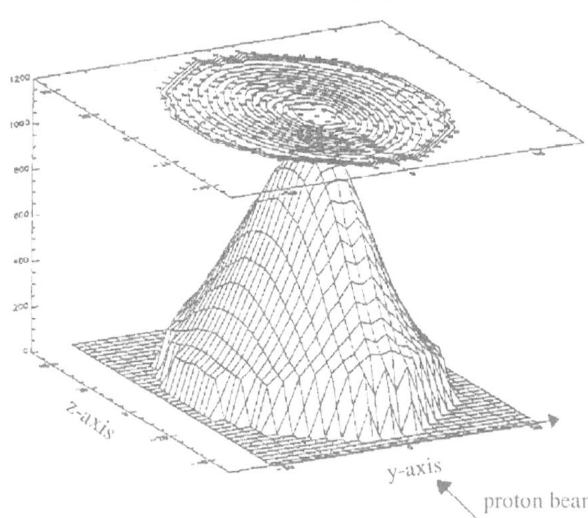

Figure 5. Fission-power distribution in Rubbia's [18] subcritical assembly driven by 3.7 GeV protons

The *static* k_{eff} of the subcritical assembly was estimated to 0.86. This is the classical eigenvalue which corresponds to the fundamental mode distribution of the neutrons, independently of the source initiating the chain reaction. It seems that Rubbia et. al. [18] used a *dynamic* k_{eff} definition, which includes the central source and therefore leads to higher values, i.e. 0.90±0.01.

The coupled *LAHET* - *MCNP* calculations [19] were carried out for a proton beam of 3.7 GeV. It produces a total energy of 67.5 GeV for each proton incident on the natural uranium target rendering a power gain of ***~18***. The graphical display of the fission power distribution shows a strong peak in the centre region where the spallation source is placed. The upper part of Figure 5 shows the projection of the power profile onto the system's mid-plane.

4. Safety Studies of Prototypical Systems

In the following safety analysis we used some typical fast reactor configurations and assumed that they would be subcritical and driven by an accelerator.

In our preliminary safety studies of fast ADSs sodium-cooled systems are analysed since codes for this kind of coolant are readily available. Contrary to sodium voiding which leads to a positive reactivity feedback, the voiding of a heavy metal coolant such as lead would cause a much smaller positive reactivity effect and therefore the following analyses for sodium-cooled ADSs tend to be pessimistic. A simplified investigation of a severe reactivity accident (RIA) in a fast ADS, which was undertaken earlier, showed that fast ADSs are rather insensitive to rapid reactivity insertions [14]. The early phase of Loss-of-Flow (LOF) and Loss-Off-Heat-Sink (LOHS) accidents was also investigated previously [7] for a fast sodium cooled ADS and showed that these types of accidents lead to sodium boiling. In the present study [30, 31], LOF and reactivity accidents in such a system were investigated beyond sodium boiling, fuel pin ruptures and up to and including molten fuel motion. Another aspect of this study was the investigation of the beneficial effect of shutting off the proton beam at a certain time during the accident. These investigations are being continued with the analysis of LOHS accidents. Also lead cooling is being considered and preliminary results will be presented.

For the present investigations the EAC-2 fast reactor accident code [29] was modified by inserting a neutron source into the point kinetics module. The system under consideration is a 800 MWe sodium-cooled reactor design which was used in the European WAC benchmark calculations. This core is set up with 10 calculational channels and in the LOF accident calculations the pump coast down has a halving time of 12 s.

In the LOF calculations presented here constant negative reactivities of -3 and -10$ were inserted in order to simulate different subcriticalities. The case with -10 $ (see Figure 6) showed a power peak with a maximum of 1.5 times nominal power and a half width of 1 s. The power rises due to sodium voiding and it decreases due to molten fuel dispersal in a few calculational channels. In the case of the -3$ subcriticality the power peak reached 5 times nominal. For comparison, the regular reactor LOF case was run which leads to a small width prompt critical power peak with a maximum of 1800 times nominal for the very conservative pin failure conditions used. This large power peak leads to the failure of all pins, whereas the mild, but longer lasting, power increases in

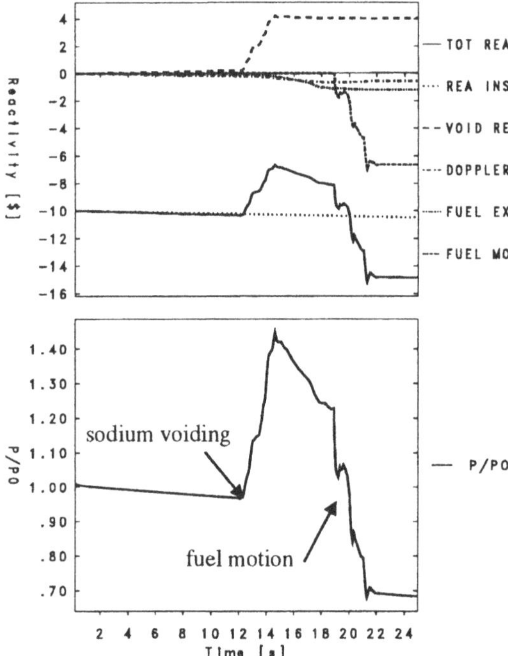

Figure 6. Reactivity and power history of LOF; Na-cooled 10$ subcritical and spallation source "on".

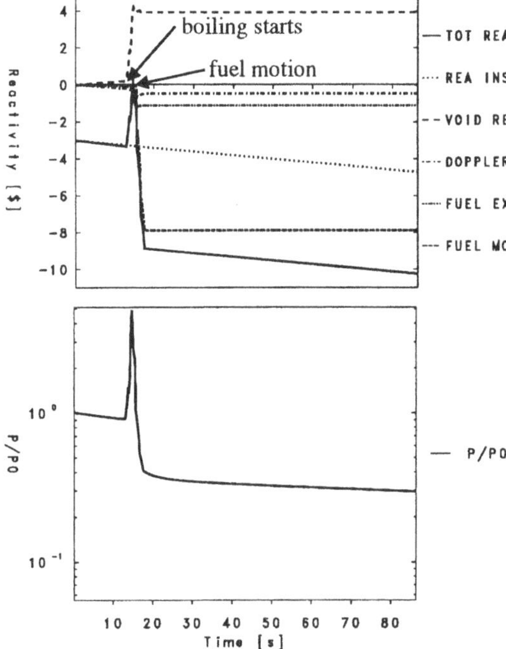

Figure 7. Reactivity and power history of LOF; Na-cooled 3$ subcritical and spallation source "on".

the ADSs lead to pin failures in only a few of the ten calculational channels. However, in the ADS cases (in which the spallation source is assumed to remain operating) the power remains at a significant fraction of nominal after the power peak. Since the pumps are further coasting down, the whole core will eventually melt if the proton beam is not shut-off. The fact that the power of an ADS (with the accelerator not shut-down) is rather insensitive to relatively large positive and negative reactivity changes has also been recognised earlier for a thermal ADS in which a salt/fuel mixture is circulated.

The ADS LOF cases were rerun with an assumed shut-off of the proton beam shortly after the sodium voiding had started. The calculations show that the power is shut down to decay heat levels in a very short time. It is shown that even if some sodium boiling and voiding has already occurred when the beam is shut-off, a complete recovery of the cooling flow can take place. Concerning the shutting-off of an ADS through interruption of the beam or switching-off the accelerator, one can argue that this can be more easily achieved than the mechanical insertion of control rods in a regular reactor.

Reactivity accidents with ramp rates of 170, 6, and 0.1 $/s and a total insertion of 3$ were investigated. These calculations confirmed that the fast insertion of a reactivity whose total insertion is not greater than the subcriticality of the system leads to no power peaks but a longer term increased power level of

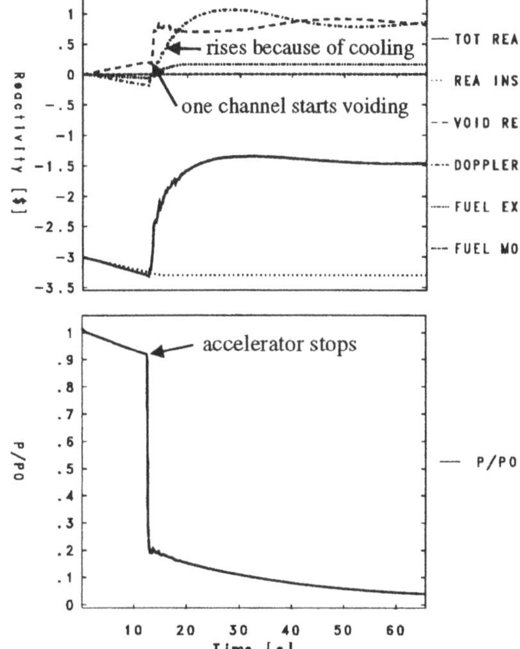

Figure 8. Reactivity and power history of LOF, Na-cooled; 3$ subcritical and spallation source switched-off after boiling.

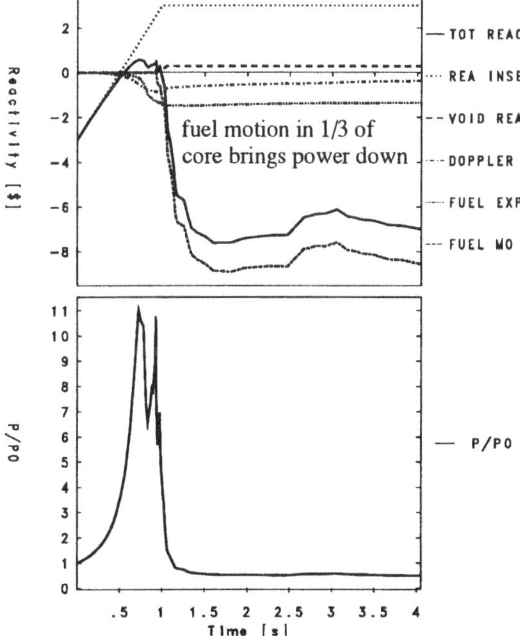

Figure 9. Reactivity and power history of LOF, Pb-cooled; 3$ subcritical, 6$ inserted in 1 s and spallation source "on".

around 1.3 times nominal. This may lead to some incoherent pin failures and fuel fragment sweep-out which would reduce the power further. The slow ramp rate of 0.1 $/s, however, led to a long-term overpower condition of around 1.6 times nominal which leads to an increased number of pin failures. To avoid such failures during a reactivity accident would either require a lower subcriticality of the ADS or the shutting-off of the proton beam.

In conclusion it can be said that cooling failure and reactivity accidents in an ADS will not cause power bursts - but if the accelerator beam is not switched off in cooling failure accidents, a slow meltdown will occur. In reactivity accidents there will also be some core damage if the accelerator beam is not switched off or interrupted. Therefore, fail-proof devices for shutting off the spallation source are indispensable for avoiding an overheating of the ADS in cooling failure, and to a lesser degree, for reactivity accidents.

In a first more detailed ADS design by C. Rubbia [16], natural circulation lead cooling in a tall vessel is proposed and only shutdown rods are included. This would make LOF accidents impossible and reactivity accidents, which are not very critical for an ADS, highly unlikely. LOHS accidents would still be conceivable. However, the passive proton beam interruption device proposed in the Rubbia design, which is based on the rise and overflow of an overheated coolant, would lead to the beam shut-off and shutdown of the ADS. The decay heat would be removed by natural air draft outside the guard vessel. To investigate that these passive devices work reliably; that lead freezing at very low power does not represent a problem; and that re-criticalities would not be a serious problem if a core melt occurred, are important research items to be investigated for a heavy metal-cooled system.

Conclusions

W-Pu and first cycle *R-Pu* in the form of MOX-fuel can to a large extent be fissioned in existing LWRs thereby exploiting the enormous energy content of the plutonium. The "rest" plutonium, consisting of several times recycled highly degraded *Pu* containing large fractions of even numbered isotopes, could be transmuted in fast ADSs. Excess neutrons can be used to breed ^{233}U in a thorium matrix. If, as expected in the near future, high-grade uranium ores become exhausted, enrichment plants (and the associated large quantities of depleted uranium) could be phased out and replaced by ADSs to "convert" spent fuel plutonium into ^{233}U, through a thorium cycle, for use in LWRs. Simultaneously, the ADS would transmute long-lived fission products such as technetium and iodine.

Within this context, therefore, the ADS is consided primarily as a replacement for enrichment plant, rather than the power reactor. More exactly, the ADS would combine the roles of a waste burner and an enrichment plant. A detailed study is required to evaluate this ADS / power reactor synergy.

6. References

1. Andriamonje S., Angelopoulos, Rubbia C., et al.(1995) Experimental study of the phenomenology of spallation neutrons in a large lead block, *CERN/SPSLC 95-17, SPSLC/ P291,*
2. Bonnaure, P., Mandrillon P., Rief H. and Takahashi H. (1986) Actinide transmutation by spallation in the light of recent cyclotron development, *4th Int. Conf. on Emerging Energy Systems*, Madrid.
3. Briesmeister, J.F. (1988) MCNP 4A - A general Monte Carlo N-Particle Transport Code, LA-12625-M, Los Alamos National Laboratory.
4. Croff, A.G. (1983) OREGEN2: A versatile computer code for calculating the nuclide composition and characteristics of nuclear materials, *Nucl. Tech.* **62,** pp 335-353. For the thorium calculations, the library pwrd5d33 lib for ThO_2 enriched with recycled , denatured ^{233}U was used. For the inert matrix calculations, the library prwus.lib for standard 3.2% enriched uranium PWR was used.
5. Eschbach, E.A., (1977) Crossed Progeny and some other nonstandard fuel cycles, Trans. Am. Nucl. Soc. **27**, p. 430.
6. Liebert W., (1995) Proposal for a comprehensive cutoff convention, Workshop "Fissile Materials and Tritium - How to verify a comprehensive production cutoff and safeguard all stocks", Geneva.
7. Lypsch, F. and Hill, R.N. (1994) Development and analysis of a metal-fuelled accelerator-driven burner, *Int. Conf. on Accelerator-Driven Transmutation Technologies*, Las Vegas.
8. Magill, J., O'Caroll, C.,Gerontopoulos, P., Richter, K. and van Geel, J. (1995) Advantages and limitations of thorium fuelled energy amplifiers, in unconventional options for plutonium disposition, *Proc. of a Tech. Committee Meeting in Obninsk*, Russian Federation, IAEA-TECDOC-840, pp. 81-95.
9. Magill, J. Matzke, Hj., Nicolaou, G., Peerani, P. and Van Geel, J, (1995) *IAEA Technical Committee Meeting on Recycling of Pu and U in Water Reactor Fuels*, Windermermere.
10. Magill, J., Peerani, P., Matzke, Hj. and Van Geel, J, (1995) *IAEA Technical Committee Meeting on Advanced Fuels with Reduced Actinide Generation*, Vienna.
11. Magill, J., Peerani, P., Van Geel, J., Landgren, A. and Liljenzin, J.-O., (1996) Inherent limitations in toxicity reduction associated with fast energy amplifiers, *Proc. of the Second Int. Conf. on Accelerator-*

Driven Transmutation Technologies and applications, Kalmar.
12. Matzke, Hj.. and Van Geel, J, (1996) in E. Merz and C.E. Walter (eds), *Proc. Nato Advanced Research Workshop "Disposal of Weapons Plutonium", Kluver Acad. Publ. Dordrecht*, p. 93.
13. Prael R.E. and Lichtenstein H. (1989) User guide to LCS: the LAHET code system, Los Alamos National Laboratory, USA, *Report Na LA-UR-89-3014.*
14. Rief, H. and Takahashi, H. (1994) Safety and control of accelerator-driven subcritical systems, Invited paper, *Int. Conf. on Accelerator-Driven Transmutation Technologies*, Las Vegas.
15. Ronen, Y. and Carmona, S., (1980) A symbiotic water breeder reactor system, *Nuc. Sci. Eng.,74, 84.*
16. Rubbia, C. et al., (1995) Conceptual design of a fast neutron operated high power Energy Amplifier, *CERN/AT/95-44* (ET).
17. Rubbia, C. et al., (1995) A realistic plutonium elimination scheme with fast Energy Amplifiers and thorium-plutonium fuel, *CERN / AT/95-53* (ET).
18. Rubio J.A. (1994), Personal communication, CERN, Switzerland .
19. Sarkar P.K. and Matthes W., (1994) New methods in high level radioactive waste management, *JRC Annual Report 94*, pp 180, EUR16251 EN.
20. International Commission for Radiological Protection (1994) Dose coefficients for intake of radionuclides by Workers, *Annals of the ICRP publication 68*, Pergamon Press.
21. Takahashi, H. and Rief, H. (1992) A critical review of accelerator-based transmutation systems, Invited paper: *OECD-Nuclear Energy Specialists' Meeting on "Accelerator-based Transmutation"*, Würenlingen.
22. Takahashi, H. and Rief, H, (1992) The energy requirement for transmuting fission products, *OECD/NEA Second General Meeting of the International Information Exchange Programme on Actinide and fission Product Separation and Transmutation*, ANL.
23. Takahashi, H. and Rief, H. (1992) The energy requirement for transmuting and isolating fission products, *ANS/ENS International Meeting*, Chicago.
24. Takahashi, H. and Chen, X. (1996) A new fuel cycle using the accelerator, The 4th Int. Conf. of Nucl. Engi. 1996 ASME/JSME, New Orleans (ICONS IV).
25. Takahashi, H. and Chen, X. et al. (1996) The evaluation of radiation damage to the target material due to the injection of medium- and high-energy protons, to be published.
26. Takahashi, H. (1996) Studies of the accelerator-driven subcritical fast reactor at Brookhaven National Laboratory, to be published.
27. Takahashi, H., Chen, X. Rief, H. and Wider, H., (1996) Determination of the subcriticality of an accelerator-driven fast reactor, to be published.
28. Uranium 1993: Resources, Production and Demand, OECD Nucl. Energy Agency and IAEA, ISBN 92-64-14019-0
29. Wider H.U. et al., (1990) The European accident code-2: overview and status, *Proceedings of the 1990 International Fast Reactor Safety Meeting*, Snowbird, Utah, USA.
30. Wider H.U., (1996) Severe accident studies in fast accelerator-driven systems, *Proc. of the 2nd Int. Conf. on Accelerator-Driven Transmutation Technologies and Applications*, Kalmar, Sweden.
31. Wider H.U., (1996) Basic safety features of ADS, *Status of the Accelerator-Driven Systems (ADS), IAEA Report*, Chapter E., to be published.

ADVANCED CANDU SYSTEMS FOR PLUTONIUM DESTRUCTION

P.G. BOCZAR, M.J.N. GAGNON, P.S.W. CHAN, R.J. ELLIS, R.A. VERRALL, A.R. DASTUR
Atomic Energy of Canada Limited (AECL)
Chalk River Laboratories
Chalk River, Ontario, Canada K0J 1P0

Abstract

High neutron economy, on-line refuelling, and a simple fuel-bundle design result in a high degree of versatility in the use of the CANDU® reactor for the disposition of weapons-derived plutonium. CANDU mixed-oxide (MOX) fuel is a near-term, technically achievable, economic option. Studies led by AECL show that four Bruce A reactors could consume 50 te of plutonium in less than 12.5 years. The symmetry in the simultaneous drawdown of excess weapons-derived plutonium from both the United States and Russia in Canada was an important consideration in the recent US Record of Decision, which includes the CANDU MOX option for further evaluation.

The CANDU versatility enables advanced options for plutonium destruction. One such option is the use of an inert matrix, non-fertile material as the carrier for weapons-derived plutonium. Mixing the plutonium with inert SiC in a standard 37-element CANDU bundle would result in destruction of 93% of the fissile plutonium (^{239}Pu and ^{241}Pu). Fuel management studies were conducted, confirming that fuelling rates and maximum powers are well within limits. Because of the very high thermal conductivity of SiC, fuel temperatures would be very low, and negligible fission-gas release is expected.

The Pu-ThO_2 cycle would also achieve a very high efficiency in plutonium destruction. With ~2.6% weapons-derived plutonium in ThO_2 in a modified CANFLEX bundle (a large central graphite displacer surrounded by 35 fuel elements in the two outer fuel rings), a burnup of 30 MW·d/kg heavy element (HE) can be achieved, and >94% of the fissile plutonium destroyed. Good neutron economy is the key to high efficiency in plutonium destruction with ThO_2. Of course, ^{233}U is produced, through neutron capture in ^{232}Th and subsequent β-decay, and partially burned in situ. This material is safeguarded in the spent fuel and is not attractive for weapons because of contamination with ^{232}U. The spent fuel could be stored until an economical and proliferation-resistant means of recycling the ^{233}U is developed.

® CANDU (CANada Deuterium Uranium) is a registered trademark of AECL.

E. R. Merz and C. E. Walter (eds.), Advanced Nuclear Systems Consuming Excess Plutonium, 163–179.
© 1997 *Kluwer Academic Publishers.*

1. Introduction

1.1 CANDU FUEL CYCLE FLEXIBILITY

The CANDU reactor has unsurpassed flexibility in its ability to accommodate different fuels and fuel cycles. This flexibility is a result of the following key features of the reactor.

- High neutron economy, that is due to the use of heavy water as coolant and moderator, the use of low-neutron-absorbing structural materials, and on-power refuelling. It also permits the use of fuel with low fissile content (with the capability of exploiting unique, proliferation-resistant fuel cycles), and enables the initial fissile content to be burned down to low levels.
- On-line refuelling provides flexibility in fuel management. One can vary the number and type of bundles added to a channel, the location of the channel to be refuelled, the frequency of refuelling, and even the axial location along the channel where the new fuel bundles are inserted. Both axial and radial power distributions can thus be shaped and controlled, as can the amount of reactivity added to the reactor during refuelling.
- The simple fuel bundle design facilitates optimization of the fuel composition. Fuel type and enrichment can be varied from ring to ring to achieve design objectives, such as tailoring reactivity coefficients, or minimizing linear element ratings. The simple design and small size of the bundle make it easy and economical to fabricate, either "hands-on" or remotely, and to test advanced fuels.

1.2 CANDU PLUTONIUM-DISPOSITIONING OPTIONS

These features enable the CANDU reactor to meet a range of plutonium-dispositioning objectives, without any major changes to the reactor. This paper examines three such options.

The first option uses conventional mixed-oxide (MOX) fuel in CANDU reactors. Although not an advanced option for plutonium dispositioning, it provides a reference for the advanced options discussed and illustrates the flexibility of the reactor in accommodating a very wide variety of fuel designs and strategies. Significant work has been done on this concept since the NATO workshop on conventional MOX fuel for dispositioning excess weapons-derived plutonium in November 1994. [1] The objective of the MOX option in plutonium dispositioning is to convert the plutonium into spent fuel, while generating useful electricity. The spent MOX fuel has similar proliferation-resistant characteristics as spent UO_2 fuel, whether CANDU or PWR. The barriers to subsequent diversion include the intense radioactivity of the spent fuel; the chemical form (oxide); dilution of the plutonium, both in the fresh fuel, and even more so in the spent fuel; degradation of the plutonium isotopic vector; and the physical form and weight of the fuel bundles (spent CANDU fuel bundles are stored in sealed assemblies).

Since the MOX option converts the plutonium to a form that has no more proliferation risk than the very much larger quantity of spent fuel from civilian reactors, there is no short-term urgency to further reduce the proliferation risk of the spent MOX fuel.

The second option that will be discussed in this paper, and the first "advanced" CANDU option, is plutonium "annihilation" in an inert matrix. Here, the objective is to destroy plutonium, without at the same time creating new plutonium. Hence an inert matrix material is used as a carrier of the plutonium. The CANDU reactor can achieve a very high efficiency of plutonium destruction in a once-through cycle without reprocessing and with no significant changes to the reactor system.

The third option that will be discussed is another MOX option, where the plutonium is mixed with ThO_2, rather than with UO_2. This option, too, achieves a very high degree of plutonium destruction. It has the additional benefit of creating a reserve of ^{233}U (created through neutron capture in ^{232}Th and subsequent β-decay), which can be safeguarded in the spent fuel until such time in the future when it is economical to recover and recycle the ^{233}U, using proliferation-resistant technology. This option has a high plutonium destruction efficiency, and the highest energy yield potential of all the options considered.

The truly remarkable fuel-cycle flexibility of the CANDU design enables these diverse plutonium management options to be achieved in an existing CANDU reactor, with no significant changes to the reactor.

1.3 MEASURES OF MERIT

This paper uses several "measures of merit" to characterize these options.

1. *Net Pu-Destruction Efficiency* (%): Net *Total* Pu destroyed, as a fraction of the *Total* initial Pu in the fresh fuel

2. *Net Fissile Pu-Destruction Efficiency* (%): Net *Fissile* Pu destroyed, as a fraction of the initial *Fissile* Pu in the fresh fuel

3. *Pu-Disposition Rate* (Mg Pu/GW_e·a): the amount of Pu used or dispositioned (i.e., in the fresh fuel) to produce 1 GW_e·a of energy

$$\sim \frac{\text{[Pu-fraction in the fresh fuel] (te Pu / te HE) * 365 d/a * (1000 MW/GW)}}{\text{Burnup (MW}\cdot\text{d}_{th}\text{/te HE) * }\eta\text{ (0.3)}}$$

4. *Energy Produced* (GW_e·a/ Mg Pu): inverse of above.

Table 1 summarizes the three CANDU plutonium management options in terms of these measures of merit. For the MOX option, two variants are considered.

2. CANDU MOX

2.1 The 1994 CANDU MOX STUDY

The objective of the MOX strategy is not to *destroy* the plutonium, but to convert it to a form that has a high degree of diversion resistance through the characteristics of spent fuel, while producing electricity. The important considerations are the timeliness of the deployment option, the plutonium disposition rate, and the economics. Of the four parameters defined in Table 1, the Pu-disposition rate is the most relevant for this objective.

In 1994, the United States Department of Energy (USDOE) commissioned an AECL-led team to examine the use of CANDU reactors for dispositioning excess weapons-derived plutonium. The target plutonium disposition rate was 50 te Pu metal in 25 years. A detailed assessment was performed, including technical, economic, safety, licensing, safeguards, security, MOX fuel fabrication, transportation, and eventual disposal of the spent fuel. The team concluded that use of two of the four 825 MW(e) reactors at the Bruce A station near Kincardine, Ontario, could achieve this. Using all four Bruce A units, 50 te of Pu could be utilized in less than 12.5 years.

CANDU MOX fuel fabrication would take place close to the source of the weapons-derived plutonium, and only finished MOX bundles would be transported to Canada.

The reference fuel uses the standard 37-element geometry and is designed to perform within the operating and safety envelopes for natural-uranium fuel. Depleted uranium is the matrix material throughout the bundle. In the central element, and the next ring of 6 elements, 5% dysprosium (a burnable poison) is mixed with the depleted uranium. Plutonium is confined to the outer two rings of fuel: 2.0% plutonium in the third ring of 12 elements, and 1.2% plutonium in the outer ring of 18 elements. The bundle average burnup of the reference MOX fuel is 9.7 MW·d/kg HE, compared with 8.3 MW·d/kg HE for natural-uranium fuel in Bruce A reactors. Peak element burnup is about the same as for natural uranium (about 16 MW·d/kg HE). The fresh fuel contains 232 g plutonium per bundle, of which 94% is fissile.

An advanced MOX fuel design was also conceived. This employs the CANFLEX geometry, a 43-element bundle having 2 element sizes arranged in rings of 1, 7, 14, and 21 elements. The CANFLEX bundle has 20% lower peak element ratings than the 37-element bundle operating at the same bundle power, and improved thermalhydraulic performance (6 to 8% higher critical channel power). The lower ratings facilitate achievement of higher burnup, and the advanced MOX design has a core-average burnup of 17.1 MW·d/kg HE, which results in a peak element burnup of under 30 MW·d/kg HE. These are burnups for which there is CANDU experience. The advanced MOX bundle contains 374 g plutonium in the fresh fuel. As in the reference bundle, the plutonium is confined to the outer two rings of fuel: 3.5% plutonium in ring 3, and 2.1% in ring 4, mixed with depleted uranium. The central 8 elements contain 6% dysprosium mixed with depleted uranium. There is some minor optimization of the internal element design (pellet size and shape, and clearances).

The use of a burnable poison mixed with the depleted uranium in these designs achieves a number of objectives. The excess reactivity of the fresh fuel is suppressed, reducing the power ripple during refuelling. The Pu-disposition rate is increased, since additional plutonium is needed to achieve a given burnup. Finally, the design results in negative void reactivity, which eliminates any power pulse during a postulated loss-of-coolant-accident (LOCA), thereby simplifying the safety and licensing analysis. Mixing the burnable poison with depleted uranium in the central, low-power elements avoids the issue of fuel performance for MOX fuel containing integral burnable poisons, that must be addressed by other reactor concepts.

Each Bruce A reactor would consume about 1 t of plutonium per year (assuming an 80% capacity factor), in both the reference and advanced cases. The higher burnup CANFLEX MOX bundle would require a lower MOX fuel fabrication capacity, thus lowering the mission cost.

In both cases, the initial plutonium content is reduced by about one third in the spent fuel. Table 2 summarizes the fuel isotopic composition for the reference 37-element bundle, calculated using WIMS-AECL. [2]

Fuel management is particularly simple for both the reference and advanced MOX options: bi-directional (adjacent channels are refuelled in the opposite direction), two-bundle shift fuelling in the direction of coolant flow. The resultant axial power distribution is excellent from the perspective of both fuel performance (fuel at extended burnup does not experience a power boost) and thermalhydraulics. A full MOX core can be accommodated with no changes to the reactor system, other than the provision for safe and secure storage of new fuel. Reference 3 gives more complete technical details of the 1994 USDOE study on the CANDU MOX option for dispositioning excess weapons-derived plutonium.

2.2 The 1996 CANDU MOX STUDY

The AECL-led team conducted additional studies for the USDOE in 1996, aimed at further increasing the plutonium disposition rate. A 50% increase in the plutonium disposition rate was achieved by increasing the plutonium content of the bundle. To compensate for the excess reactivity, the burnable poison content in the central elements was increased from 7% to 15%, and the purity of the coolant and moderator was downgraded from 99.75% to 97% purity.

The resultant 37-element fuel bundle has a plutonium loading of 300 g confined to the outer 2 rings (3.1% Pu in ring 3, 1.6% Pu in ring 4), 15% dysprosium in the central 7 elements, with depleted uranium as the base material throughout the bundle. The average discharge burnup is slightly greater than in the earlier study, 10 MW·d/kg HE.

The plutonium disposition rate in a Bruce A reactor is 1.5 t Pu per year per reactor (assuming an 80% capacity factor). The MOX fuel fabrication capacity requirement is 78 t per year per reactor. As with the earlier design, the reactor operates within the natural-uranium license envelope.

The modified CANFLEX advanced MOX fuel bundle has a plutonium loading of 470 g (4.6% Pu in ring 3, 2.6% Pu in ring 4), 15% dysprosium in the central 8 elements, with depleted uranium as the base material throughout the bundle. The average discharge burnup is about 17.1 MW·d/kg HE. The plutonium disposition rate in a Bruce A reactor is 1.2 t per year per reactor. The MOX fuel fabrication capacity requirement is 41 t per year per reactor.

2.3 STATUS OF CANDU MOX OPTION FOR PLUTONIUM DISPOSITIONING

One major advantage of the CANDU option is the participation of a trusted third country, Canada, that can provide security and safeguards assurances in a balanced, simultaneous drawdown of both US and Russian weapons-surplus plutonium. CANDU MOX is a low-cost, low-risk option, readily available in the near-term, which would enable a quick start to the disposition of weapons-surplus plutonium, by converting it to spent fuel. It is one of the options chosen for further evaluation by the USDOE in its recent Record of Decision.

Preparations are now in place for the first physical tests towards qualifying the use of CANDU MOX fuel, fabricated from weapons-surplus plutonium. The program consists of the fabrication of a small amount of CANDU MOX fuel in the United States and in Russia, for testing under simulated CANDU reactor conditions in AECL's NRU research reactor at the Chalk River Laboratories. Although AECL has tested MOX fuel for decades, these new tests will confirm the behaviour of fuel using weapons-grade plutonium. This type of small-scale experimental test was endorsed by the G7 leaders at the Nuclear Summit in Moscow in 1996 April.

There is real interest in Russia in the CANDU option. A joint Canada-Russia feasibility study sponsored by the Canadian government builds upon previous USDOE studies. Its aim is to establish the viability of a CANDU MOX fuel fabrication plant in Russia, addressing related safeguards and security issues. The first interim report, issued in the fall of 1996, establishes the feasibility of CANDU MOX fuel fabrication in Russia.

3. Plutonium Annihilation in CANDU

3.1 INTRODUCTION

The Pu-annihilation concept incorporates the weapons-plutonium in an inert matrix that will facilitate the burning of the plutonium without generating new plutonium — thus resulting in maximum plutonium destruction. The carrier's purpose is to dilute the plutonium for incorporation in standard CANDU fuel elements and bundles. (Without dilution, the heat generation would be so localized that it could not be removed efficiently and the plutonium would melt.) A related application is actinide burning—the destruction of actinide waste produced in normal LWR reactor operation, and concentrated by reprocessing. Long-lived, carcinogenic nuclides are the main targets for actinide burning, specifically, ^{237}Np, ^{244}Cm and ^{241}Am. This application is the main motivating force for the European interest in inert-matrix fuels.

3.2 DESIRABLE PROPERTIES OF INERT MATRIX MATERIALS

Besides neutronic considerations, the selection of suitable inert-matrix materials must consider the following materials properties:

- compatibility with coolant and clad, the latter to elevated temperatures for accident conditions;
- phase stability—changes in phase that are due to temperature changes or irradiation could degrade the fuel's performance (including possibly disintegration of the matrix material to powder);
- irradiation properties—under irradiation, the carrier material should not swell or undergo phase changes. It is known that some potential candidates, such as Al_2O_3, quickly amorphize, with associated swelling, when exposed to fission-fragment damage;
- compatibility of the fission products with the matrix carrier;
- melting temperature—a high melting temperature provides insurance during off-normal and accident scenarios;
- thermal conductivity—a high thermal conductivity, leading to lower operating temperatures, is not a prerequisite, but is a strong advantage for safety considerations. Such an advantage over UO_2 provides a strong additional motivation to compensate for the costs of moving to inert-matrix fuels, and would allow operation at higher bundle powers;
- Pu-microstructure—the formation of a solid solution or fine dispersion of the Pu-containing phase is a prerequisite.
- heat capacity—high heat capacity ensures that after-heat generation during a LOCA will not heat the fuel excessively. However, at the same time, it means that there is more stored heat for dissipation in a loss-of-coolant flow (or regulation) incident. Generally, high values of heat capacity are preferred to limit fuel temperature increases.

3.3 CANDIDATE MATRIX MATERIALS

Many candidates are being considered by various countries. The Japanese are focusing on rock-like candidates, which should be especially stable for underground disposal after irradiation. A partial list of candidates under consideration in Europe, the United States and Canada includes SiC, $MgAl_2O_4$ (spinel), $ZrSiO_4$ (zircon), ZrO_2, CeO_2, $CePO_4$ and BeO. Advantages and disadvantages of some of these materials are discussed below:

- SiC has a very high thermal conductivity and melting point, and, as a good industrial material for many applications, is being widely studied. It is generally very stable chemically and has high resistance to oxidation.
- $MgAl_2O_4$ (spinel) has a cubic microstructure, which, theoretically, suggests good irradiation properties. However, accelerator simulation tests, described below, do not confirm this expectation.
- $ZrSiO_4$ (zircon) is a naturally existing mineral, which suggests that it is stable and would form a stable waste form.
- ZrO_2 is known to have good irradiation properties (no swelling) but has poor thermal conductivity. Stabilized zirconia with burnable poisons is the main candidate of the Los Alamos National Laboratories (LANL).
- CeO_2 has a cubic microstructure, identical to that of UO_2, suggesting good irradiation properties. Again, this has not been confirmed by our accelerator simulation tests.
- BeO has very high thermal conductivity and is stable; however, it is toxic and therefore has not yet been seriously considered.

Table 3 shows thermal properties of these candidates.

3.4 ACCELERATOR SIMULATION TESTS

Because of the large number of candidates and the expense of fabricating fuel and performing irradiation tests, AECL has used the Chalk River Laboratories Tandem Accelerator to simulate irradiation damage, thus providing an initial screening of the candidates. The damage caused by fission fragments moving through the inert matrix with kinetic energy of about 70 MeV is severe (compared with neutron or γ-ray damage). Therefore, the specimens were bombarded with a beam of iodine at 70 MeV, a typical fission fragment and energy. Bombardment dose and sample temperature were varied, the former between 10^{18} ions/m^2 and 10^{20} ions/m^2, the latter between room temperature and 1200°C. Expected results were obtained for benchmark tests on Al_2O_3 and UO_2; Al_2O_3 is known to rapidly amorphize and swell from fission-fragment damage, whereas UO_2 does not amorphize or swell. The samples of $ZrSiO_4$ and $CePO_4$ showed swelling; the samples of SiC and ZrO_2 (including additives) did not show any swelling.

3.5 OTHER SIC TESTS

Sintered samples of SiC containing Al_2O_3 (a sintering aid) and CeO_2 or CeC (Ce is a non-radioactive surrogate for plutonium) were prepared with various amounts of Ce up to 30 wt %. Various sintering aids, in addition to Al_2O_3, were also tested to reduce the normally high temperatures required for sintering SiC. With these additives, sintering densities of 96% theoretical density (TD) were achieved at 1860°C and 90 to 94% TD at 1780°C. This work is on-going.

Compatibility tests of SiC with alkaline water (pH of 10.7) at 300°C were performed with good results. Similarly, compatibility tests with Zircaloy-4 were performed up to 1700°C. No interaction was seen at 1000°C; some diffusion of Si into Zircaloy was seen at 1500°C; and at 1700°C a layer of ZrC formed at the SiC/Zircaloy interface, and the remainder of the Zircaloy formed a molten Zr-Si alloy phase.

On the basis of these assessments, AECL is focusing its efforts on SiC as the most promising candidate for inert-matrix applications. Its very high thermal conductivity will result in very low temperatures, both in normal operation and in postulated accidents, with the expected benefit of low fission-gas release. Its high melting temperature is also a benefit. There do not appear to be any long-lived activation products resulting from its irradiation, and it would appear to be a stable waste form. Simulated irradiation performance, and other tests, are positive to date. Needless to say, much further work must be done to assess SiC, and other candidates. If other materials are shown to be superior, they too can be used for actinide burning or plutonium annihilation in CANDU reactors.

3.6 REACTOR PHYSICS ASSESSMENTS FOR ACTINIDE BURNING

AECL has performed reactor physics assessments of actinide burning in CANDU systems under contract to a commercial client. Although the analysis has not been as detailed as the work previously described on CANDU MOX fuel for plutonium dispositioning, these studies have nonetheless shown actinide burning in CANDU reactors to be technically feasible from the reactor physics aspect. The results are reported here because they are directly applicable to plutonium annihilation. These assessments were conducted using the WIMS-AECL [2] lattice code with the ENDF/B-V data library, and the RFSP [4] fuel management code, and are more detailed than earlier studies. [5]

For actinide burning in CANDU systems, SiC was chosen as the inert-matrix carrier, with a standard 37-element bundle. The actinide-mix consists of the ^{237}Np, ^{241}Am, ^{243}Am, and plutonium from spent PWR fuel. The actinides were mixed with SiC uniformly throughout the bundle. Because the fission energy derives almost totally from plutonium, and because there is no fertile material (either ^{238}U or ^{232}Th) present to produce additional fissile material to compensate for the loss of plutonium, the reactivity and bundle power decrease rapidly during irradiation. To compensate for what would otherwise be a very high refuelling ripple (increase in reactivity, and local power during refuelling), a burnable poison was added to the bundle. For this application, gadolinium was chosen (having a higher depletion rate than the dysprosium used for CANDU MOX fuel). The fast burnout rate of gadolinium reduces the burnup penalty.

Three actinide inventories were considered: 400, 200 and 100 g per bundle in the high-, medium-, and low-inventory cases, with total plutonium contents of 351 g, 175 g, and 89 g per bundle, respectively, and fissile plutonium loadings of 236 g, 118 g, and 59 g per bundle, respectively. The bundles had gadolinium contents of 60, 20 and 10 g per bundle, respectively. Gadolinium was confined to the innermost 7 fuel elements; in the high-inventory case, for example, the central element had 20 g gadolinium, and the remaining 40 g gadolinium was distributed uniformly over the six elements in the next ring.

Coolant void reactivity in all cases is negative, as is the power coefficient. Hence there would be no power pulse in a postulated LOCA, and the safety and licensing analyses would be greatly simplified. The fuel temperature coefficient is very slightly positive; however, this is irrelevant in the safety analysis since any increase in heat in the fuel would immediately be transferred to the coolant because of the high thermal conductivity of the SiC, thus reducing the coolant density and producing a negative reactivity feedback because of the negative void reactivity coefficient.

Table 4 gives the actinide composition for the fresh and spent fuel, for the high-inventory case: 63% of the total original actinide inventory is destroyed, as is 91% of the initial fissile plutonium inventory. The high-inventory case results in the destruction of 0.68 Mg of actinides in a CANDU 6 reactor per year (assuming an 80% capacity factor).

Detailed, realistic fuel management simulations were performed using RFSP for a standard CANDU 6 reactor. On-line refuelling enables a full core of Pu-SiC to be used in CANDU. Acceptable bundle and channel powers were obtained for all cases. A bi-directional, 2-bundle shift refuelling scheme was used in the high- and medium-inventory cases, and a 4-bundle shift refuelling scheme was used in the low-inventory case. Refuelling rates in bundles per full-power-day were 9.2, 20.4 and 52.3 for the high-, medium- and low-inventory cases, respectively. The current fuel handling system can easily meet the requirements of the high- and medium-inventory cases; modifications would be required to meet the demands of the low-inventory case.

In all cases, maximum time-average channel powers were less than 6320 kW, and maximum time-average bundle powers were less than 960 kW. Again, these are within current limits. With the high thermal conductivity of the SiC matrix, corresponding fuel temperatures would be very low, and fission-gas release is expected to be negligible. (For a fuel centreline temperature of 2000°C in UO_2, the SiC fuel temperature would be about 500°C.)

Figure 1 shows the axial bundle power profile for the highest power channel (a central channel, N-10, near adjuster rods), in the high-inventory case. The axial power distribution corresponds to an instantaneous "snapshot" at one moment in time. The axial power distribution is typical of that with enrichment in CANDU, and in fact is similar to that with MOX fuel: it peaks towards the inlet end of the channel (axial bundle position 4), thereafter decreasing towards the outlet end. Only relatively fresh fuel would experience power boosting during refuelling (in shifting from axial bundle position 1 to 3, and from position 2 to 4), and the critical channel power (the channel power at which the fuel first experiences dryout) will be higher than for a cosine-shaped axial power distribution; thus thermal margins are better.

3.7 REACTOR PHYSICS ASSESSMENTS FOR PLUTONIUM ANNIHILATION

The above results for actinide burning are directly applicable to the case of burning weapons-derived plutonium. Table 5 shows the composition of the fresh and spent fuel for the high-inventory case above, with 250 g weapons-derived plutonium in the fresh fuel (no other actinides), calculated using WIMS-AECL with the ENDF/B-V data library. Void reactivity and power coefficient are both negative. Refuelling rate, and power distributions will be very similar to the high-inventory actinide-mix case above (RFSP was not re-run).

Looking at the "measures of merit" in Table 1, plutonium annihilation in an inert matrix has high net-plutonium and fissile-plutonium destruction efficiencies. The Pu-disposition rate is comparable with the CANDU MOX options, at 1.66 Mg Pu/GW(e) a. The fission energy derived from the plutonium is also good.

4. CANDU Pu-Thorium

The final CANDU plutonium-annihilation option to be considered in this paper employs the use of ThO_2 as a carrier for the plutonium. This is a responsible, forward-looking strategy that uses plutonium to convert ^{232}Th to ^{233}U, to be used as a future energy resource. The ^{233}U would be safeguarded in the spent fuel, with all the proliferation-resistant features of spent UO_2 or MOX fuel. Moreover, the radiation fields caused by the presence of ^{232}U (which emits copious α-particles) and its daughter products (particularly ^{208}Tl, which emits a 2.6 MeV γ-ray), provide a high degree of self-protection and render ^{233}U unattractive as a weapons material. The ^{233}U could be recovered in the future using a proliferation-resistant technology, when the price of uranium is high enough to warrant its recovery. Uranium-233 is the best fissile material in a thermal reactor, having the highest η-value (fission neutrons produced per neutron absorbed). In CANDU reactors, the production of ^{233}U also maintains the option of the "self-sufficient-equilibrium thorium cycle", a long-term fuel cycle option which, in equilibrium, results in as much ^{233}U in the spent fuel as is needed in the fresh fuel. This CANDU "near-breeder" fuel cycle provides long-term assurance of fissile fuel supplies.

The assessment of Pu-ThO_2 for plutonium management in this paper was limited to reactor physics lattice calculations, using the multigroup lattice code WIMS-AECL. Actinide inventories were calculated using a fully-coupled multiregion WIMS-AECL / ORIGEN-S code package. [6,7] Reactor calculations and fuel management simulations were not performed. However, given the CANDU flexibility in fuel management, no technical feasibility issues are anticipated.

A somewhat different approach was taken in designing the Pu-ThO_2 fuel bundle for this application. To maximize the destruction of the plutonium, good neutron economy was desired. A reduction in void reactivity was also sought, to compensate for the faster dynamic behaviour of the fuel (shorter neutron lifetime, and smaller delayed-neutron fraction). To achieve these two objectives, the central elements in a CANFLEX bundle were replaced with a large central graphite displacer. Plutonium at 2.6% (354 g per bundle) was mixed with thorium in the remaining 35 elements in the outer two fuel rings of the CANFLEX bundle. Computer analyses confirmed that enrichment grading in the outer two fuel rings would result in peak element ratings that are comparable to those in a 37-element bundle with natural-uranium fuel. The resultant burnup was 30 MW·d/kg HE, a burnup for which there is CANDU experience with Pu-ThO_2 fuel. Void reactivity was 8.6 mk, which is judged to be acceptable with the current shutdown system. Computer simulations also showed that using SiC instead of graphite in the central displacer reduces the magnitude of the void reactivity somewhat.

Addition of a small amount of burnable poison to the central displacer would further reduce void reactivity, increase the plutonium loading per bundle as well as the absolute amount of plutonium destroyed, but would decrease the plutonium destruction efficiency. (The plutonium destruction efficiency would be reduced from about 77% to 71%, by poison addition that reduces void reactivity from about 8.6 mk to zero.)

Table 6 shows the composition of the fresh and spent fuel: 77% of the total plutonium is destroyed, and 94% of the fissile plutonium, a destruction efficiency which is similar to that for plutonium annihilation in an inert matrix. Fissile ^{233}U (including its parent ^{233}Pa) is produced to the extent of 168 g, which can be recovered and recycled in a proliferation-resistant fashion. (For example, the ^{233}U, remaining plutonium, and ^{232}Th could be co-extracted, without separating the ^{233}U, and fresh weapons-derived plutonium added to maintain burnup, and further destroy plutonium stockpiles). Looking at the “measures of merit” in Table 1, this option achieves high plutonium destruction efficiency and has the highest energy yield, even without recycling the ^{233}U. Recycling the ^{233}U would increase the energy yield many fold.

5. Summary

The CANDU system provides unsurpassed flexibility for plutonium management through high neutron economy, on-line refuelling, and a simple, economical fuel-bundle design.

TABLE 1. Summary of CANDU plutonium management options

	Pu/U MOX (1)	Pu/U MOX (2)	Pu-inert matrix (annihilation)	Pu/Th
Net Pu-Destruction Efficiency (%)	34	23	82	77
Net Fissile Pu-Destruction Efficiency (%)	58	41	93	94
Pu-Disposition Rate (Mg Pu/GW_e·a)	1.56	2.22	1.66	1.04
Energy Produced (GW_e·a/ Mg Pu)	0.64	0.45	0.61	0.96 (>>1 with ^{233}U recycle)

TABLE 2. CANDU MOX: Fuel composition for reference bundle (g/bundle)

(37-element bundle, 1994 study)

Nuclide	0 MW·d/kg	9.7 MW·d/kg
^{239}Pu	218	79
^{240}Pu	14	58
^{241}Pu	0.3	13
^{242}Pu	0.05	4
Total	232	153

TABLE 3. Thermal properties of candidate inert-matrix materials

Material	Melting Temp. (°C)	Thermal Conductivity 100°C	1000°C	Heat Capacity ($J/cm^3 \cdot K$)
ZrO_2	2715	1.9	2.3	2.6
BeO	2530	220	20	3.1
$MgAl_2O_4$	2135	10.7		
CeO_2	2600	10.9		2.6
SiC[a]	2700	61.6	27	2.2
SiC[b]	2700	77.5	50	2.2
Si	1410	108		1.7
UO_2	2878	8.8	3.2	2.6

[a] Sintered α-phase
[b] Sintered β-phase

TABLE 4. Fuel composition: Actinide-burning in an inert matrix (g/bundle)

Nuclide	0 MW·d/kg	582 MW·d/kg
^{239}Pu	205	5.7
^{240}Pu	96	57
^{241}Pu	31	15
^{242}Pu	19	42
Total Pu	351	125
^{237}Np	20	10
^{241}Am	20	2
^{243}Am	3.5	8.7

Total of 400 g actinide mix per bundle

TABLE 5. Fuel composition: Pu-annihilation in an inert matrix (g/bundle)

Nuclide	0 MWd·/kg	733 MW·d/kg
^{239}Pu	235	7
^{240}Pu	14	23
^{241}Pu	0.9	10
^{242}Pu	0.07	10
Total Pu	250	46

TABLE 6. Fuel composition - Pu/ThO_2 (g/bundle)

Isotope	0 MW·d/kg	30 MW·d/kg
Pu-238	0.2	0.3
Pu-239	331.3	5.3
Pu-240	21.2	45.2
Pu-241	1.4	14.5
Pu-242	0.2	15.9
Total Pu	354.3	81.2
U-233 + Pa-233	0	167.7
Np-237	0	0.004
Am-241	0	0.59
Am-243		2.21
Cm-242		0.26
Cm-244		0.42

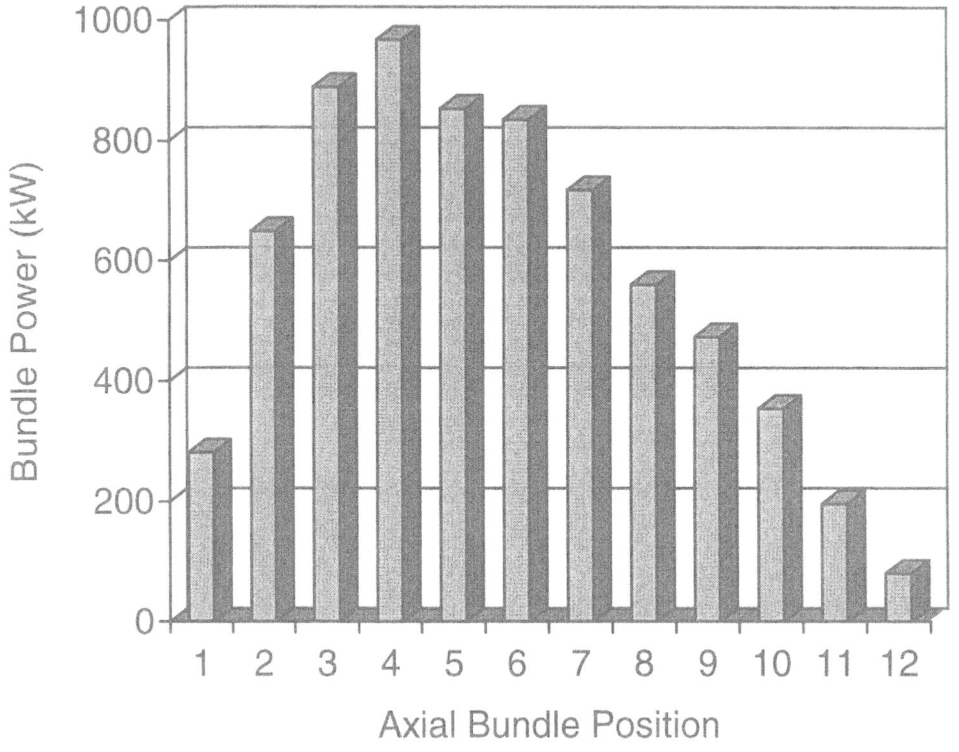

Refuelling direction: →
Maximum Channel Power, 6858 kW, channel N-10

FIGURE 1: Typical axial bundle power profile, Channel N-10, actinide mix

CANDU MOX offers a timely, technically achievable, cost-effective option for dispositioning weapons-derived plutonium. The use of the Bruce A reactors in Canada for the simultaneous drawdown of plutonium from both the United States and Russia offers an attractive symmetry and is actively being considered.

If the objective in plutonium management is annihilation, then CANDU can achieve this using an inert-matrix fuel. On-line refuelling enables reactivity to be added incrementally to compensate for the rapid depletion of the plutonium, allowing a full core of inert-matrix—plutonium fuel. High neutron economy enables a large fraction of the plutonium to be destroyed. A detailed reactor assessment including realistic fuel management simulations indicates that this is achievable in existing CANDU reactors. More than 93% of the fissile plutonium can be destroyed in a single pass.

The Pu-ThO_2 cycle offers a high plutonium destruction efficiency, while creating a stockpile of ^{233}U that is safeguarded in the spent fuel for future recovery using proliferation-resistant technology. It is a forward-looking option that maximizes the energy potential from the plutonium.

Hence CANDU offers immediate, plus evolutionary approaches to plutonium management.

6. References

1. Boczar, P.G., Kupca, S., Fehrenbach, P.J., and Dastur, A.R. (1994) Plutonium Burning in CANDU, presented at the NATO Advanced Research Workshop on Mixed Oxide Fuel (MOX) Exploitation and Destruction in Power Reactors, Obninsk, Russia, 1994 October 16-19.

2. Donnelly, J.V. (1986) WIMS-CRNL --A User's Manual for the Chalk River Version of WIMS, Atomic Energy of Canada Limited Report, AECL-8955.

3. Boczar, P.G., Hopkins, J.R., Feinroth, H. and Luxat, J.C. (1995) Plutonium Dispositioning in CANDU, presented at the IAEA Technical Meeting, Recycling of Plutonium and Uranium in Water Reactor Fuels, Windermere, U.K., 1995 July 3-7. Also Atomic Energy of Canada Limited Report, AECL-11429.

4. Rouben, B. (1996) Overview of Current RFSP-Code Capabilities for CANDU Core Analysis, Atomic Energy of Canada Limited Report, AECL-11407.

5. Meneley, D.A., Dastur, A.R., Verrall, R.A., Lucuta, P.G., and Andrews, H.R. (1994) Annihilation of Plutonium in CANDU Reactors, presented at the NATO Advanced Research Workshop on Mixed Oxide Fuel (MOX) Exploitation and Destruction in Power Reactors, Obninsk, Russia, 1994 October 16-19.

6. Hermann, O.W., and Westfall, R.M. (1993) ORIGEN-S - SCALE System Module to Calculate Fuel Depletion, Actinide Transmutation, Fission Product Buildup and Decay, and Associated Radiation Source Terms, in *SCALE: A Modular Code System for Performing Standardized Computer Analyses for Licensing Evaluations*, NUREG/CR-0200, Rev.4 (ORNL/NUREG/CSD-2/R4), Vol.II, Part I, (Draft November 1993).

7. Gauld, I.C. (1997) A Coupled Two-Dimensional WIMS-AECL and ORIGEN-S Depletion Analysis Code System: I. Program Abstract, II. Users Manual, and III. Programmers Manual, internal Atomic Energy of Canada Limited Report.

Advanced Systems for Pu Utilization

M. Salvatores, M. Delpech
Nuclear Reactor Directorate, CEA-Cadarache, FRANCE

NATO Advanced Research Workshop
"Advanced Nuclear Systems consuming Excess Plutonium"
13-16 October 1996, Moscow, Russia

1. INTRODUCTION

The aim of the present paper is not a policy contribution to the debate of Pu utilisation, whatever its origin. Our more modest goal is to review technically potential ways to use Pu on large scale. This has also been the objective of numerous publications, conferences and seminars.

A few recent publications give a wide overview of several aspects of this complex issue. The GLOBAL'95 conference held in Versailles in 1995, had a number of sessions on Pu utilisation, both from the strategic and the technical point of view, and both civil and weapon-grade Pu issues were treated [1]. Unconventional options for Pu disposition have been discussed in a IAEA meeting [2]. The physics of Pu recycling in fission reactors has been reviewed in detail in a OECD-NEA report, prepared by a specific Working Party set up the NEA Nuclear Science Committee [3]. As far as France, ref. 4 gives an overview from the French national point of view.

In the present paper a survey will be given both of "evolutionary" systems and of more revolutionary concepts. In fact, it is widely recognized that fission reactors offer several effective ways to use Plutonium :

- Thermal neutron reactors (MOX-fuelled PWRs with different moderator-to-fuel ratios V_m/V_f).
- Fast reactors, MOX-fuelled.
- Pu w/o U fuels in both thermal and fast neutron reactors.
- Accelerator-driven sub-critical fission reactors.

For all these options, feasibility studies have been performed in many laboratories. However, besides core related studies, the impact on the fuel cycle (e.g. waste production) has to be evaluated.

E. R. Merz and C. E. Walter (eds.), Advanced Nuclear Systems Consuming Excess Plutonium, 181–202.
© 1997 *Kluwer Academic Publishers.*

Studies are underway in France on all these options :

- system studies,
- experimental studies (fuels and targets development and irradiation, experimental physics validation in critical assemblies).

This last point is relevant, to judge of the maturity of a particular option.
Finally, options for Pu use in CANDU or HTR will not explicitly treated here.

As far as weapon-Pu, cooperative efforts are underway to investigate strategies for its utilisation in standard reactors (see for example the Russian-French AIDA-MOX program, documented in Ref. 1).These programs are not explicitly mentioned, since most of the technical solution proposed, fall in a wider range of solutions, valid for any kind of Pu.

2. 100 % MOX fuel in PWRs

It is worth to remind here that MOX recycling in LWRs is successfully performed in a number of countries [5]. For example, in France seven 900 MWe PWRs are actually loaded with MOX fuel assemblies (~ 30 % of the core) and the COGEMA fuel fabrication plant MELOX is reaching full capacity in Marcoule.

Feasibility studies to extend Pu recycling to 100 % MOX-fuelled cores (core performances, reactivity coefficient limitations etc) do not indicate impossibilities. Appropriate core designs can be defined. However, it is known that two major drawbacks have to be considered : Pu vector degradation during multirecycling (limit on void coefficient) and Minor Actinide (MA) build-up. These two facts are simply related to the physics properties of actinides, and in particular to the defavorable fission-to-capture cross-section ratios in thermal neutron spectra.

Examples of Pu consumption and MA build-up are given in table I.

The results presented in this table, are obtained using a "mixing" hypothesis on the Pu recycling. In fact a strict "self-recycling" will lead to unrealistic high Pu content in the fuel at the second recycling. The hypothesis used here is to mix first-generation Pu to second (and successive) generation Pu, in the ratio about 3:1.

It is also to be noted, that we have chosen to show results corresponding to a burn-up of 55 GWd/t, since the economic objective of long irradiation times has an impact on the evolution of the needed Pu content for successive recyclings. Data in table I, do not show feasibility-related parameters. We refer to e.g. reference 3 for a detailed discussion of these aspects. However, it can be reminded here that the feasibility of two recyclings, in particular using high V_m/V_f ratios, could be safely envisaged, even taking into account uncertainties. Moreover, experimental programs (like the EPICURE program [6] at CEA-Cadarache) have provided essential experimental evidence to validate MOX physics.

TABLE I

Masses (kg/TWhe) in different types of PWRs at the end of each recycle and after 3 years cooling (burn-up = 55 GWd/t)

	Pu	MA	MA/Pu
PWR-UOX	29.3	3.78	0.13
PWR-MOX (V_m/V_f = 4)			
Cycle 1	- 71	8.7	- 0.12
Cycle 2	- 79	12.9	- 0.16
Cycle 3	- 85	15.0	- 0.18
PWR-MOX (V_m/V_f = 2)			
Cycle 1	- 64	16.4	- 0.26
Cycle 2	- 69	19.3	- 0.28
Cycle 3	- 73	21.7	- 0.29
PWR-MOX (V_m/V_f = 1.1)			
Cycle 1	- 54	21.8	- 0.40
Cycle 2	- 56	22.3	- 0.40
Cycle 3	- 58	23.0	- 0.40

The effectiveness of the use of high V_m/V_f ratios is clearly shown by the ratio of (Pu consumption)/(MA production). Present CEA studies indicate that a MOX core with V_m/V_f = 4 would offer the best performances and should be feasible.

As far as minor actinide production, table II gives the details related to the global data of table I. It is evident that, even in the most favorable case (i.e. V_m/V_f = 4), the minor actinide build-up is very significant after two recyclings (despite the fact that a favorable "mixing" approach to fuel loading has been used). This means that, even with an optimized V_m/V_f ratio, indefinite recycling can hardly be envisaged.

TABLE II

Minor Actinide production (kg/TWhe) at the end of cycle and after 3 years cooling (burn-up = 55 GWd/t)

	PWR-MOX 100 %								
V_m/V_f	4			2			1.1		
Number of recyclings	1	2	3	1	2	3	1	2	3
^{237}Np	0.2	0.2	0.2	0.4	0.4	0.3	0.4	0.4	0.4
^{241}Am	2.2	3.7	4.4	6.7	8.4	9.7	10.7	11.1	11.5
^{243}Am	4.1	6.0	7.1	5.6	6.5	7.3	6.5	6.6	6.9
Cm	2.2	3	3.3	3.7	4	4.2	3.9	3.9	4
Total MA	8.7	12.9	15	16.4	19.3	21.7	21.8	22.3	23.0

Table III shows void reactivity effects and soluble Boron efficiency during Pu recycling vs V_m/V_f ratios. A positive void effect is unacceptable as well as a small Boron efficiency, which indicates potential problems to ensure control of the core (the usual value for Boron Efficiency is about - 9 pcm/ppm). Thus, Pu recycling is difficult in PWR and not acceptable for usual V_m/V_f ratios.

Finally, it should be mentioned that mixed MOX/enriched-U fuels can be envisaged for PWRs. In that case, limitations related to multirecycling can probably be reduced, with the obvious drawback to generalize MOX recycling to all the fuel of all a nuclear power park, which can have severe economic penalties [9].

TABLE III

Reactivity coefficients vs V_m/V_f ratios

V_m/V_f	2		4	
	Cycle 1	Pu recycling Equilibrium state	Cycle 1	Pu recycling Equilibrium state
Void effect (% of Keff)	- 13	+ 20	- 48	- 9
Soluble Boron worth (pcm/ppm)	- 3	- 2	- 9	- 6

3. MOX FUEL IN FAST REACTORS (CAPRA-TYPE)

In recent years, significant work has been devoted to assess the characteristics of a fast neutron Pu burner in the frame of the CAPRA program [7]. Most results have been documented in the open literature and a detailed feasibility study has been completed (documented elsewhere [8]). Some characteristics of the reference CAPRA core (based on oxide fuel at high Pu content) are given in table IV and a core lay-out is given in fig. 1.

An extensive experimental program is underway to validate the concept :

- fuel and S/A concept (in particular at SUPERPHENIX) (see for example fig. 2),
- core physics characteristics (CIRANO program at the MASURCA critical facility).

TABLE IV

Characteristics of the Reference CAPRA Oxide Core (04/94)

Feed Pu enrichments (Pu/(U+Pu+Am))	%	C1 = 43.0 C2 = 44.7
Fuel residence time	efpd	855
Cycle length Sub-cycle length Load factor	efpd efpd	285 142.5 0.80
Reactivity loss over cycle	pcm	~ 9000
Pu consumption[(a)] MA production	kg/TWeh kg/TWeh	- 74.2 9.7
EOL peak damage EOL peak burn-up	dpaNRT Fe % h.a.	124 20
Max linear rating	W/cm	504
EOC Sodium void reactivity EOC Doppler constant	pcm pcm	1564 -455
βeff	pcm	324

(a) Pu of core loading : from 2nd recycling in a PWR-MOX at high burn-up.

FIGURE 1
REFERENCE CAPRA CORE CHARACTERISTICS

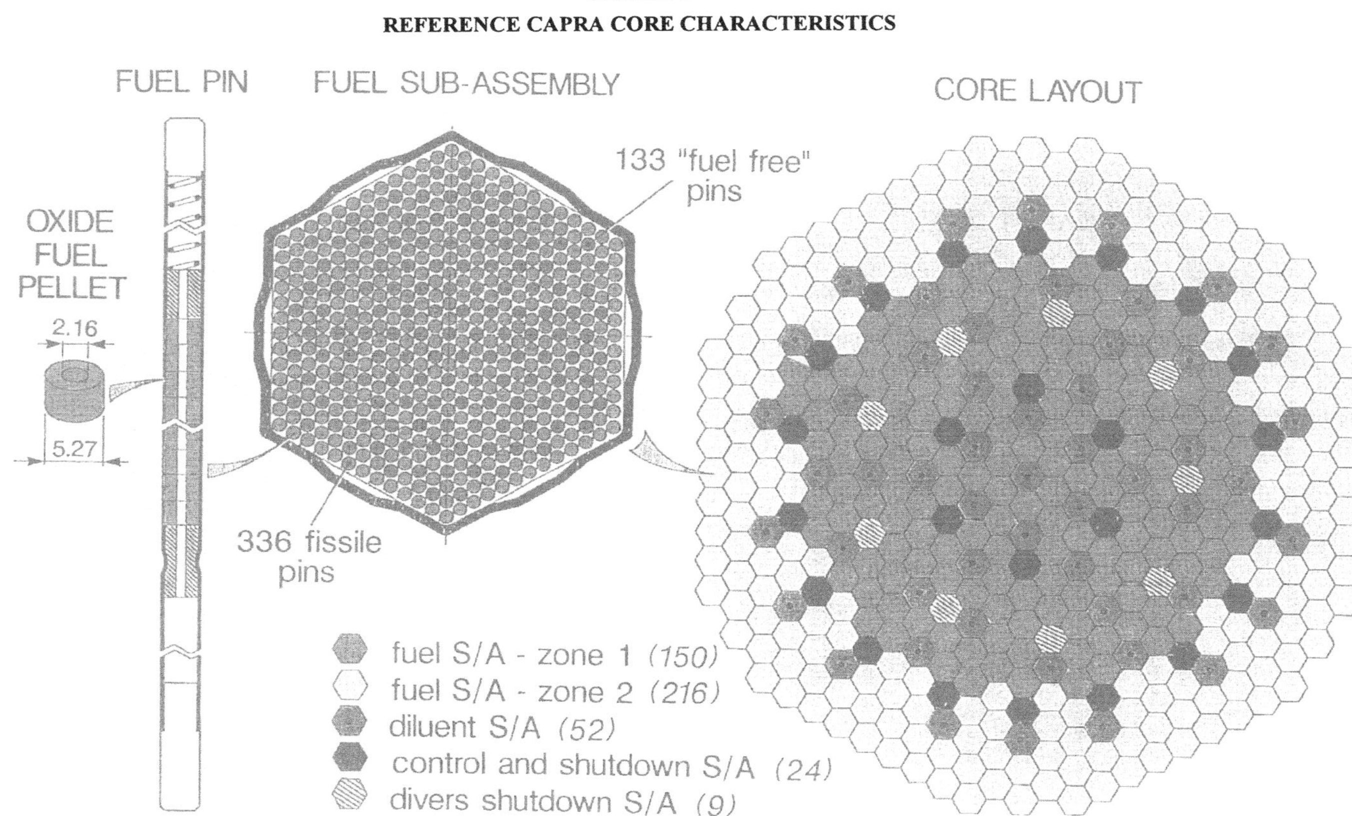

FIGURE 2

CAPRA1A & CAPRA1B SUBASSEMBLIES TO BE IRRADIATED IN SUPERPHENIX

- ♦ Fuel : $(U,Pu)O_2$ 31 % Pu Pin dimension : 5.42 × 1.7 mm
- ♦ Main objectives :
  - Use of different Plutonium qualities
    - CAPRA1A : "standard" Pu coming from PWR uranium fuels
    - CAPRA1B : "degraded" Pu coming from MOX fuels (second generation)
  - Test the heterogeneous bundle
    - large number of pins (397)
    - heterogeneous bundle

○ fissile pins

● dummy pins (filled with F17)

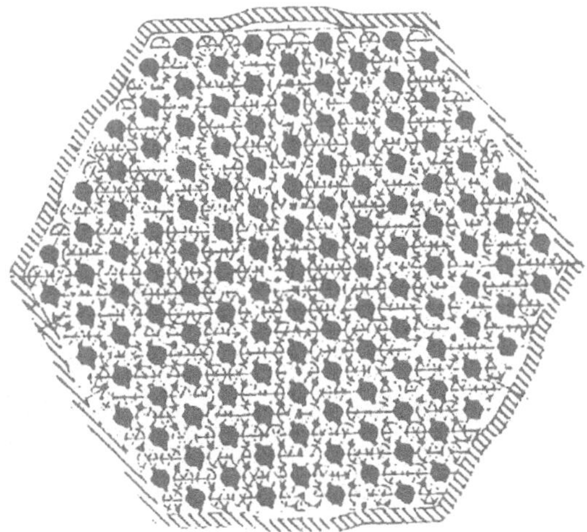

The indications coming from the CAPRA studies underline the potential of fast reactors to provided the best required tool for several options to control Pu stocks, to increase or to stabilize them and even to decrease them over realistic periods of time. The use of a "reversible" core concept is then relevant for industrial, long-term applications.

Moreover, the FR option provides a unique capability to accommodate Pu vectors over an extremely large range. This point will be further illustrated later on.

In this respect, the experimental validation underway in the frame of the CAPRA program has a particular value.

4. SCENARIO STUDIES AND IMPACT ON THE FUEL CYCLE

Reactor concepts and (multi)-recycling options such as the ones indicated previously should be considered in the frame of a "realistic" reactor park, to evaluate the consequences in terms of mass-flow and, among others, in terms of potential radiotoxicity in a waste repository/storage.

Scenario studies have been performed, to establish the fuel cycle main parameters at equilibrium [9]. Two significant scenarios are recalled here, namely :

- Pu multirecycling in a HM-PWR-MOX (100 %) with $V_m/V_f = 4$ (HM : High Moderation).
- Pu multirecycling in a CAPRA-type fast reactor after two recyclings in a PWR-MOX (standard $V_m/V_f = 2$).

Figures 3 and 4 give the annual mass flows related to these scenarios, which define the needs for reprocessing and fabrication of MOX fuels, and the potential inventories in a waste repository (taking for Pu losses a value of 0.1 %, presently validated at the industrial level).

As far as the potential source of radiotoxicity in a repository, one can evaluate the consequence of Pu recycling according to the two previous scenarios and to intercompare its time evolution to that associated to the fuel unloaded by the same PWR park and directly disposed, without reprocessing.

FIGURE 3

PARK OF PWRs (60 GWe - 400 TWhe/year)

EQUILIBRIUM STATE

BURN-UP : 55 GWD/t

SCENARIO # 1 : RECYCLE OF PU IN HIGH MODERATED REACTOR (Rmod = 4)

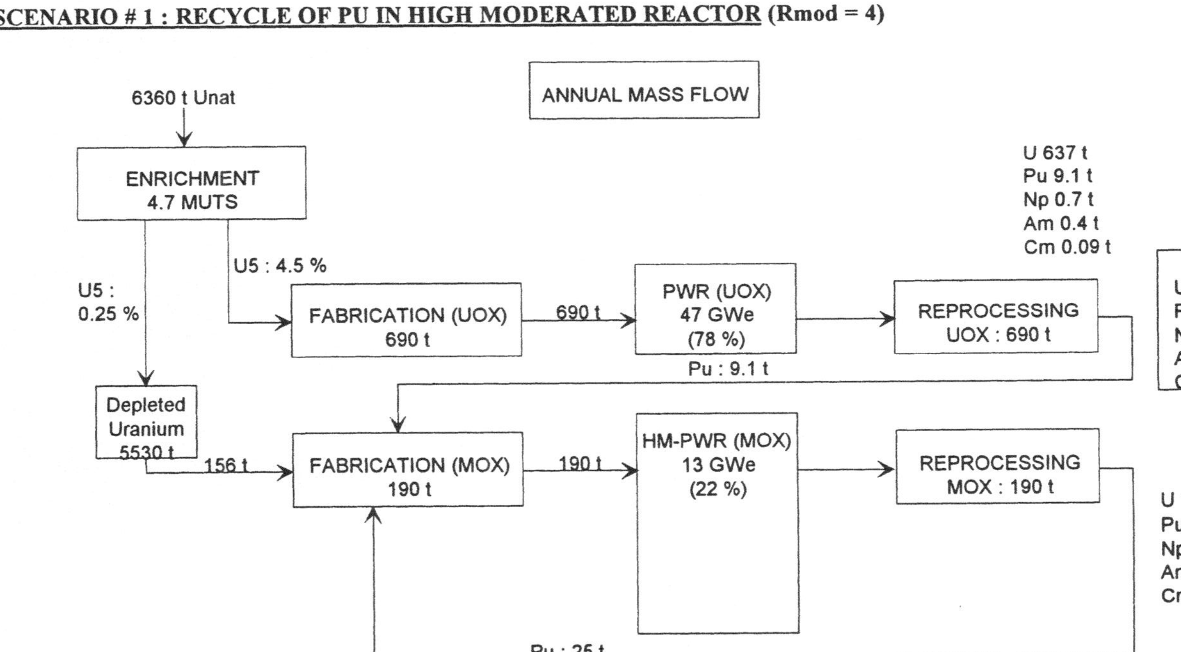

FIGURE 4

PARK PWR + CAPRA (60 GWe - 400 TWhe/year)
EQUILIBRIUM STATE
BURN-UP PWR = 55 GWD/t
BURN-UP CAPRA = 140 GWD/t

<u>**SCENARIO # 2 : RECYCLE OF Pu : twice in PWR and after in CAPRA**</u>

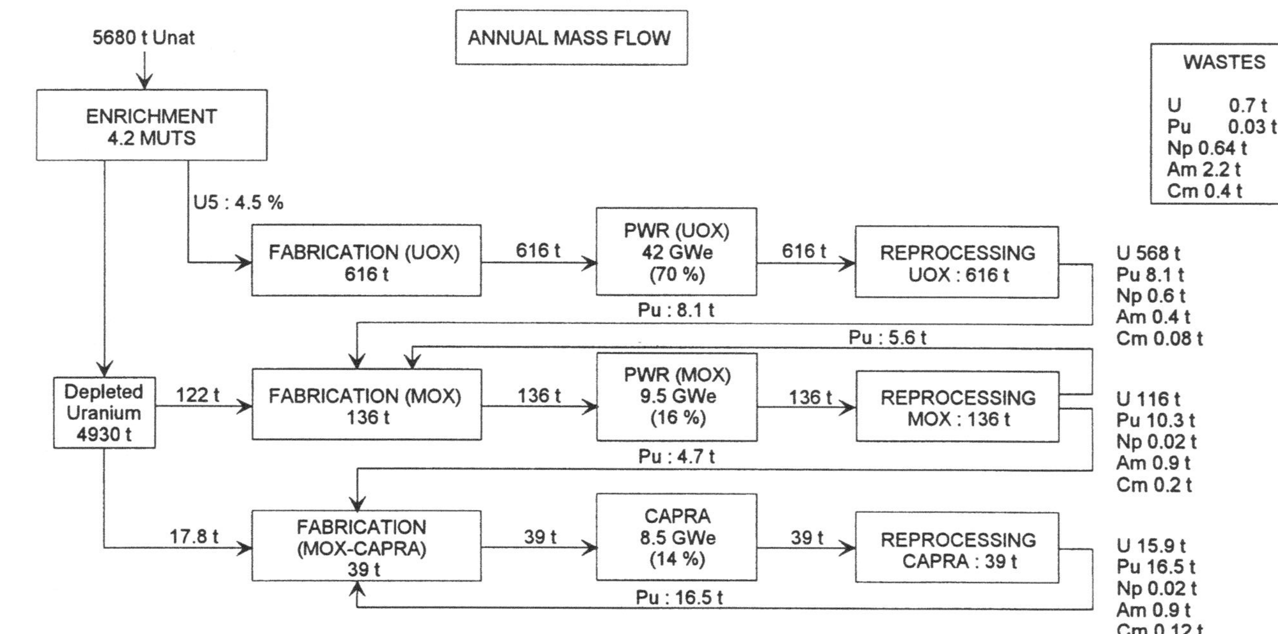

FIGURE 5
POTENTIAL RADIOTOXICITY EVOLUTION WITH TIME

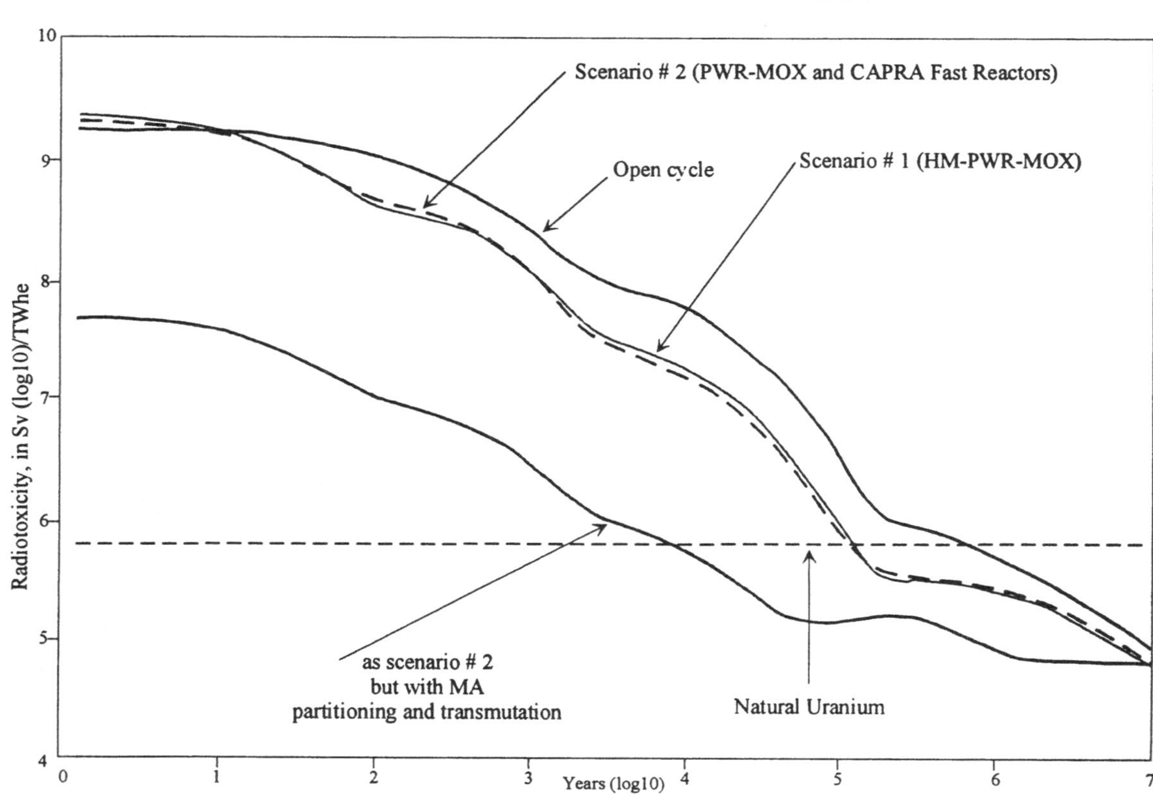

The results shown in figure 5, indicate that Pu recycling do reduce radiotoxicity in the wastes in both scenarios (by approximately a factor 3 to 5, all over the time scale). Obviously, to reach higher reduction factors, one has to envisage to separate minor actinides (mainly Am) from the waste and transmute them. If this is done in the same CAPRA-type reactor (scenario # 2) and using a decontamination factor for minor actinides of the order of 1 %, one obtains a reduction in radiotoxicity of a factor 50 to 100 (see figure 5 and reference 10).

5. BEYOND MOX FUELS

In the previous paragraphs, we have given a quick survey of advanced, but evolutionary concepts to deal with Pu, its utilisation and consumption. However, to enhance the Pu consumption, on obvious option has been explored, namely the use of U-free (Pu w/o U) fuels both in LWRs and in Fast reactors.

The support for Pu can be an inert material (oxides like CeO_2 or MgO, metals etc ...).

A special case is the use of Thorium instead of Uranium (e.g. in $(Th,Pu)O_2$ fuels).

Studies have been performed in several laboratories on PWR-type reactors, which show that one can in principle approach the theoretical Pu consumption limits (~ 110 ÷ 130 Kg/TWhe). Of course these studies are often of a conceptual nature, and most work is needed to demonstrate their feasibility and the associated economics.

Four examples of this type of studies, performed in France, will be quickly reviewed in what follows.

5.1 Pu w/o U in a PWR of the N4-type (1450 MWe)

The first example considers a standard large-size PWR, in which a theoretical U-free fuel is introduced. The main goal is to assess the Pu consumption rate and the quality of the Pu-vector at the end of irradiation (cycle length : 1200 EFPD). Two cases have been considered - Case 1 : the initial Pu is a first generation Pu from standard PWRs - Case 2 : Weapon Pu is loaded. The results are given in table V.

TABLE V

U-free Pu fuel in a standard PWR

Case 1 - Pu from PWR-UOX (burn-up : 33 GWd/t)

Mass of Pu in the reactor (t)	Cycle length (EFPD)	Fraction of initial Pu consumed	Fraction of total fissions	Pu consumption (kg/TWhe)	Pu vector Pu8/Pu9/Pu0/Pu1/Pu2 At beginning (% at)	At the end (% at)
6.1	1200	75 %	70 %	109	0/58/24/13/5	1/3/36/18/42
Case 2 - Weapon Pu						
6.1	1200	72 %	71 %	105	0/93/6/0.8/0.2	0/20/44/23/13

Most work is of course needed to assess the feasibility of these concepts, but this is beyond the scope of the present paper. Here we are mostly interested to look to some simple indications on the possible consequences of such an option, and in particular to point out the significant degradation of the Pu vector.

5.2 An alternative approach : the APA (Advanced Plutonium fuel Assembly) concept

At CEA, a theoretical study has been performed [11], to look at an alternative approach to use Pu in specialized assemblies of a standard PWR. The type of assembly is indicated in fig. 6.

The work reported in [11] indicates the advantages of the APA concept : high burn-ups and increased local moderator-to-fuel ratio enables to consume 60 % of second-generation Pu, while the minor actinide production represents only 8 % of this figure. Full core studies and thermohydraulic studies, indicate acceptable overall core behaviour. However, no specific technology-related program is underway at present to test such concept.

5.3 Pu w/o U in a CAPRA-type fast reactor

An overall assessment [12] of the possibility to conceive a large U-free core has been performed (fuel design, physics, core management, transient analysis and the energetic potential in a whole core accident). The general design remains the reference CAPRA 4/94 one with diluent S/As and an heterogeneous fuel bundle with fuel-free pins. Even if it is clear that

FIGURE 6

The APA (Advanced Plutonium fuel Assembly) concept (from reference 11)

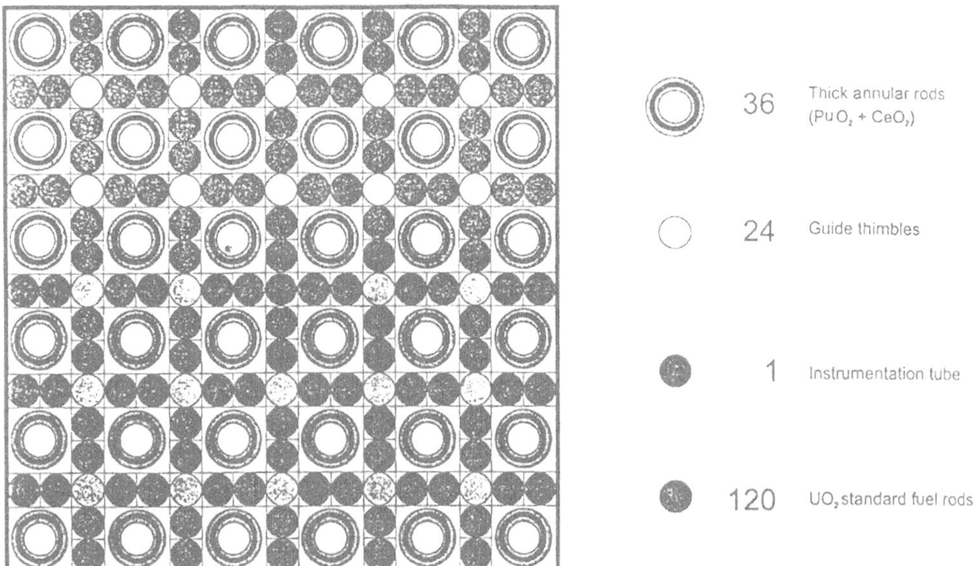

High local V_m/V_f (= 6) - High burn-up

Pu consumption - 76.5 Kg/TWhe

MA production + 9.0 KG/TWhe

Final Pu vector (Global) (% at) Pu8/Pu9/Pu0/Pu1/Pu2 : 5/17/31/18/29

no definitive conclusions could be drawn from that work, the possibility to conceive a large U-free core seems demonstrated. The more noticeable trends to be underlined are :

- the interest to specialise U-free cores in the use of degraded Pu qualities : physics studies demonstrated that a near conventional Doppler effect could be recovered when associating such Pu isotopics with the introduction of a $^{11}B_4C$ moderator in the empty pins of the fuel S/A (enhancement of Pu 240 and Pu 242 contributions),

- the choice of a nitride fuel, in a form that still remains widely open, seems to be a necessity if one remains in a logic of a PUREX type reprocessing,

- an innovative annular fuel element (spherical particles of pure PuN are poured in the annular space between an external cladding and an inner tube) must be seen as one possibility knowing that much more conventional fuel concepts can also be considered. Nevertheless, the essential part of the conclusions remains valid and a generic character. The innovative fuel element has the merit to show that, when considering U-free fuels, the introduction of an inert matrix is not an absolute necessity and that solutions can also be found in S/A and fuel pin designs.

The main characteristics of a CAPRA core with Pu w/o U fuel are given in table VI, for two different Pu vectors of the initial loading.

The excellent capability to use highly degraded Pu vectors is clearly indicated.

It is worth to remind that the options for a U-free fuel are examined also in relation with parallel studies performed at CEA in the frame of the SPIN program, to provided solution to the recycling of Am in reactors according to the so-called "heterogeneous" mode (i.e. an Americium compound, such AmO_2 or Am_2O_3, on an "inert" matrix). Inert matrix irradiation programs are underway (see fig. 7 and ref. 13) in particular for promising solutions like MgO or spinel $MgAl_2O_4$.

TABLE VI

Pu w/o U in a CAPRA-type fast reactor

Fuel Management (d)	6 x 170	6 x 170
Initial Pu inventory (kg)	10960	14520
Moderator	yes ($^{11}B_4C$)	yes ($^{11}B_4C$)
Pu quality	*	**
Pu burning (kg/TWhe)	116	124
MA production (kg/TWhe)	14	22
Burn-up (at %)	45	35
Dose (dpaNRT)	136	142
EOC sodium void worth (pcm)	1030	2000
EOC Doppler constant (pcm)	- 385	- 530

* Equilibrium Pu from MOX-CAPRA multirecycling (3/33/40/8/16)
** Equilibrium Pu from (Pu W/oU) - CAPRA multirecycling (3/17/51/8/21)

⇒ Doppler from even Pu isotopes and use of moderator
Excellent capability to use strongly degraded Pu vectors

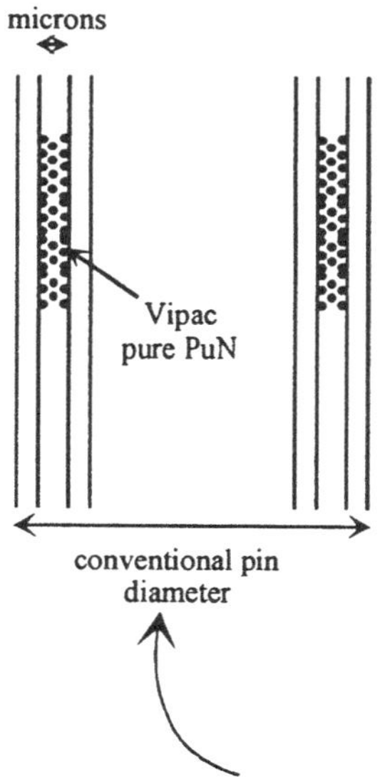

Schematic view of a fuel concept using a pure Pu compound (PuN)

FIGURE 7

MATINA1 EXPERIMENT IN PHENIX

- Objectives :

  Test and selection of best matrices for :
  - for Transmutation of MA (mainly Am) in heterogeneous mode,
  - for non-fissile pins in CAPRA-type S/As
  - for the option "Pu w/o U"

- Composition of the experimental device (under irradiation)

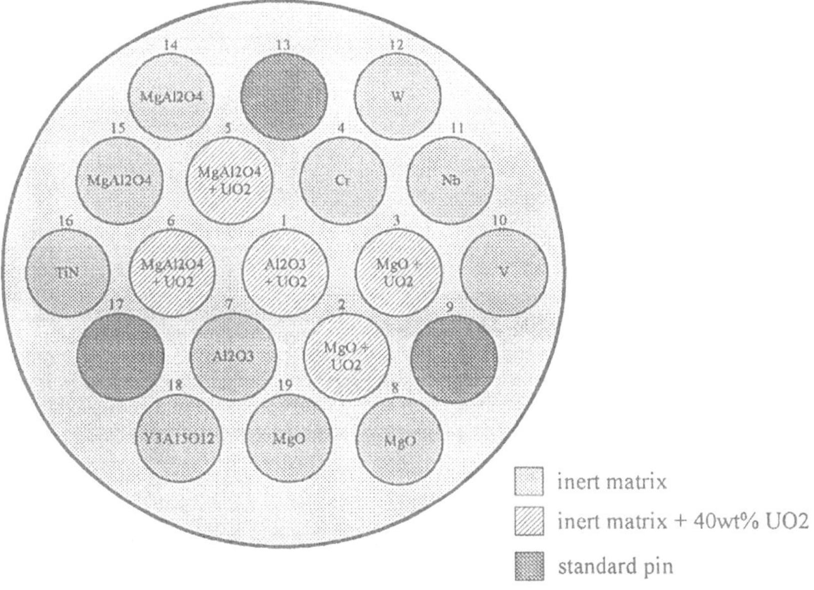

5.4 Use of Th-matrix (Oxide fuel)

The use of Th as a candidate "matrix" for U-free fuels has been advocated. Besides essential arguments related to feasibility, economics and strategic issues (outside the scope of the present review), one can assess the performance of $(Th,Pu)O_2$ fuels in terms of Pu consumptions. Two examples will be given in what follows.

First, if this fuel is used in a CAPRA core (for a burn-up of 150 GWd/t) one can compare $(U,Pu)O_2$ and $(Th,Pu)O_2$ solutions, as it is done in table VII [14].

TABLE VII

$(Th,Pu)O_2$ and $(U,Pu)O_2$ - fuelled CAPRA cores comparison

	$(Th,Pu)O_2$	CAPRA 4/94
Pu enrichment inner/outer core	21.7/26.4	43.0/44.7
Initial reactivity (pcm)	1880	9000
Residence time (EFPD)	1184	3 × 285
Pu loaded (Kg)	8177	9234
ΔPu (Kg/TWhe)	- 82.0[a]	- 74.2
ΔMA (Kg/TWhe)	+ 6.6	+ 9.7
EOC Sodium void reactivity (pcm)	1970	1564
EOC Doppler constant (pcm)	- 800	- 455

(a) Uranium build-up : 1876 Kg U-233 (+ Pa 233), 125 Kg U-234, 9 Kg U-235, 1 Kg U-236 (total 59 Kg/TWhe)

A second example is the investigation of $(Th,Pu)O_2$ fuel a standard PWR (900 MWe - class, managed by third-of-core with 287 EFPD cycles and average discharge burn-up of 33 GWd/t). Mass inventory data are given in table VIII [15].

TABLE VIII

900 MWe PWR with $(Th,Pu)O_2$ fuel (Burn-up : 33 GWd/t)

Total Pu %	Fissile Pu % (mass)	Total Pu consumption (Kg/TWhe)	U-233 production (Kg/TWhe)	Minor Actinide build-up (Kg/TWhe)
6.5	4.6	113	46	9.9

The potential of the Th use are confirmed by the results of table VI and VII. Fuel cycle (fabrication, reprocessing) issues should be considered.

The feasibility of U free fuels is enhanced by using Thorium (see Table IX) which improve the void negative effect.

TABLE IX

Void effect for Pu w/o U fuels in a PWR with standard V_m/V_f ratio

	PuO_2 + inert matrix	$PuO_2 + ThO_2$
Void effect (%) at BOL	+ 3	- 32

6. ROLE OF ACCELERATOR DRIVEN (ADS) SYSTEMS

Both critical reactors and ADS have potentially the same maximum Pu consumption rate, if the subcritical core in the ADS is comparable (e.g. neutron spectrum) to the corresponding critical one. Moreover, it is well understood that, due to physics arguments, when using Pu, there is no need for an external excess of neutrons, (see e.g. ref. 16).

However, subcriticality can be desirable in terms of margins to avoid reactivity accidents, in some specific cases, when the use of unconventional fuels is proposed. This can have relevance if one wants to maximise Pu consumption, whatever Pu vector is considered (e.g. in the case of U-free fuels) and one has to cope with the decrease of β_{eff} and Doppler constant.

Moreover, an external source can allow to realize a longer core life-time (i.e. low reactivity swing during the cycle). This has been pointed out elsewhere [17], where this approach has been compared for both Thorium and Uranium fuels.

The most interesting case is in fact the U-Pu cycle : in fact to have $\Delta\rho$/cycle $\simeq 0$ (corresponding to a Pu enrichment at equilibrium of 13.5 %), one should design a subcritical core with Keff $\simeq$ 0.97.

This means that for an ADS fed by a ~ 1 GeV proton beam, one would need a beam current of $i_p \simeq 27$ mA (in the case of a subcritical reactor supplying 1500 MWth), if the spallation process in the target produces ~ 30 ÷ 40 neutrons/proton.

For these conditions, the fraction of the energy produced in the subcritical core which is used to feed the accelerator, is approximately 10 %.

Many studies have been made to use ADS for burning Pu, and several possible core lay-outs have been proposed.

However, it seems that the first step towards a full feasibility demonstration of the potential of ADS, should be a limited scale (in terms of power) demonstration of the ADS component couplings, and the operational demonstration within a range of significant values of the most important parameters (e.g. $i_p \simeq 2 \div 5$ mA for 1 GeV protons or, alternatively $i_p \simeq 5 \div 15$ mA for $0.2 \div 0.3$ GeV protons ; power in the subcritical core : $20 \div 40$ MWth ; Keff : $0.92 \div 0.97$). Studies in this direction are underway at CEA and an ADS concept based on a subcritical ($K_{eff} \simeq 0.95$) core with $(Pu,U)O_2$ fuel (45 % Pu content), Na coolant, producing 40 MWth is under study (the HADRON concept). Neutrons are supplied to this core by spallation of ~ 1 GeV protons ($i_p \simeq 2$ mA) over a liquid metal target (an alternative with a solid W target is also provided). Variants with $(Pu, Th)O_2$ fuel and lead coolant are also investigated.

7. CONCLUDING REMARKS

The use of Plutonium in fission reactors, critical or subcritical, can be envisaged according to a number of options, both in thermal or in fast neutron spectrum cores. The physics phenomena are generally well known, even if uncertainties (e.g. nuclear data uncertainties) play a significant role in defining limitations in extending conceptual designs beyond present knowledge.

From a technology and feasibility point of view, the use of MOX fuel in PWRs is an industrial reality, which includes core operation, fuel fabrication and performance, fuel reprocessing. The use of MOX in fast reactors has also been demonstrated.

To increase Pu consumption in medium-term scenarios of nuclear power utilisation, the use of CAPRA-type fast reactors can represent a realistic option, and large scale experimental demonstrations are underway.

More unconventional core and fuels can be envisaged, but the feasibility is still far from being demonstrated.

Finally, accelerator-driven system can be envisaged, even if their use seems to be more attractive in the frame of advanced waste management options, for the transmutation of radioactive elements, such as Am and Cm, which can be unlikely candidates for fuels of standard critical fission reactors [18].

8. REFERENCES

1. Proceedings of the GLOBAL'95 Conference on Emerging Nuclear Fuel Cycles, Versailles, Sept. 1995.

2. Proc. IAEA Technical Committee Meeting on "Unconventional options for Pu disposition", IAEA-TECDOC-840, Vienna (1995).

3. "OECD/NEA Working Party in the Physics of Pu Recycling - Final Report" 6 Volumes, OECD-NEA, 1995.

4. B. Barré "A French view on Nuclear Energy and Fast Neutron Reactors", prepared for n° 100 issue of the PNC Technical Review (1996).

5. OECD/NEA Report on the Management of Plutonium, to be published.

6. J.P. CHAUVIN et al., "EPICURE : Synthesis of an Experimental Program Devoted to the Validation of MOX recycling", paper E-133, Proc. Int. Conf. PHYSOR'96, Mito, Japan, Sept. 1996.

7. The proceedings of a series of seminars give a detailed picture of the CAPRA studies and its international framework : 1st CAPRA Seminar, Cadarache, March 1994 ; 2nd CAPRA Seminar, Karlsruhe, Sept. 1994 ; 3rd CAPRA seminar, Lancaster, Nov. 1995.

8. A. Languille et al. "CAPRA core studies - The oxide reference option", p. 874-881, Proc. Int. Conf. GLOBAL'95, Versailles, Sept. 1995.

9. M. Delpech et al. "Scenarios of Plutonium and Minor Actinide Management at equilibrium" Proc. NEA - Information Exchange Meeting on Partitioning and Transmutation - Mito (Japan), Sept. 1996.

10. J. Tommasi et al., Nuclear Technology, 111, 133 (1995).

11. A. Puill, J. Bergeron "APA : An Advanced Concept for Using Plutonium in PWRs", Nuclear Technology, to be published.

12. J. Rouault et al. "The design of U-free large fast reactor cores. The CAPRA programme trends". Proc. Int. Conf. PHYSOR'96. Mito (Japan), Sept. 1996.

13. C. Prunier et al. "The CEA SPIN Program : Minor Actinide Fuel and Target aspects" Proc. Int. Conf. GLOBAL'95, Versailles, Sept. 1995.

14. J. Tommasi, Private Communication.

15. A. Puill, Private Communication.

16. M. Salvatores et al, Nucl. Sci. Eng. 116 (1994) p. 1-18.

17. M. Salvatores et al. "The potential of Accelerator-Driven systems for Transmutation or Power Production using Thorium or Uranium Fuel Cycles" to be published.

18. M. Salvatores, J.P. Schapira, H. Mouney "French Programs for Advanced Waste Management Options". Proc. 2nd Int. Conf. on Accelerator-Driven Transmutation Technologies, Kalmar, Sweden, June 1996.

BURNING OF WEAPON-GRADE PLUTONIUM IN VVER AND HTGR REACTORS

N.N.Ponomarev-Stepnoy, E.S. Glushkov, I.K.Levina
Russin Research Center "Kurchatov Institute", Moscow, Russia

1.BACKGROUND

The implementation of the nuclear disarmament program poses the following problems: the cut-off of the further production of weapon-grade plutonium and the disposition of disarmament plutonium.

In solving these problems we take into consideration, along with non-proliferation safeguards,

- the structure, state and capabilities of the existing nuclear-industrial complex and
- the state and prospects of nuclear power.

Nuclear-industrial complex of Russia

The Russian nuclear-industrial complex was created to provide the country with nuclear weapon. The complex integrates the whole chain of technologically related enterprises, including mining and processing of uranium ores, enrichment facilities, plants for manufacturing of fuel elements, commercial nuclear reactors, radiochemical plants for reprocessing of spent fuel, plutonium handling installations and other components of the nuclear fuel cycle.

In Russia the fuel cycle is oriented to the chemical reprocessing of spent NPP fuel. The RT-1 plant can reprocess 440 tons of spent VVER-440 fuel per year, which is sufficient to serve the domestic and foreign NPPs equipped with VVER-440s. The end products of reprocessing are:

- uranium containing 2 to 4% of U-235 which is re-enriched with highly enriched uranium obtained in reprocessing spent fuel from naval reactors and used as RBMK fuel;
- plutonium (plutonium dioxide) which is enclosed in special packages and transported to a storage facility. As a result, about 30 t of plutonium has been accumulated to date; this quantity will achieve 47 t by the year 2005.

The RT-2 plant for reprocessing of spent VVER-1000 fuel was planned and is partially constructed. According to the project the plant must produce uranium dioxide pellets for VVER-1000 fuel elements and plutonium oxide powder for mixed uranium-plutonium MOX fuel for the VVER-1000 reactor. Closing the VVER-440 and VVER-1000 fuel cycles requires to complete the construction of the RT-2 plant and a complex for production of mixed uranium-plutonium fuel.

Nuclear power in Russia and its fuel cycle

In Russia nuclear power develops on the basis of the existing nuclear industrial complex.

In the 1970s-1980s the rates of NPP construction in Russia were very high; about 50% increase in the total NPP capacities of the country was planned by the end of the century. The Chernobyl accident frustrated these plans.

At present Russia has nine operating NPPs totalling 21242 MW with thirteen VVER units, eleven RBMK units and one BN unit.

The power strategy of Russia up to the year 2000 assumes, as a main task, the backbiting of existing NPPs. The commissioning of new NPP capacities is planned during the decades that follow. The increase of nuclear capacities in the North-West, the Central area and the Far East in the period of up to 2010 is under review now. At the same time it is proposed to create, by 2030, the preconditions for a considerable increase of nuclear's contribution to the country's electricity output.

For the present-day scale of nuclear power there is no problem in its fuel supply. Russia's nuclear power with its open or semi-closed fuel cycle will be sufficiently provided with fuel from the available reserves of raw materials within the next few decades. An inevitable absolute growth of the nuclear capacities and an increase of their contribution to the total fuel balance as well as the extension of regions and the increasing number of countries using nuclear energy are expected in the future. Then, although it is hard to predict exactly when there comes that time, nuclear power will encounter the problem of nuclear fuel resources, unless it is on the point of recycling nuclear fuel.

E. R. Merz and C. E. Walter (eds.), Advanced Nuclear Systems Consuming Excess Plutonium, 203–217.
© 1997 *Kluwer Academic Publishers.*

In Russia the nuclear fuel cycle strategy was established at the stages of formation and intense development of the domestic nuclear power. The two-component nuclear power based on thermal reactors and breeder reactors was proposed. This structure assumes a closed fuel cycle: chemical reprocessing of spent fuel with the reuse of both recovered uranium and recovered plutonium.

Following this concept, Russia has gained some experience in reprocessing of spent fuel with the recovery of plutonium and its recycling in reactors. the greatest efforts were made on the utilization of plutonium in fast reactors. works on the use of MOX fuel in VVERs were also conducted. The slowdown in the rates of nuclear power development and a surplus in uranium fuel lessened the efforts on development of mixed fuel for VVERs. Nevertheless, pilot fuel elements from mixed uranium-plutonium oxide were developed and put through preliminary rests; moreover, the physical basis for VVER cores with the mixed fuel was elaborated.

2. CUT-OFF OF WEAPON-GRADE PLUTONIUM BUILD-UP

The RF and US Governments agreed to decommission reactors producing weapon plutonium until the year 2000. Three of all these reactors that ever operated in both countries are in operation now: two in Seversk and one in Zheleznogorsk. The other plutonium production reactors were shut down or decommissioned.

To shut down the reactors in Seversk and Zheleznogorsk it is necessary to create alternative sources of heat and electricity which are generated now by these reactors. As the joint Russian-American estimations showed, this problem is unlikely to be solved until 2000. This has brought up a proposal about the conversion of the cores in the operating plutonium reactors, so as to change over them from the two-purpose (power + Pu generation) regime to the single purpose (power generation) one. Such a solution

- ensures the cut-off of production of weapon plutonium in the shortest possible time,
- resolves the problem of power supply in Seversk and Tomsk until the decommissioning of the plutonium reactors,
- solves the employment problem for the personnel of the production complexes,
- enhances the safety,
- reduces by an order of magnitude the fuel inventory in the cycle and thereby can improve the economic indices of the operation of these reactors.

A fuel in the form of enriched uranium dioxide in an aluminium matrix will be used in the case of the conversion of the reactors. Therefore, the problem of burning of weapon-grade uranium will be also partially solved. The question about burning of weapon-grade plutonium will be studied at the next stages.

The project "Conversion" is being implemented by the joint Russian-American efforts. The feasibility study has been made; the stage of working projecting, research and development has been prepared, after which it will be possible to proceed to the implementation.

The joint work showed

- the technical feasibility of the conversion of the operating two-purpose plutonium reactors to the power regime without production of plutonium,
- the enhancement of the safety and, hence, the reduction of the risk of operation of these reactors after their conversion,
- the conversion of the plutonium reactors will allow the termination of plutonium build-up before the year 2000,
- the operation of the reactors in the power regime is ensured up to the year 2005 and, as a result, the conversion appears to be more economically efficient than the creation of alternative energy sources.

New compensating capacities must be developed and put into operation by the moment of decommissioning of the converted reactors. In deciding between the nuclear and fossil options for the compensating sources of energy, preference should be given to the nuclear ones. The main argument for the nuclear sources is the solution of the employment problem for the reactor specialists living in Zheleznogorsk and Seversk.

3. BURNING OF PLUTONIUM IN REACTORS

Among various options for the disposition of disarmament plutonium Russia favours the burning of the plutonium in power reactors. This is explained by the desire to use the high energy value of plutonium and the capabilities of the existing nuclear-industrial complex. Under the conditions of

Russia it is difficult to find arguments for some other options for the disposition of weapon-grade plutonium, especially as Russia has the stockpiles of reactor plutonium from reprocessing of spent fuel.

In Russia various variants of burning of weapon-plutonium in power reactors: VVERs, BNs, HTGRs are under development. Moreover, Russian specialists in collaboration with AECL specialists are examining the possibility of plutonium burning in CANDU reactors. This approach corresponds formally to the term "spent fuel standard" accepted in the US. However, in some cases the Russian variant of weapon plutonium burning in reactors involves the possibility of spent fuel reprocessing and the recycling of nuclear fuel, including plutonium. First of all, this is true for the BN reactor, but the possibility for recycling of spent VVER fuel is not ruled out.

This report considers the possibilities of weapon plutonium burning in VVER and HTGR reactors. The data presented will be a help in estimating the degree of readiness to the implementation as well as its possible time and cost. These estimates are required for the comparison and the choice of one or other variants of weapon plutonium burning.

VVER-1000.

The utilization of plutonium in VVER-1000 reactors is one of the most realistic problems, because reactors of this type are in operation in Russia and abroad. The calculation investigations into the possibility to use weapon-grade plutonium in VVER-1000s have been conducted in the Russian Research Centre "Kurchatov Institute" for some years. Below is given in brief the results of the analysis of the basic problems concerning plutonium burning and some neutron-physical characteristics of the VVER-1000 cores loaded with mixed uranium-plutonium fuel (MOX fuel).

Under consideration is the simplest method of involving plutonium in the VVER-1000 fuel cycle - the direct replacement of a part of uranium fuel by MOX fuel with no essential changes in the core structure and the unit operating conditions.

The physical features of the core with uranium-plutonium fuel (reduction in the fraction of delayed neutrons and in the efficiency of absorbers used) pose restrictions on the number of MOX fuel assemblies in the VVER-1000 of existing design. Their fraction is 1/3 in most of the cycles considered. The example of the arrangement of the core one-third loaded with weapon plutonium and the basic characteristics of the fuel cycle are presented in Fig.1 and Table 1, respectively. The table comprises also the data on the core containing power plutonium of the following isotope composition: 1.3% Pu-238, 57.7% Pu-239, 27.1% Pu-240, 8.6% Pu-241 and 5.3% Pu-242. As can be seen from the results, the variant with the substitution of one third of uranium fuel assemblies by plutonium ones decreases the efficiency of weapon plutonium burning. An increase in the efficiency can be achieved by increasing the fraction of MOX fuel assemblies up to 100%. However, this can require to change the design of control rods and to introduce burnable poison into fuel elements.

As follows from the calculation investigations, the VVER-1000 cores containing both uranium and MOX fuel assemblies are characterized by an increased nonuniformity in the distribution of energy release from one fuel element to another. The reason is a considerable difference in the neutron-physical properties of the uranium and plutonium fuel assemblies. This leads to the necessity of profiling the MOX fuel assemblies either by fuel elements with different plutonium contents (see Fig.2) or by plutonium and uranium fuel elements (see Fig.3).

The reduced fraction of delayed neutrons and the lower efficiency of absorbers in the MOX-fueled VVER-1000 reactor impart an particular urgency to the investigation of the safety problems. The transients qualifying as a reactivity accidents were assessed in the RRC KI. The processes initiated by the accidental withdrawal of a control rod were considered. It was shown that the basic criteria for the safe VVER-1000 operation are met in the cases considered.

The question about the sufficiency of the efficiency of the VVER-1000 emergency protection system in the case of the one-third loading of the core by plutonium fuel is considered in the context of the work under way now on modernization of the VVER-1000 uranium fuel cycle. The modernized fuel cycle will involve zirconium guide channels and spacers, burnable integrated in fuel and core arrangements with a reduced leakage of neutrons. The investigations having been made showed that the modernized cycle allows an about 20-25% increase in the efficiency of emergency protection for the three-year cycle as compared with the VVER-1000 cycles in use now. A further increase in the efficiency of the control rods of the control and protection system approximately by 20% can be achieved by increasing the diameter and the B-10 content of the absorbing rod with the corresponding increase in the guide channel diameter. Thus, one might expect that a certain decrease in the efficiency of emergency protection in VVER-1000s resulting from the substitution of plutonium fuel assemblies for a third of the core can be compensated by the above methods. The greater number of MOX fuel assemblies (more than a third of all the fuel assemblies in the core) will require evidently to change the design of the VVER-1000 reactor, in particular, to increase the number of mechanically-driven control rods in the control and protection system.

The physical features of the mixed fuel and the ensuring problem of the representative prediction of the core characteristics point out the need for the upgrading and verification of the codes for the neutron-physical calculations of the VVER-1000 cores with the mixed fuel. To obtain the experimental data for the verification of the codes the critical experiments with MOX fuel should be conducted. Under review now is the possibility of conducting such experiments in the critical facilities of the RRC "Kurchatov Institute", at the Institute of Atomic Energy in Rzez (Czechia), in the BFS facility of the FEI in Obninsk.

Alongside with the substantiation of the neutron parameters of the cores with uranium-plutonium fuel for the licensing purposes it is necessary to substantiate the serviceability of MOX fuel elements. The loop tests of plutonium fuel elements for VVER-1000s are in progress now at the NIIAR (Dimitrovgrad). A program for development and trial operation of three pilot fuel elements containing weapon plutonium in the VVER-1000 reactor of the Balakovo NPP has been drawn up. The possibility of the trial operation of these fuel elements was justified by the neutron-physical calculations performed at the RRC KI.

All these efforts are parts of the total program of work on weapon plutonium burning in VVERs. The substantiation of the decision about the loading of one-third of the VVER-1000 core by plutonium fuel assemblies requires to make a complex of calculations and experiments, including, in particular,

- conducting of the critical experiments with uranium-plutonium fuel in Russia,
- upgrading and verification of the codes for the calculation of the neutron-physical characteristics of the VVER-1000 cores with plutonium fuel (comparison with the results of the precision calculations made by Russian and foreign codes, analysis of the results of the Russian and foreign critical experiments with plutonium fuel),
- complex calculations for the substantiation of the safety with particular emphasis on reactivity accidents,
- trial operation of plutonium fuel assemblies in operating VVER-1000s,
- review of the Western experience in the use of MOX fuel in PWRs.

A considerable volume of works should be carried out to substantiate the production of MOX fuel by the enterprises of the Russian nuclear complex.

On the operation of RRC KI specialists, there are grounds to believe that the trial operation of three MOX fuel assemblies in a VVER-1000 reactor might be started until the year 2000 and that the pilot loading of a third of the core in an operating VVER-1000 reactor by MOX fuel assemblies might be made until 2004.

High temperature gas-cooled reactor

In November 1993 the Minatom addressed to the US Government a proposal on the joint US-Russian Programme on development of a GT-MHR reactor plant which could be efficiently used for burning weapons-grade plutonium. The first GT-MHR reactors for weapons-grade plutonium burning were suggested to be constructed in Seversk.

At present joint Russia-USA development of the Conceptual GT-MHR Project is under way, with the financial backing given by the General Atomics (USA) and Minatom (RF). Recently Framatome (France) has joint the Programme. The Conceptual Project is planned to be completed in October 1997. The Project's objectives are listed in Table 2.

In the reactor of this type weapons-grade plutonium is used in the form of undiluted plutonium dioxide. The employment of microspheres with multilayer coatings of pyrocarbon and silicon carbide allows a high burnup of plutonium: up to 90% of the initial Pu-239 charge, to be reached in a once-through reactor cycle. Plutonium contained in the spent fuel is of no interest for weapon production. The ceramic structure of the spent fuel and the multilayer coating properties make possible its long-term disposal in geological formations without reprocessing.

Thus, the concept of weapons-grade plutonium disposition using the GT MHR includes the following stages (Fig.4):

- reprocessing of weapons-grade plutonium into another form, namely, coated particles, which practically rules out the possibility of its use for military purposes;
- burning weapons-grade plutonium in the GT-MHR reactors with efficient generation of useful power;
- long-term disposal of the spent fuel in geological formations.

The GT-MHR concept is based on the use of the core with the graphite moderator and helium coolant as well as of the fuel in the form of microspheres with multilayer pyrocarbon and silicon carbide coatings. The GR-MHR consists of a reactor enclosed into a steel high pressure vessel incorporated with the power conversion system (Fig.5) using gas turbine cycle (Fig.6). The core (Fig.7) consists of hexagonal graphite prism fuel elements based on coated particles (Fig.8). The GT-MHR is characterized by enhanced safety and high efficiency (up to 50%), which make it more economically advantageous comparing with other reactor types.

The main GT-MHR characteristics are given in Table 3.

In the GT-MHR design the quantity of useful power generated per a gram of burnt plutonium for single fuel irradiation run is larger than in any other reactor system. This can be seen in Table 4.

The foundation of the GT-MHR Project is both the solutions proven in the development and operation of foreign high temperature reactors as well as longer than 30 year Russian experience with designing the HTGR reactors.

The microsphere technology and design have been proved by multiple reactor tests of the fuel with uranium dioxide carried out in Russia, the long-term (over 20 years) operation of the AVR reactor, experiments for the substantiation of the DRAGON Project, and experiments performed in the USA. The suggested modes of the GT-MHR fuel use compared with the results of this type fuel tests and performance are listed in Table 5.

The discharged fuel elements are suitable for direct disposal in stable geological formations. The protective coatings of microspheres prevent radioisotope spreading into the environment. The leaching tests have demonstrated a higher ability of coated fuels to retain wastes as compared with the method using borosilicate glass.

Russia has evaluated alternative variants for reprocessing 50 tons of plutonium using the GT-MHR. One of these variants suggests the construction of three plants, each containing four reactor modules: in Seversk, Krasnoyarsk, and at the Mayak Combine.

The first reactor module may be put into operation in 8.5 years after the decision is made, the last one in six years after than date. These plants will reprocess 50 tons of plutonium during 25-30 years after the decision is made.

The problem of speeding up the weapons grade conversion into forms which make difficult its military application can be solved by increasing the rate of production of Pu microspheres with multilayer protective coatings. The use of multilayer coatings would increase the resistance of starting plutonium to proliferation and diversions. Production of Pu microfuel could be organized in Russia in the nearest future.

The location of a fuel production facilitiy and a GT-MHR plants in Seversk having a well organized infrastructure for weapons grade plutonium handling and security reduces the proliferation risk in shipment of the latter. The same can be said about the facilities in Krasnoyarsk and at Mayak.

According to the estimates (Table 6) the total capital expenditures for development and construction of a plant with four reactor modules as well as fuel production facilities and long-term spent fuel repositories in Seversk will be compensated for profits to be received from selling the electrical power. Minor expenditures for GT-MHR fuel production are accounted for by the use of undiluted plutonium fuel as this requires smaller volume of fuel production comparing with the MOX fuel for other reactors.

The high probability that the public would admit the HTGR alternative for WGPu utilization is not only determined by such key factors as a high level of inherent safety, not melting core, no need in the evacuation of the population in any credible accidental situations, but also by a high degree of burnup of the initially charged plutonium, which is higher than the "spent fuel standard".

Thus the GT-MHR can be efficiently used simultaneously for the production of useful power and for surplus weapons-grade plutonium disposition. Due to these properties the GT-MHR can be used with the uranium (thorium) fuel for commercial purposes.

Table 1

Characteristics of the VVER-100 fuel cycles using plutonium (stationary refueling)

Characteristics	Uranium fuel cycle	33% MOX (power plutonium)	33% MOX (weapons- grade plutonium)
Fuel charge,t (heavy metals)	65.4	65.4	65.4
Material of spacers and control rod guide tubes	stainless steel	zirconium	zirconium
Number of charged fuel assemblies (MOX FA)	54(-)	55(18)	55(18)
Average initial isotope content in MOX FA, kg/t			
Pu	-	60	34.4
Pu	-	40	32.4
U-235	43.1	2.0	2.0
Type of refueling scheme	out-in-in	out-in-in	out-in-in
Core life-time, eff.day	297	300	301
Minimum core subcriticality, %(coolant temperature 280°C, all) control rods in the lower position, except the most efficient one, xenon poisoning)	-	3.2	3.6
Average burnup of discharge FA, MW day/kg.U	41.1	40.8	41.1
Plutonium content averaged over all discharged FA,kg/t			
Pu	10.0	19.4	14.1
Pu	6.8	11.2	8.7
Plutonium content averaged over MOX FA discharged,kg/t			
Pu	-	38.4	19.0
Pu	-	21.1	11.0
Plutonium annually charged into the reactor, kg/yr			
Pu	-	415	247
Pu		289	233
Plutonium annually discharged from the reactor, kg/yr			
Pu	217	436	306
Pu	148	261	189

Table 2

GT-MHR objectives

High efficient electricity generation.

Weapons-grade plutonium utilization:
- Reprocessing of Pu into coated microspheres
- Pu burning without recycle into GR-MHR
- Disposal of spent fuel in geological formations without reprocessing.

High safety level ensuring core not melting due to inherent characteristics and use of passive systems.

Ability to use various fuels (uranium, plutonium, thorium, etc.) without recycle.

Commercial power production on the basis of uranium fuel.

Table 3

Main characteristics of high-temperature GT-MHR reactor

Characteristic	Unit
Thermal power, MW	600
Efficiency of electrical power production, %	up to 50
Helium coolant temperature (inlet/outlet),°C	490/850
Helium pressure, MPa	7
Core diameter (inner/outer), m	2.96/4.84
Outer diameter of radial reflector, m	7
Core height, m	8
Fresh Pu charge, kg	750
Pu-235 burnout degree, %	90
Refueling multipleness	3
Charged Pu, kg/yr	250
Pu-239 enrichment of charged Pu, %	94
Discharged Pu, kg/yr	70
Pu-238 enrichment of discharged Pu, %	30
Quantity of destroyed weapons-grade plutonium for 60 year per one unit, t	15

Table 4

Comparison of Pu uses in power reactors of various types

	GT-MHR(four units)	VVER-1000 (with MOX charge in 1/3 core)	Fast sodium reactor BN-800
Thermal power, GW	4×0.6	3.2	2.1
Net efficiency, %	47	31.2	37.2
Electrical power, GW	1.15	1.0	0.8
Electrical power production GWe/yr from 50 t of weapons-grade Pu	46	25	17.2
Fraction of destroyed Pu-239% (without recycle)	90	50	13

Table 5

Summarized data on irradiation of TRISO microspheres comparing with the requirements to GT-MHR-fuel

Programme	Description of fuel	Maximum fluence of fast neutrons $10^{25} n/m^2$ (E > 0.18)	Maximum burnup	Temperature region during irradiation, °C
GT-MHR requirements	TRISO-covered PuO_2 fuel particles in compacts	4	800 MWday/kg	up to 1300
Russia	TRISO-covered UO2 fuel particles in spherical elements	2-3	up to 20% FIMA	500-1650
USA	TRISO-covered UC_2 fuel particles (93% enrichment) in compacts	12	up to 78% FIMA	900-1550
	TRISO-covered UCO fuel particles (20% enrichment) in compacts	7.8	up to 22% FIMA	1100-1250
	TRISO-covered PuO_2 fuel par ticles in compacts	2.2	737 MWday/kg	up to 1440
Germany	TRISO-covered UO_2 fuel particles (10% enrichment) in spheres and compacts	< 6.2	up to 14.9% FIMA	1100
Japan	TRISO-covered UO_2 fuel particles (4-10% enrichment) in compacts	up to2.8	up to 9.4% FIMA	1300
DRAGON	TRISO-covered PuO_2 fuel particles, mixed graphite in compacts	1.5	747 MWday/kg	up to 1275

Table 6

Total investment expenditures and techno-economic indices

		Sum, million US doll.
1.	R & d activities	
1.1.	On development and implementation of HTGR and other NPP equipment	275
1.2.	On development of fuel manufacturing technology and facilities	97
2.	Construction of 12 HTGR power units	2640
3.	Construction of fuel manufacturing complex	58.0
4.	Construction of a repository for long-term WGPu-SNF storage	50.5
	Total expenditures prior to HTGR putting into operation	3120.5
	Integral operating costs for storage maintenance	6781.4
	Electrical power cost for 1000 kWh	21.4$
	Expenditures for conversion of 50t weapons-grade plutonium	-

Fig.1 Fuel assemblies loading scheme of a core of WWER-1000.
Equilibrium fuel cycle (1/3 of weapon plutonium fuel assemblies)

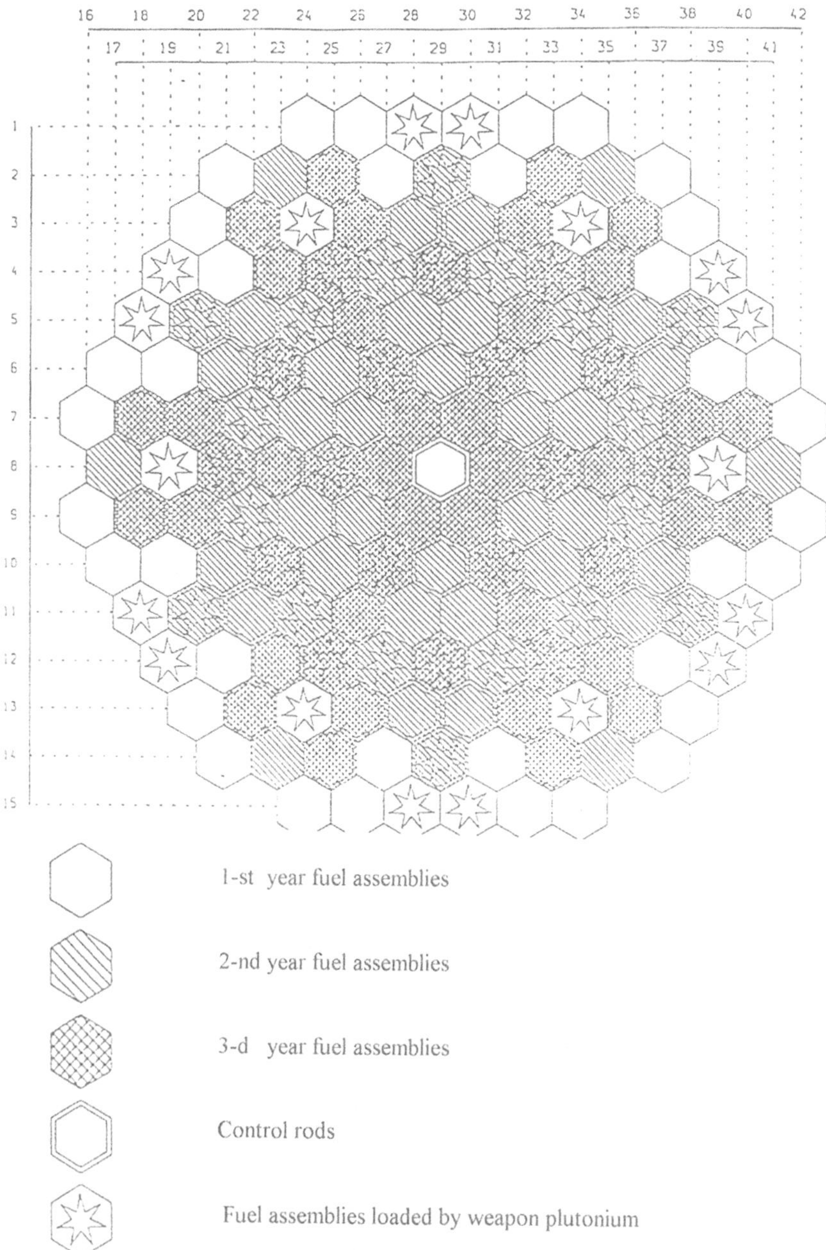

Fig.2 Fuel assembly loaded by weapon plutonium.

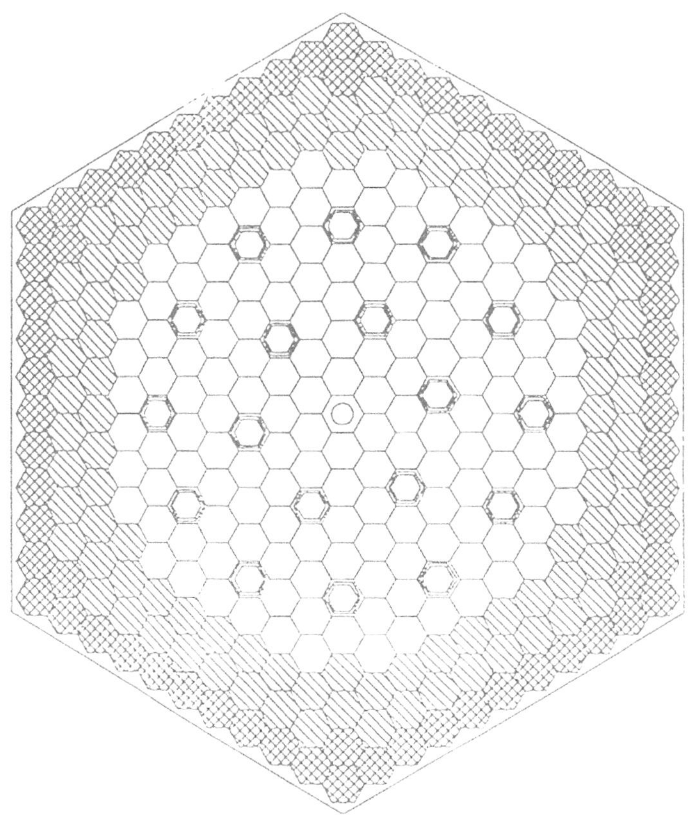

Fuel assembly 4.4 of fission plutonium

Fuel assembly 2.4 of fission plutonium

Fuel assembly 2.0 of fission plutonium

Central channel

Structural channel

Fig.3 Fuel assembly loaded by weapon plutonium fuel rods and uranium fuel rods

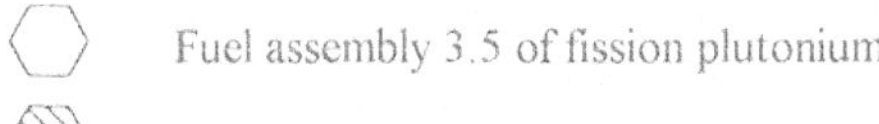

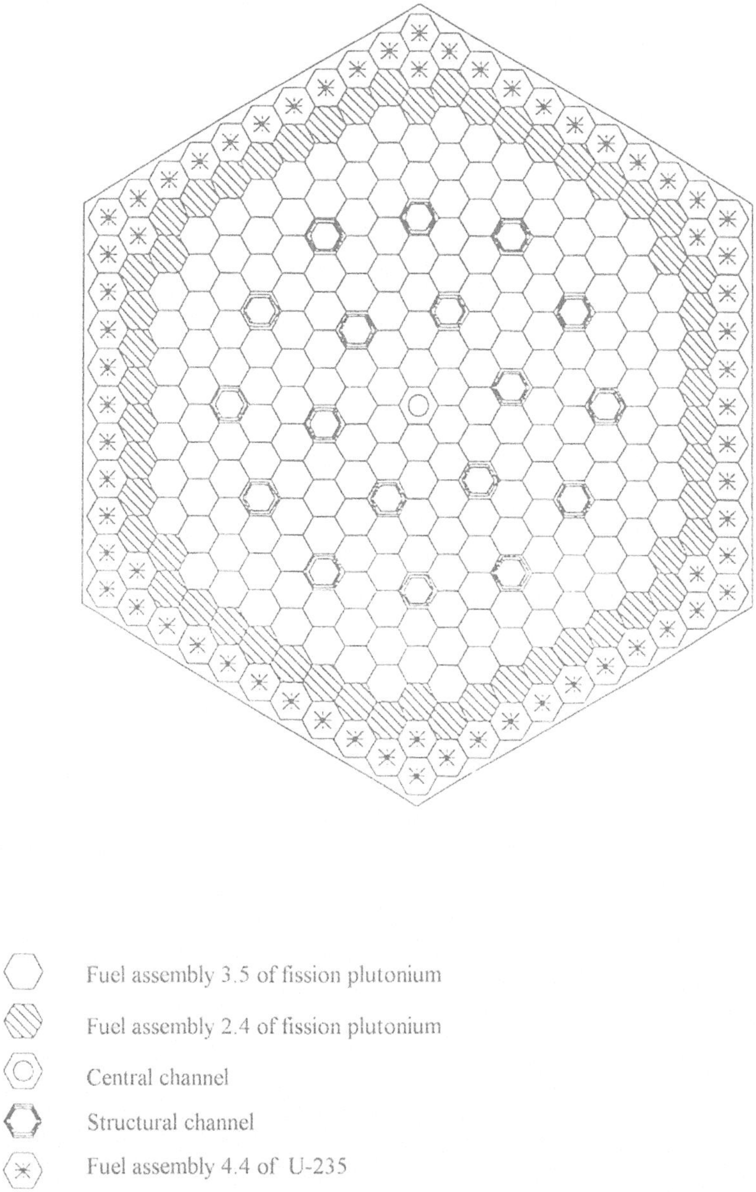

Fuel assembly 3.5 of fission plutonium

Fuel assembly 2.4 of fission plutonium

Central channel

Structural channel

Fuel assembly 4.4 of U-235

Fig. 4: DISPOSITION OF WGPu

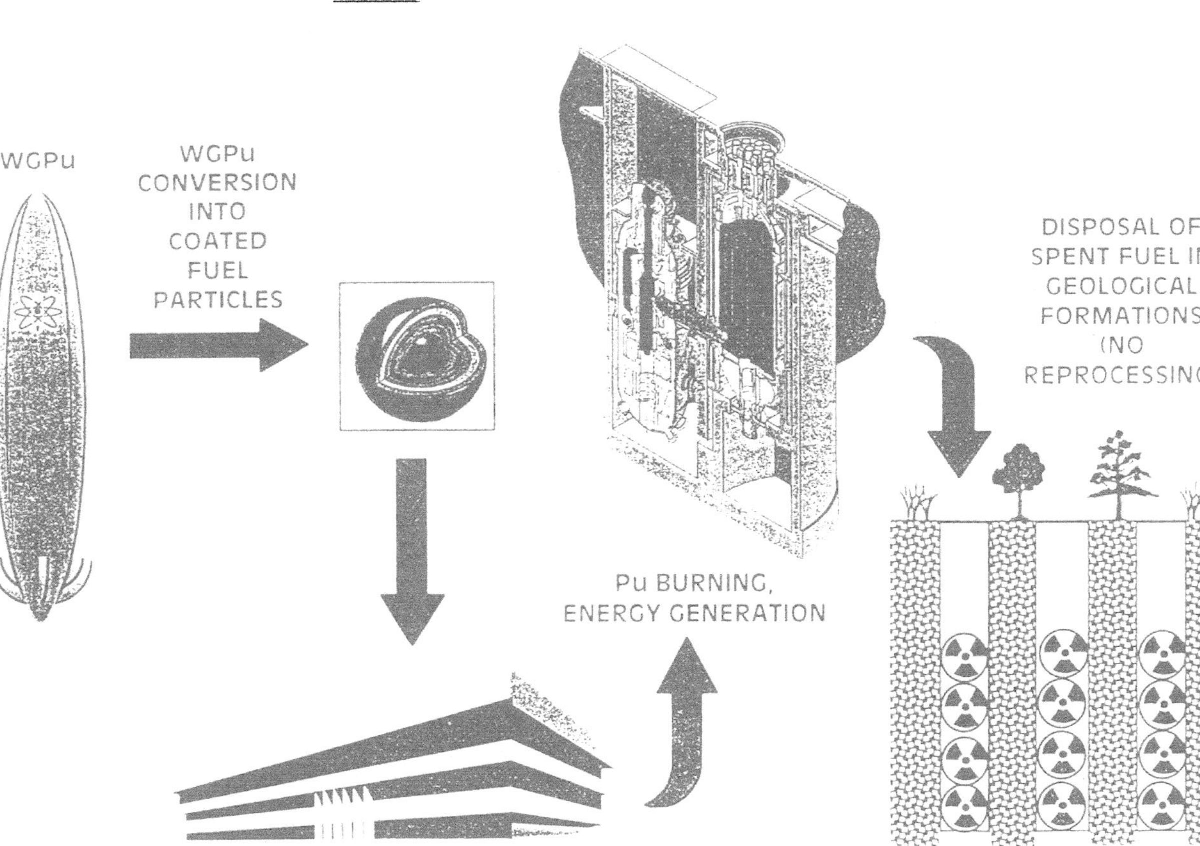

Fig. 6: GT-MHR POWER CONVERSION PROCESS FLOW DIAGRAM

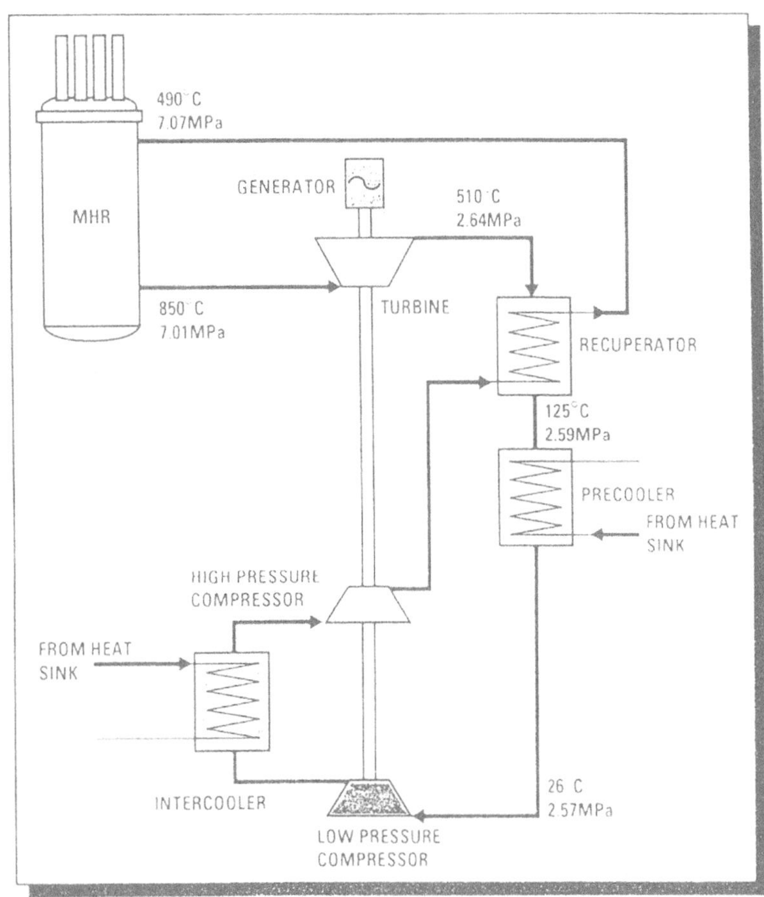

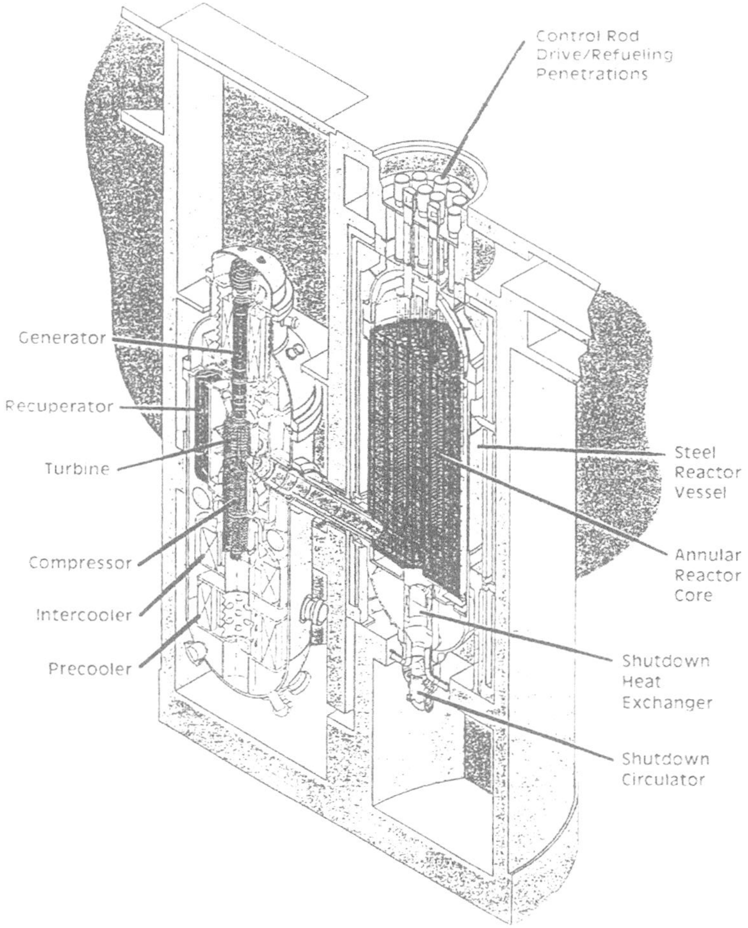

Fig. 5: GT-MHR MODULE ARRANGEMENT

FIGURE 7

GT-MHR CORE LAYOUT

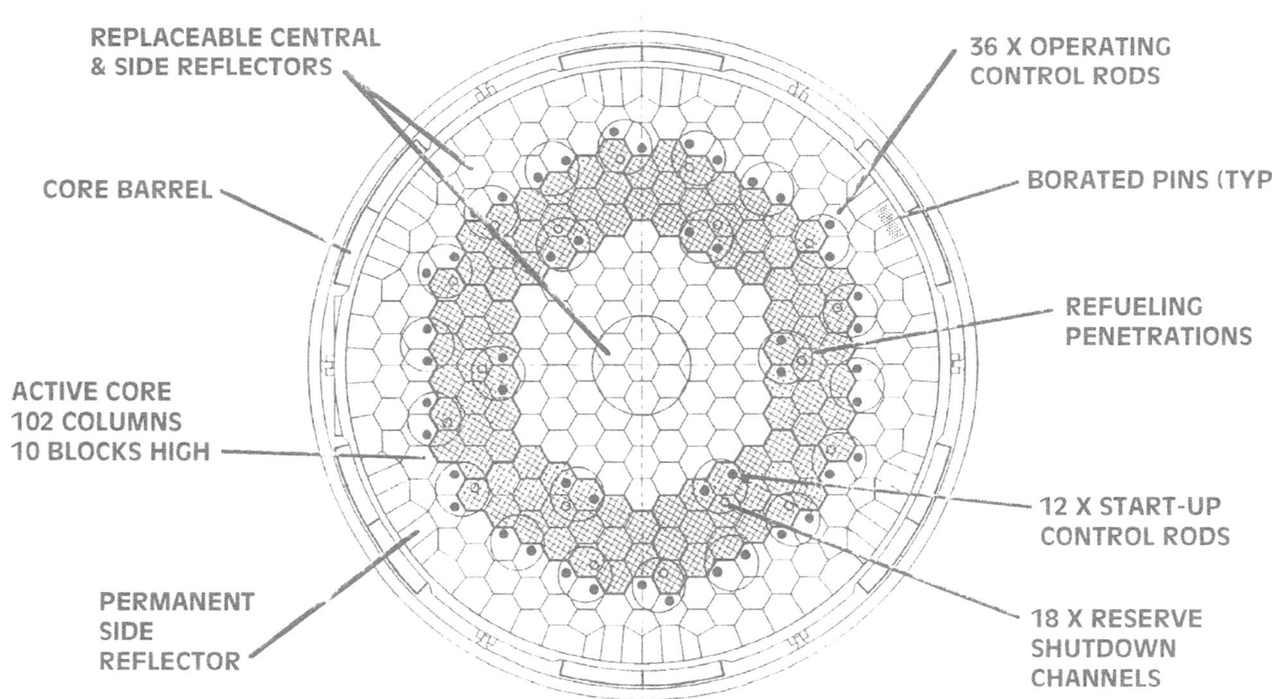
REPLACEABLE CENTRAL
& SIDE REFLECTORS
CORE BARREL
ACTIVE CORE
102 COLUMNS
10 BLOCKS HIGH
PERMANENT
SIDE
REFLECTOR
36 X OPERATING
CONTROL RODS
BORATED PINS (TYP)
REFUELING
PENETRATIONS
12 X START-UP
CONTROL RODS
18 X RESERVE
SHUTDOWN
CHANNELS

UTILIZATION OF EXCESS PU IN MOLTEN SALT REACTORS

Alekseev P.N., Ignatiev V.V., Prusakov V.N., Ponomarev-Stepnoy N.N., Stukalov V.A., Subbotin S.A.
Russian Research Center "Kurchatov Institute", Moscow

For the large scale and long range implementation of nuclear power (NP) it should meet the following requirements; cost effectiveness, sufficiency of resources; safety; acceptable environmental impact and nonproliferation.

Two first requirements can be well solved by using a two component NP involving thermal and fast solid fuel reactors. This NP structure would permit the natural resources (uranium and/or thorium) to be much more effectively used, uranium production to be reduced, and, hence, the radon penetration into the biosphere would be sufficiently lower.

The ways of reaching the required breeding level, enhancing the safety of nuclear power units, and reducing the capital investments are currently known, what is needed is time and funds for their realization. By the time when the society realizes the need for NP development the NP two-component structure will be practically ready although much will have to be done for optimization both of the nuclear power units (NPU) and of the NP structure including fuel cycle facilities.

Difficulties may be encountered with the public attitude to the level of environmental impact which will be in the main determined by the quantities of radionuclides both in the fuel cycle (uranium, plutonium) and in the radwaste disposals (minor actinides, fission products), and of those globally dispersed (carbon-14, tritium, krypton-85); and with the public attitude to the question of nonproliferation of nuclear materials, which will be in the main determined by the availability of pure Pu in fuel cycle.

It can be shown right now that the risk of short-lived radionuclides (for example, iodine-131) and of such as strontium-90 and cesium-137 can be reduced to the permissible level due to ensurance of the required safety level at NPPs, disposals, fuel cycle facilities; ways of decreasing releases of carbon, tritium, krypton to the air can be seen. This can be demonstrated in practice or using indirect evidences (for example, persistence of some types of barriers for hundreds or thousands of years). But it is difficult to show that long-lived actinides and fission products could be reliably disposed for some millions of years.

Of course, search for the ways of reliable radwaste disposal should not be abandoned but, on the other hand, the possibility of using actinides for energy production and, thus, for closing the fuel cycle not only with uranium and plutonium but also with minor actinides, should also be considered. The problem of transmutation of hazardous long-lived fission products is awaiting its solution as well. In this case a more sophisticated NP structure would be needed, that could be achieved both by using additional technological procedures for fuel reprocessing and production, or more sophisticated NPU or by designing new NPU. It may happen that the use of minor actinides and fission products in the thermal and fast solid fuel reactors will require more sophisticated reactor designs, cast doubt on their safety, demand development of new types of fuel. Therefore in this case it is reasonable to consider the possibility of developing the three-component NP structure which would use, along with the thermal and fast reactors, the reactors-burners (R-B) for incineration of minor actinides and transmutation of some fission products. On the first stage of NP development some part of weapon grade Pu could be used as neutron source in R-B

More attractive in such multicomponent system with taking into account nonproliferation restriction we consider the following scheme of nuclear fuel cycle with molten-salt-burner reactor MSR (see Fig.1):

- discharged fuel goes through dry gas-fluoride processing;
- uranium and, possibly, a part of plutonium are recycled as a fuel in thermal and fast solid-fuel reactors;
- another part of plutonium together with all minor actinides and some fission products are incinerated in a burner reactor; stable and short-lived fission products are removed from the burner reactor by the separation systems of the reactor itself. These problems can be solved by chemical and physical methods. Among the last ones the thermodiffusion and molecular centrifugation.

E. R. Merz and C. E. Walter (eds.), Advanced Nuclear Systems Consuming Excess Plutonium, 219–224.
© 1997 *Kluwer Academic Publishers.*

- upon its additional purification the remaining plutonium can be used in thermal and fast reactors;
- the stable and short-lived fission products are directed to an interim storage facility from where, unless find application in processes or medicine, they are sent to final disposal.

The advantages of the MSR as a burner reactor follow not only from a possibility for its combination with the gas-fluoride technique of fuel reprocessing, which is low-cost and produces a small quantity of waste, but also from its capability to use fuel of any nuclide composition without Pu purification from MA and FP. The MSRs have the flexibility to utilize any fissile fuel in continuous operation with no special modification the core as demonstrated during MSRE operation for U^3, U^5 and Pu.

The MSRs further require a minimum of special of fuel preparation and can tolerate denaturing and dilution of the fuel. Moreover, the systems of this reactor could be simultaneously used as the components of the external fuel cycle.

In addition, there is a possibility to eliminate reactivity accidents in the reactor, providing it is operated under subcritical conditions with acceptable, economical external neutron sources. The subcritical mode could be important for the MSR burner because of the some decrease of the delayed neutrons due to convection and composition of the fuel.

Development of some components of such a nuclear power system that will be acceptable for large-scale implementation in the long-term perspective could be launched just now, because this will promote the efficient resolution of the following immediate challenges:

- the gas-fluoride technique will help to reprocess efficiently spent fuel from power reactors, including RBMKs, which allows the expenses for this process to be covered at the cost of recycling of uranium, whereas plutonium together with minor actinides and fission products would be sent to interim storage (for 30-50 years) in a nonweapon-grade form until the price of uranium increases so that plutonium recycling becomes economical. Pu from RBMK is not suitable for thermal power reactors and could be used for denaturating of weapon grade Pu.
- the mixture of 100 t low grade RBMK plutonium (9/40/41/42≈0,49/0,35/0,10/0,06) with 100 t of weapon-grade Pu will give to NP 200 t of efficient for energy production Pu (9.40/41/42≈ 0,71/0,21/0,05/0,03)

The advantages of chemical Dry and Pirochemical Processing Method and molecular centrifugation or thermal diffusion separation over the aqueous processes are:

a) compactness due to the absence of moderator;

b) simplicity of the schemes and equipment used;

c) acceptance for short cooling times;

d) small fuel inventories out of reactor;

e) no radiation damage to process solvents;

f) solid process requiring minimum treatment for packaging, shipment, and storage;

g) no "naked" Pu in reprocessing.

The dry fluoride method can be applied in various versions depending on the agent used, choice of the scheme of separation and cleaning of the components, etc. The proposed version of the method (see Fig. 2) provides uranium/plutonium separation already in the fluoridation phase due to the use of chlorine trifluoride (ClF_3). The separation is accomplished in a flame apparatus at a temperature higher than 1000°C. UF_6 obtained from uranium fluoridation is then removed together with fission products forming volatile fluorides (mainly group 5-7 elements). Plutonium, in the form of PuF_4, and a part of fission products forming non-volatile fluorides (group 1-4 elements) as well as americium, curium and a insignificant quantities of residual uranium (about 0.1%) fall as a solid precipitate on the apparatus bottom.

UF_6 is subjected to cleaning from fission products by distillation under pressure in the packed towers, and from neptunium - by chemisorption on fluoride magnesium, and then is transferred for reenrichment.

In future PuF_4 can be either cleaned from fission products in the cycle "fluoridation to PuF_2 - thermal decomposition", or converted into PuO_2 by pyrohydrolysis and then passed for intermediate

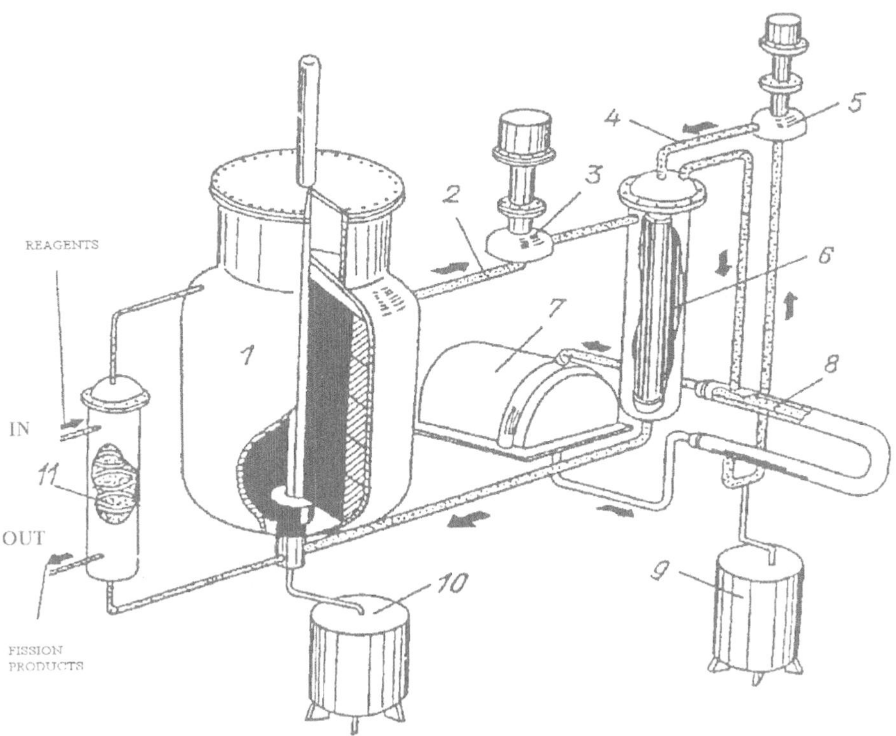

Fig. 1. Molten Salt Reactor

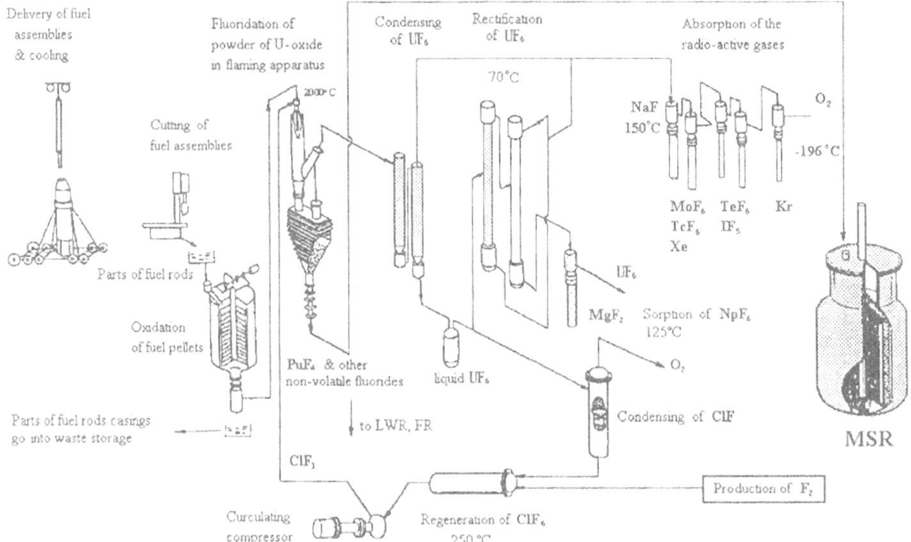

Fig. 2. Gas fluoride reprocessing of spent nuclear fuel of power reactor

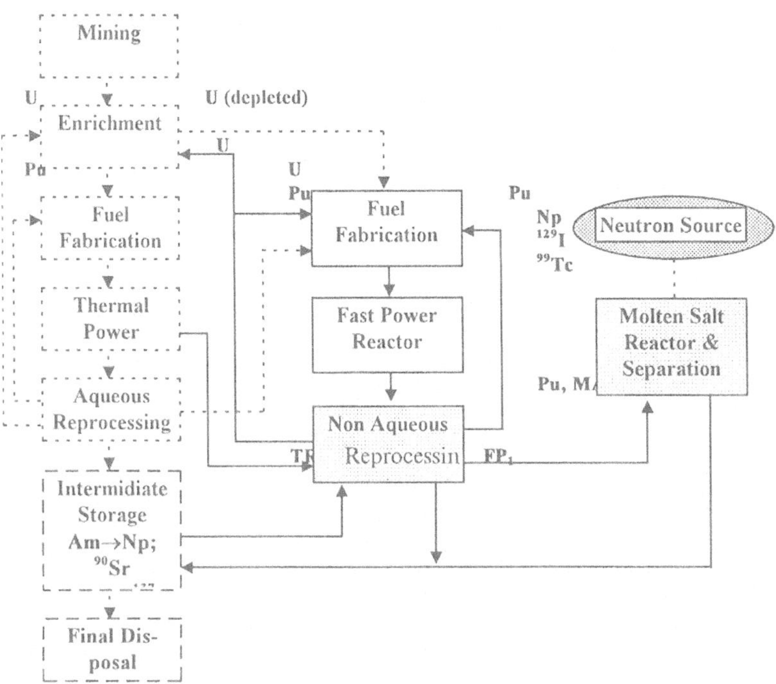

Fig. 3. Multicomponent Nuclear Energy System with closed fuel cycle for all actinides and dangerous long-life fission products. FP_1, FP_2 represent short-lived and long-lived FPs, respectively

long-term storage, or used directly as the nuclear fuel in the molten salt reactor. In the latter case there is no need to clean plutonium from fission products and minor actinides.

The fluorides of fission products, which are separated from UF_6 during the distillation, are subjected to pyrohydrolysis to convert them into the oxide form, then they are vitrificated and, after cooling, passed to controlled storages for final disposal. Fluorine is recycled.

The basis which will be used to carry out the processing task rely on:

- well established theoretical data;
- experimental results of the MSR program (MSRE, MSBR, DMSR, etc.);
- the experimental results for fluoride volatilization Process.

Development efforts on the MSR concept are of the same age as nuclear power as a whole. The first experimental MSR began to operate in the USA in 1951. It was developed in the framework of an ARE (Aircraft Reactor Experiment) program. An MSRE reactor with a thermal capacity of 8 MW was constructed in the USA in 1965. The basic purpose of the MSRE was to check to working capacity of the individual design units, to perfect the fuel and coolant technology, and to study the dynamics of this type of reactors. In the course of the four-year campaign the rector operated at first, on U-235 and then was converted to U-233.

Extraction of U-235 from the fuel composition was made by fluoridation.

The main result of the above experiment was a convincing demonstration that power MSRs with circulating fuel where feasible with the limits of the state of the art at the time. Since that time the stages of extensive development of MSR problems in the USA gave way to the periods of stagnation or even stoppage of funding. It should be noted that the periods of more active interest in MSRs took place when serious difficulties and new problems appeared on the traditional path of nuclear power development and were apprehended by the public.

Now the future the fuel cycle of the MSR burner should be aimed at solving the following tasks:

- introduction of highly radioactive fissile materials (Pu, NP, Am, Cm) as well as some fission products into the fuel cycle;
- minimizing the volume of radwaste generated by the fuel cycle as a whole and its individual stages. The waste must be in a compacted form suitable for storage and disposal in small volumes;
- reprocessing of fuel and its recycling in nuclear power installations must be realized with as few operations as possible and a minimum loss of radionuclides to be incinerated beyond the protective barriers of the closed fuel cycle.

CONCLUSIONS

1. There is a possibility to update the NP fuel cycle, which permits the NP safety, ecology, cost effectiveness and sufficient resources supply and non proliferation to be ensured for the long perspective.
2. A multi-component structure (fig. 3) of the fuel cycle may be laid into the nuclear power involving:
    - advanced light-water reactors;
    - advanced fast reactors;
    - dry fluoride reprocessing of spent fuel;
    - molten salt "burners" with a neutron source;
    - molten salt physical and electrochemical separation of actinides and fission products.
3. In such multycomponent system weapon grade Pu cold be converted into the non weapon grade form without degradation of its potential properties as nuclear fuel with delayed utilization without reduction of economical efficiency of nuclear energy system. Weapon grade Pu even in denaturated form is valuable neutron source for utilization of minor actinides and for incineration of dangerous long-life fission products.

ACCELERATOR-DRIVEN SYSTEMS - A NEW PERSPECTIVE ON FISSION ENERGY

WACLAW GUDOWSKI
Royal Institute of Technology,
S-100 44 Stockholm, Sweden

ABSTRACT

In last few years Accelerator-Driven Systems (ADS) have been a subject of intensive research in many countries. This paper presents some basic facts about accelerator-driven transmutation with emphasis on Pu-incineration and makes a general overview of some of the research activities in ADS. Possible impact of accelerators on nuclear energy production and transmutation of nuclear wastes is preliminary assessed.

1. Introduction

The end of the Cold War prompted and intensified the direct contacts between Russian weapon scientists and their Western colleagues which consequently lead to development of some research activities which otherwise were hindered by the political constraints. Accelerator-Driven Systems (ADS) became one of these extensive activities which got a special acceleration from the new cooperation possibilities. In July 1991 the first meeting between the Russian and Western accelerator and reactor experts was devoted to accelerator-driven systems and took place in Saltsjöbaden in Sweden [1]. This workshop stimulated further successful international cooperation in the field. First International Conference on Accelerator-Driven Transmutation Technologies and Applications took place in Las Vegas (USA) July 25-19 1994 and gathered about 180 participants. Next conference, Second International Conference on Accelerator-Driven Transmutation Technologies and Applications in Kalmar (Sweden) 3-7 June 1996 became a forum for more 200 participants from 24 countries discussing vividly the most important issues in accelerator, reactor and transmutation technologies. Next ADTT Conference will take place in Czech Republic in 1998.

The idea to use an accelerator as an intense, external neutron source (through the spallation processes) to assist the nuclear reactors is not very new. The first practical attempts to promote accelerators to generate potential neutron sources were made in the late 1940's by E.O. Lawrence [2] in the United States, and W.N. Semenov in the USSR. The first application proposed was the production of fissile material in the frame the MTA project at the Lawrence Livermore Radiation Laboratory [3]. This project was abandoned in 1952 when high grade Uranium ores were discovered in the United States. The Canadian team at

E. R. Merz and C. E. Walter (eds.), Advanced Nuclear Systems Consuming Excess Plutonium, 225–235.
© 1997 *Kluwer Academic Publishers.*

Chalk River always has been strong proponent of such a producer of fissile material which could be used in conjunction with a conversion-efficient CANDU reactor [4]. The concept of the accelerator breeder also has been studied by Russian scientists. Under the guidance of V.I. Goldanski, R.G. Vassylkov made a neutron yield experiment in depleted Uranium blocks using the accelerator at Dubna [5].

In last few years hybrid systems were proposed for different purposes. ADS on fast neutrons for the incineration of higher actinides was proposed at Brookhaven National Laboratory (PHOENIX-project) and now is carried out in Japan as a part of OMEGA-programme. Los Alamos National Laboratory has developed several ideas - under the common name: Accelerator-Driven Transmutation Technologies (ADTT) - to use the hybrid system on thermal neutrons with a linear accelerator for incineration of Plutonium and higher actinides as well as for transmutation of some fission products in order to reduce effectively long-term radioactivity of nuclear waste. Three years ago Carlo Rubbia and his European group at CERN proposed different hybrid system based on cyclotron or race-track like combination of linac (LINCYC) with a bending section to produce nuclear energy or to transmute Pu and minor actinides and/or some fission products using the Thorium fuel cycle. Some basic preliminary experiments were already performed by the CERN-group.

Fig. 1 presents in a very short and concise way the history of the development of hybrid system ideas through the diagram of the accelerators proposed for Accelerator-Driven Systems. On X-axis is the energy of proton beam, on Y - the current required for the estimated performance. In it quite clear that the first proposed accelerators required an extremely high current intensity at the modes energy below 1 GeV. Current projects are more focused on higher energies and lower currents which reflects the technological constraints for accelerator construction.

The high power accelerator technology required for ADS has been under continuous development for the past decades. Linear accelerators, linacs, have been developed into highly reliable and efficient research (and military) tools. There is confidence in that a high power (200 mA, 1.6 GeV), continuous wave (CW) accelerator can be built at a reasonable cost (In October 1995 the US DOE committed to the demonstration of the accelerator technology for application to tritium production). On the other hand the technology the of circular proton accelerators, such as the segmented cyclotron or synchrotron recently improved so that a proton beam of 10-15 mA is achievable. The cyclotron does not require a large physical area and has some other benefits compared to linac. In recent evaluations it was found that the most efficient operation current for a cyclotron-type machine would be ~10 - 20 mA and for a linear accelerator ~100 mA.

It is important to note that only LANSCE at Los Alamos National Laboratory and SINQ at Paul Scherrer Institute (Würenlingen/Villingen) have existing accelerator facilities, which could accommodate a low power demonstration experiment for the accelerator-driven system.

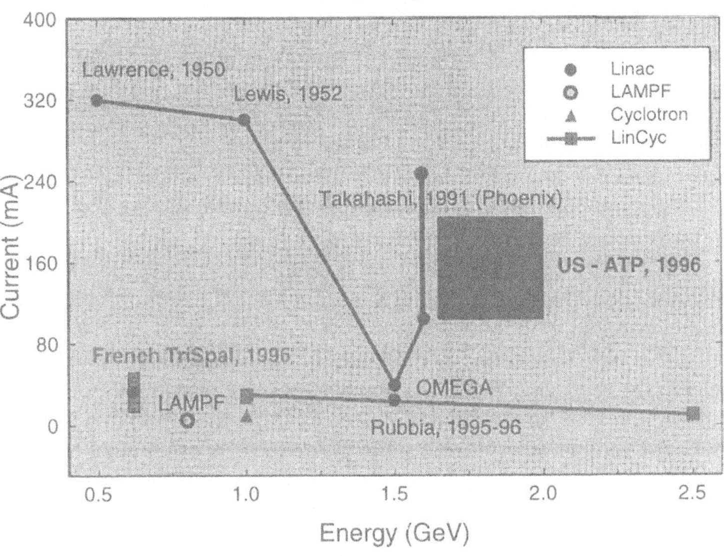

Fig. 1. Accelerator proposed during the years for Accelerator-Driven Systems. Accelerator parameters given as a function of particle energy and particle current.
ATP - Accelerator-Production of Tritium, American Project for Tritium Production,
TriSpal - Tritium-Spallation, French project for Tritium production.

2. ADS and Critical Reactors.

Accelerator-driven Systems are often presented and "marketed" as an ultimate solution for the future nuclear power system including an alternative approach for the nuclear waste management. The proponents of these systems claim that ADS can address most of the important concerns of today nuclear technology. What are those concerns? With a certain simplification one can say that there are three main reasons why nuclear technology is perceived in many countries as a dangerous and even unacceptable technology:

1. Criticality concern:

   Operation of the critical reactors relies on delayed neutrons and their life-time, so reactor fuel has to be well characterized;

   Run-away risks - reactor can explode (Chernobyl fear);

   No passive mechanisms for criticality controls, however, some feedbacks work really well (like boiling of the water in LWRs)

2. Decay heat problems:
   Reactor core can melt-down (TMI + Chinese syndrom);
   Solid fuel put severe constraints on the cooling system and lack of passive cooling systems;
   Economy drives reactor towards high power.

3. Nuclear waste handling and proliferation concerns:
   Reactors produce long-lived waste (mainly actinides) and geological storage has low public acceptance due to the unimaginable time scales;
   Pu in wastes can be a serious proliferation risk.

Fig. 2 shows risk perception related to the nuclear technology as perceived by so called general public and experts.

It can be seen that there is a significant difference in risk perception between experts and general public. For the general public - the potential worst case scenario accidents have the decisive role in perception of the risks, it can explain the high risk perception for criticality accidents and for the geological repositories. For experts - daily operational problems and real technical difficulties have the biggest impact on risk perceptions.

Risc perception for accidents related to nuclear power

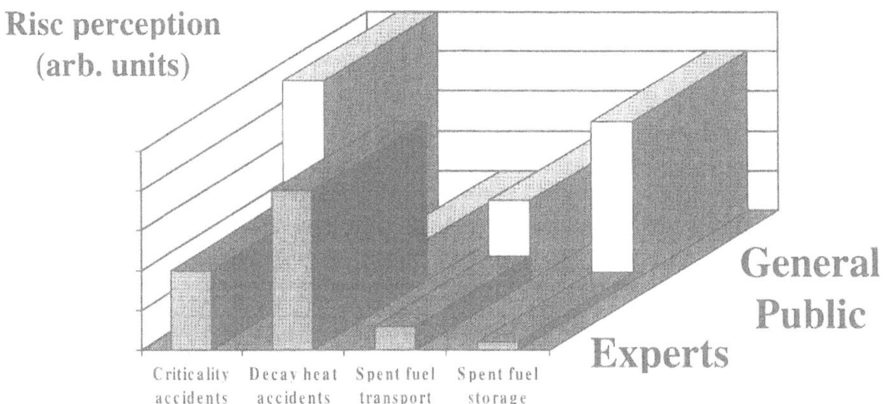

Type of accidents

Fig. 2. Risk perception of the hazards related to nuclear power. Generalized picture. Transportation of nuclear fuel/waste does not create great concerns in Sweden, in other countries perception is very different (Germany, USA).

In this context ADS can play a unique role for the future of nuclear technology, because ADS addresses the concerns which are perceived as serious by the general public. ADS offer solution for the safety issues associated with criticality: ADS operates in non self-sustained chain-reaction mode and therefore minimizes the criticality concern. ADS is operated in a subcritical mode and stays subcritical, regardless of the accelerator being on or off. The accelerator provides a convenient control mechanism for subcritical systems than that provided by control rods in critical reactors, and subcriticality itself adds an extra level of operational safety concerning criticality accidents. A subcritical system driven with an accelerator decouples the external neutron source (spallation neutrons) from the fissile fuel (fission neutrons). Accelerator driven systems can in principle work without safe-shutdown mechanisms (like control rods) and being independent from the delayed fission neutrons can accept fuels that would not be acceptable in critical systems. In other words the exact characteristic of the fissionable materials is not necessary. Also separation of the long-lived nuclei from the high-level waste and transmuting them into short-lived or non-radioactive wastes would ease a lot of constraints for geological repositories, may significantly reduce their costs and increase the public acceptance.

Moreover, the possibility of burning Plutonium in these systems, including the option of very deep burning of weapon-grade or reactor-grade Plutonium enhance the potential non-proliferation role of ADS.

There are not straight forward differences in managing of the decay-heat in ADS compare to the conventional reactors. However the new options of non-solid form of fuel (like liquid or quasi liquid fuel) together with subcriticality open the new possibilities for designing passively safe systems or systems with significantly enhanced safety with regard to decay-heat problems.

3. Transmutation and Spallation

The basic process of accelerator-driven nuclear systems is nuclear transmutation. The nuclear transmutation was first demonstrated by Rutherford in 1919, who transmuted ^{14}N to ^{17}O using energetic α-particles. I. Curie and F. Joliot produced the first artificial radioactivity in 1933 using α-particles from naturally radioactive isotopes to transmute boron and aluminum into radioactive nitrogen and oxygen. It was not possible to extend this type of transmutation to heavier elements as long as the only available charged particles were the α-particles from natural radioactivity, because the Coulomb barriers surrounding heavy nuclei are too great to permit the entry of such particles into atomic nuclei. Transmutation processes can be induced efficiently - as shown on Fig.3 - mainly by photons, neutrons and charged particles. However, due to small cross-sections, the use of photons is very limited; charged particles bombarding directly the transmutation target are also of very limited use due to the Coulomb barrier. Neutrons are by far the most efficient transmutation tool. The conventional nuclear reactors, specially those types with a good neutron economy, can be used as a neutron source for transmutations purposes. Accelerator-Driven Systems are also one of the options. High power accelerators coupled with the spallation process can be used to produce large numbers of neutrons, thus providing an alternative method to the use of nuclear reactors for this purpose. Spallation offers exciting new possibilities for generating intense neutron fluxes for a variety of purposes.

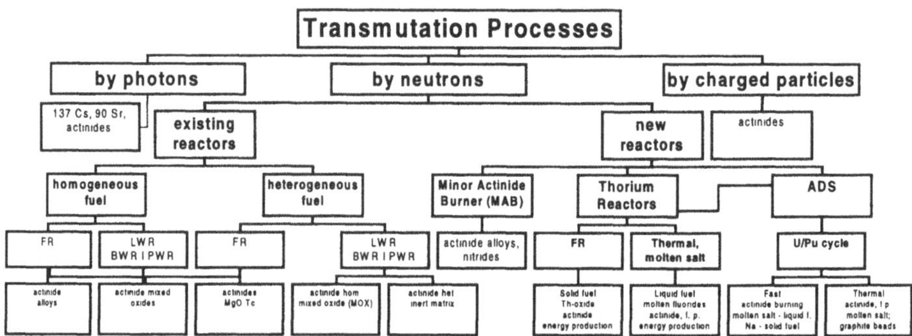

Fig. 3 Classification of the transmutation processes

Spallation - Fig. 4 - refers to nuclear reactions that occur when energetic particles (e.g. protons, deuterons, neutrons, pions, muons, etc.) interact with atomic nucleus - the target nucleus. In this context "energetic" means kinetic energies larger than about 100 MeV per nucleon. At such energies the nuclear reactions do not proceed through the formation of a compound nucleus. The initial collision between the incident projectile and the target nucleus leads to a series of direct reactions whereby individual nucleons or small groups of nucleons are ejected. At energies above a few GeV per nucleon, fragmentation of the nucleus can also occur. After the initial phase, the nucleus is left in en excited state and subsequently relaxes to its ground state by "evaporating" nucleons, mostly neutrons [6].

Some target materials, such as ^{232}Th and ^{238}U, can also be fissioned by lower energy (~1 MeV to ~20 MeV) neutrons. Spallation, high energy fission and low energy fission produce different nuclear debris (spallation and fission products).

The function of the spallation target in the ADS is to convert the incident high energy particle beam to low energy neutrons. These requirements can be summarized as:

1. Compact size to enable good coupling to the surrounding blanket,
2. High power operation, on the order of 10 to 100 MW,
3. High neutron production efficiency,
4. Reliable and low maintenance operation,
5. Safe and low hazard operation,
6. Small contribution to the waste stream.

It is believed today that molten lead or molten lead-bismuth eutectic (LBE) are the best target choices, meeting most of the requirements above. A problem with LBE, however, is the production of radioactive and highly mobile polonium isotopes, especially ^{210}Po, from high energy proton and neutron reactions on lead and bismuth. This will be a concern because the polonium in the LBE will be rapidly released at operating temperatures. Pure lead, on the other hand, has a lower polonium production, but higher operating temperatures. Further assessments are needed in order to make a choice between these target materials.

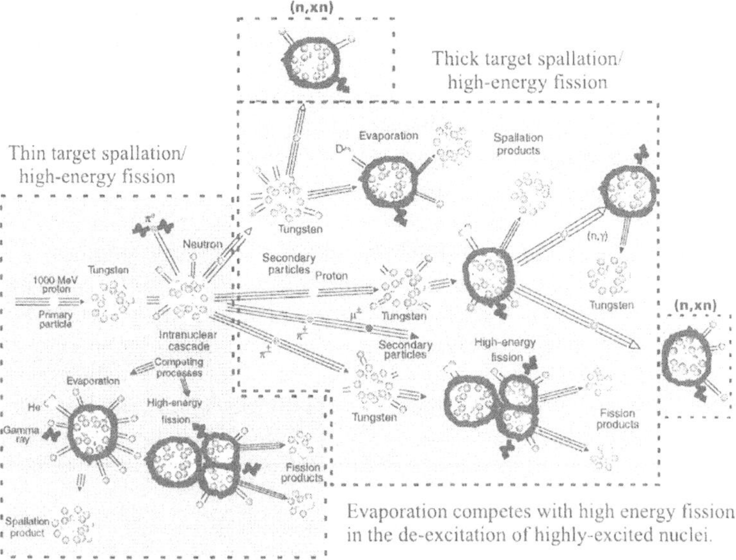

Fig. 4. Illustration of high-energy processes in a spallation thick and thin targets [6].

Fig. 5 presents spallation neutron production in different target materials as a function of the incident proton energy. Target diameter 25 cm, length 100 cm. The linear equation for the neutron production (upper left corner) is valid for energy range 0.8 < E < 4 Gev [7]

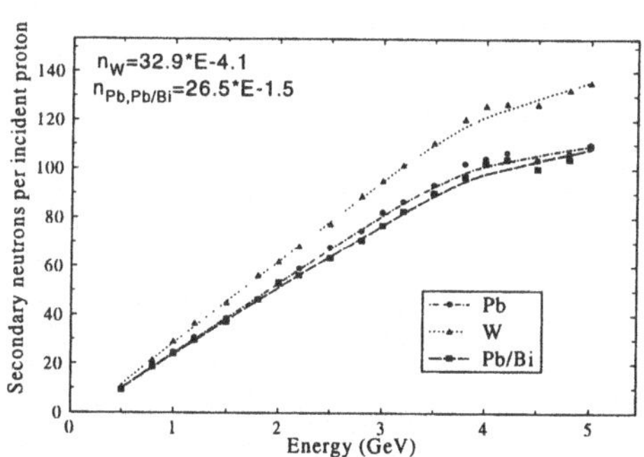

Fig. 5 Spallation neutron production in different target materials as a function of the incident proton energy. Target diameter 25 cm, length 100 cm. The linear equation for the neutron production (upper left corner) is valid for energy range 0.8 < E < 4 Gev

The thermal cross-sections for transmuting MA and FP are larger than the fast neutron cross-sections. However, the thermal neutron cross-sections of the transmutation products are also large. It is therefore desirable in thermal systems to remove the products in order to reduce unproductive losses of neutrons. The capture of fast neutrons by the fission products and by the structural material is smaller than for thermal neutrons and, from the point of neutron economy, a fast reactor is better than a thermal reactor. Also, one would like to take advantage of the high η-values for ^{239}Pu and the other actinides to produce extra neutrons by high-energy fission for use in transmutation of the long-lived fission products.

The Thorium-Uranium fuel cycle is very attractive for future ADS. It has at least two advantages over the traditional Uranium-Plutonium cycle used in most nuclear reactors:

1. The Thorium-Uranium cycle produces a smaller amount of higher actinides than the Uranium-Plutonium cycle, because of the small capture to fission ratio in ^{233}U and because of the presence of two other fissionable isotopes of Uranium (^{235}U and ^{237}U) in the chain leading to Plutonium and the other heavier actinides
2. Even if ^{233}U is a good nuclear weapon material, the Thorium-Uranium cycle is perceived as safer than the Uranium-Plutonium cycle from a nuclear weapons proliferation standpoint, because of the presence of the hard-γ emitter ^{208}Tl in decay chain of ^{232}U which is a minor product of the cycle, and because it is imagined that an isotopic dilution of ^{233}U with depleted or natural Uranium in the feed, or in the start-up fuel, would make ^{233}U difficult to extract in pure form.

The very generic scheme of the Accelerator-Driven System is presented on Fig. 6.

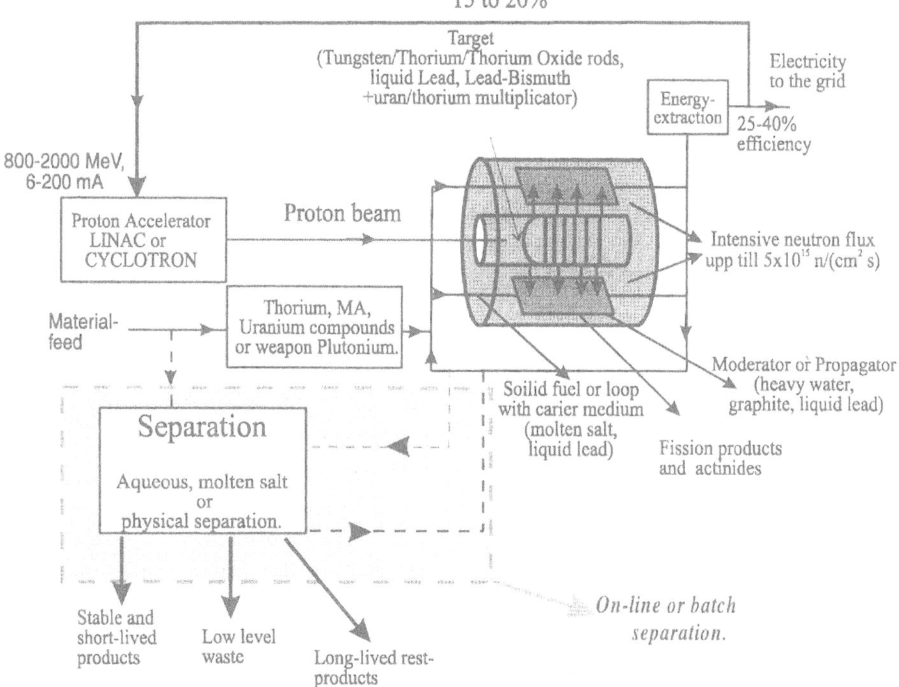

Fig. 6. A generic scheme of the Accelerator-Driven System.

3. Review of the existing projects

Fig. 7 shows a classification of existing ADS concepts according to their physical features and final objectives. The classification is based on neutron energy spectrum, fuel form (solid, liquid), fuel cycle and coolant/moderator type, and objectives for the system. ADS systems - like reactors - can be designed to work in two different neutron spectrum modes - on fast or on thermal neutrons. The CERN-group headed by C. Rubbia investigates the possibilities (CERN application to EC, 1995) to design a system which will exploit the neutron cross-section resonances in what could be classified as a "resonance neutron" mode. Both, fast and thermal systems are considered for solid and liquid fuels. Even quasi-liquid fuel has been proposed based on the particle fuel (pebble bed) concept developed by BNL.

The objective for some nuclear transmutation systems is to transmute existing nuclear wastes from light water reactors, mainly Pu and minor actinides, with or without concurrent energy production. These projects can be classified as an attempt to close the LWR-fuel cycle. Other systems are designed to take advantage of the Thorium fuel cycle for energy production. As can be seen in Fig. 6 most concepts are based on linear accelerators, however the new ideas of constructing high current cyclotron presented by CERN-group [8]can make the cyclotron option much more attractive economically.

4. Conclusions

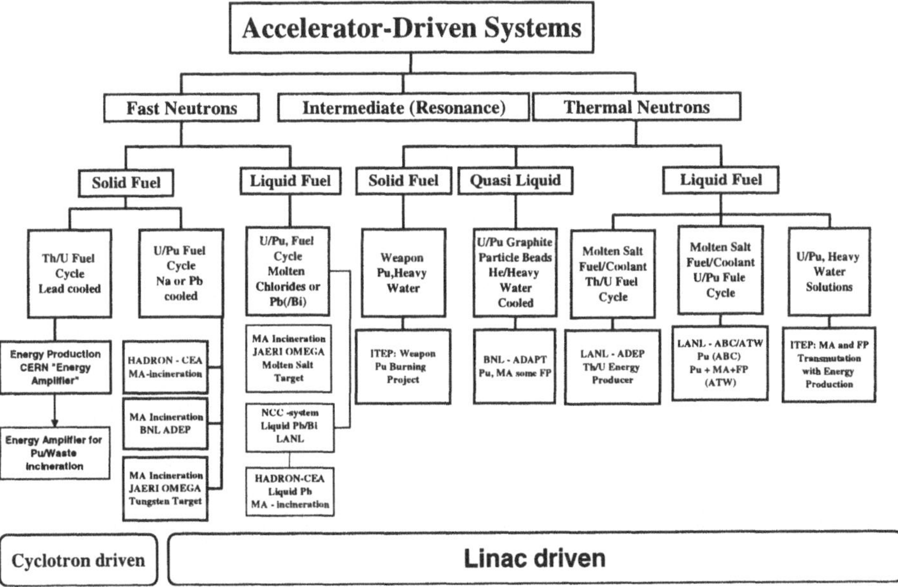

Fig. 7. Classification of existing ADS concepts.

A brief summary of technical disadvantages of the different ADS from the Swedish perception point of view is presented in Table I.

TABLE I. TECHNICAL DISADVANTAGES/ADVANTAGES AND PUBLIC PERCEPTION FOR ADS

ADS type	**Technical Disadvantages/**Advantages	**Public Perception (Swedish perspective)**
Solid fuel and thermal neutrons	Flux/power density distribution make system unfeasible. Serious safety concerns with "beam-on" power transients(e.g. unprotected loss of flow accidents)	Similar to LWR (subcriticality advantages important)
Aqueous system with liquid/quasi-liquid fuel	Pressurized system can not meet the "blow-down" criterium(Low pressure drives the economy to the bottom).Radiolysis of the water creates serious problems	Too risky, no technology developed, difficult safety in depth!
Solid fuel and fast neutrons	Very high inventory/burnup ratio (e.g. 3160 /250 for Japanese minor actinide burner). Liquid sodium cooling (with all the odds). Very powerful accelerator (BNL Phoenix) if electrical energy is to be produced. Radiation damage problems.	Low acceptance for fast neutron spectrum - but subcriticality can balance this. No experiences in fast reactors, sodium questionable. TOO LARGE INVENTORY
Liquid fuel and fast neutrons	Even worse inventory / burnup ratio - 4.5% (e.g. 5460 /250 for Japanese actinide burner) Problems with chloride salts (corrosion)	As above. Can liquid lead help?
Solid fuel and liquid lead (Thorium based)	Good ratio: transmutation rate/inventory (in equilibrium), transmutation of some FP (Tc, I) possibility of inherent safety features. **LEAD TECHNOLOGY to be developed.**	Appealing physical parameters and passive safety potential. Lead technology not mastered, low acceptance for fast spectrum.
Liquid fuel and thermal neutrons	High thermal neutron flux - radiation damage and activation problems. Separation chemistry on- or quasi on-line. Difficult safety in depth.	Low inventory, low pressure and high transmutation rate are attractive. No technology developed for liquid fuel.

Finally one can conclude that ADS can complete and broader the nuclear power option, there are specific objectives, mainly in fuel inventory toxicity reduction, nuclear waste toxicity reduction and LLFP incineration. Most of them can be addressed successfully with

ADS:

- by minimizing the core inventory and allowing high neutron fluxes and "exotic" (not well behaving) fuels (thermal systems but not exclusively),
- thorium breeding without excess breeding or the use of enriched material.

Different paths exist for ADS development :

- utilization of existing spent LWR fuel should be of interest from the environmental conservation point of view, could be a joint effort for weapon-grade and reactor-grade Pu burning,
- Thorium fuel cycle should be the strategic objective for the future nuclear power (both reactor and ADS), it requires, however, 20-30 years of an intensive development.

ADS will not eliminate completely the necessity of the geological repository, it can make it cheaper, more acceptable and safer. Could pay for transmutation costs itself through energy production. Moreover ADS open new possibilities for the nuclear technology and can contribute to the environmental restoration

Future of ADS and the nuclear technology in general will be probably based not on its technological feasibility, which can be mastered today but it will be rather a moral choice for the society. To make this choice, all the alternative technologies have to be put "on equal footing". ***Today all the technologies for energy production carry global threats.***

Acknowledgments

This paper is partially based on materials prepared for the IAEA "Accelerator-Driven Systems - Status Report" which is to be published in 1997.

References

[1] International Specialist Meeting on "Accelerator-Driven Transmutation Technology for Radwaste and Other Applications", Saltsjöbaden June 24-28, 1991.

[2] E.O. Lawrence and N.E. Edlefsen, "On the Production of High Speed Protons", Science, LXXII, NO. 1867, 376, October 10, 1939.

[3] C.M. van Atta, "A Brief History of MTA Project", ERDA Information Meeting on Accelerator Breeding , January 19-19 (1977).

[4] G.A. Bartholomew and P.R. Tunnicliffe, Eds, "The AECL Study for an Intense Neutron generator", Atomic Energy of Canada Limited, Report No. AECL-2600 (1966).

[5] R.G. Vassylkov, V.I. Goldanski et al., Atomnaya Energiya 48, 329 (1978).

[6] G.J. Russel, E.J. Pitcher and L.L. Daemen, "Introduction to Spallation Physics and Spallation Target Design", International Conference on Accelerator Driven Transmutation Technologies and Applications, Las Vegas, NV, July 25-19 1994.

[7] J. Carlsson, MsC thesis, Department of Nuclear and Reactor Physics, Royal Institute of Technology, January 1997.

[8] C. Rubbia, Los Alamos ADTT Review, November 1996.

On Solving the Fissionable Materials Non-Proliferation Problem in the Closed Uranium-Thorium Cycle.

V.E.Marshalkin, V.M Povyshev, Yu.A.Trumev.
RFNC VNIIEF

Abstract.

The solution of the fissionable materials and nuclear technologies non-proliferation problem has been analyzed in the closed Uranium-Thorium cycle.

It has been shown that using thorium instead of ^{238}U limits the actinides generation by uranium isotopes, and absence of uranium fuel surplus, active ^{233}U, ^{235}U isotopes dilution by ^{234}U and ^{236}U isotopes together with radiological danger of ^{232}U decay products , create technological barriers for using nuclear fuel materials for making explosive devices. Fissionable materials accumulated (weapon and power-plant ones) can naturally and efficiently be utilized in the uranium-thorium cycle.

E. R. Merz and C. E. Walter (eds.), Advanced Nuclear Systems Consuming Excess Plutonium, 237–257.
© 1997 *Kluwer Academic Publishers.*

Introduction.

The disarmament process that started in the latest years supposes appearing about 50 tones of excessive for military purposes plutonium of weapon quality in the USA and about the same quality of it in Russia /1/. The quantity of highly enriched uranium released in this process is an order greater and amounts to 500 tones of surplus weapon quality uranium in each of the countries. Because of the fact that manufacturing an explosive nuclear device supposes using several kilograms of plutonium and several times greater quantity of weapon uranium, this quantity of weapon materials is sufficient for manufacturing dozens of thousands of explosive devices. That's why the disarmament process has qualitatively complicated ensuring fissionable materials and nuclear technologies non-proliferation and activated research for solving this problem.

It is a well-known fact /1/ that primitive nuclear devices with energy emission of about 1 trothyl equivalent kiloton, can be manufactured of exhaust nuclear fuel plutonium from ordinary power reactors. The world nuclear power industry has already generated about 1000 tones of such plutonium and continues generating it. At present time from these 1000 tones about 86 tones of plutonium was extracted from the irradiated fuel, extraction rate exceeds the rate of its using, and by the beginning of the next century , the supplies of such plutonium will amount to 110-170 tones.

At present time it is accepted that the main barrier for nuclear weapon propagation is limitation of access to fissionable materials. That's why the work on studying perfection of fissionable materials recording and monitoring are carried on extensively, and apparently, on realization of these achievements during storage. We see this way as a necessary stage , but with poor prospective in terms of solving the non-proliferation problem because of intensive plutonium and other actinides generation as by-products in nuclear power industry and their accumulation.

The principal solutions for limitation of access to fissionable materials for using them in weapon production are: first, absence of these materials in storage, and second, unfitness of nuclear fuel components for use in manufacturing explosive devices. Both these conditions are implemented during thermal-neutron reactor operation in the closed uranium-thorium cycle. Besides, in thorium fuel, both weapon plutonium and uranium, and power-plant plutonium and other actinides from expend nuclear fuel can be burned by fission reaction in the most efficient way.

1. Generation of actinides and their burning by fission reaction on reproducing isotopes ^{238}U and ^{232}Th.

On the scheme 1 and the scheme 2 isotope accumulation chains in a thermal reactor have been shown. In each cell together with isotope's chemical symbol its half life , decay type, values of fission cross-section in a thermal point and fission resonant integral in resonance area (first column), values of neutron capture cross-section in a thermal point and resonant integral in resonance area (second column) are given.

The correlation between decay constants $\lambda^i_{dec} = \ln 2 / T^i_{1/2}$ and burn-out constants $\lambda^i = F\sigma^i_c$ for the i-isotope determines the correlation between

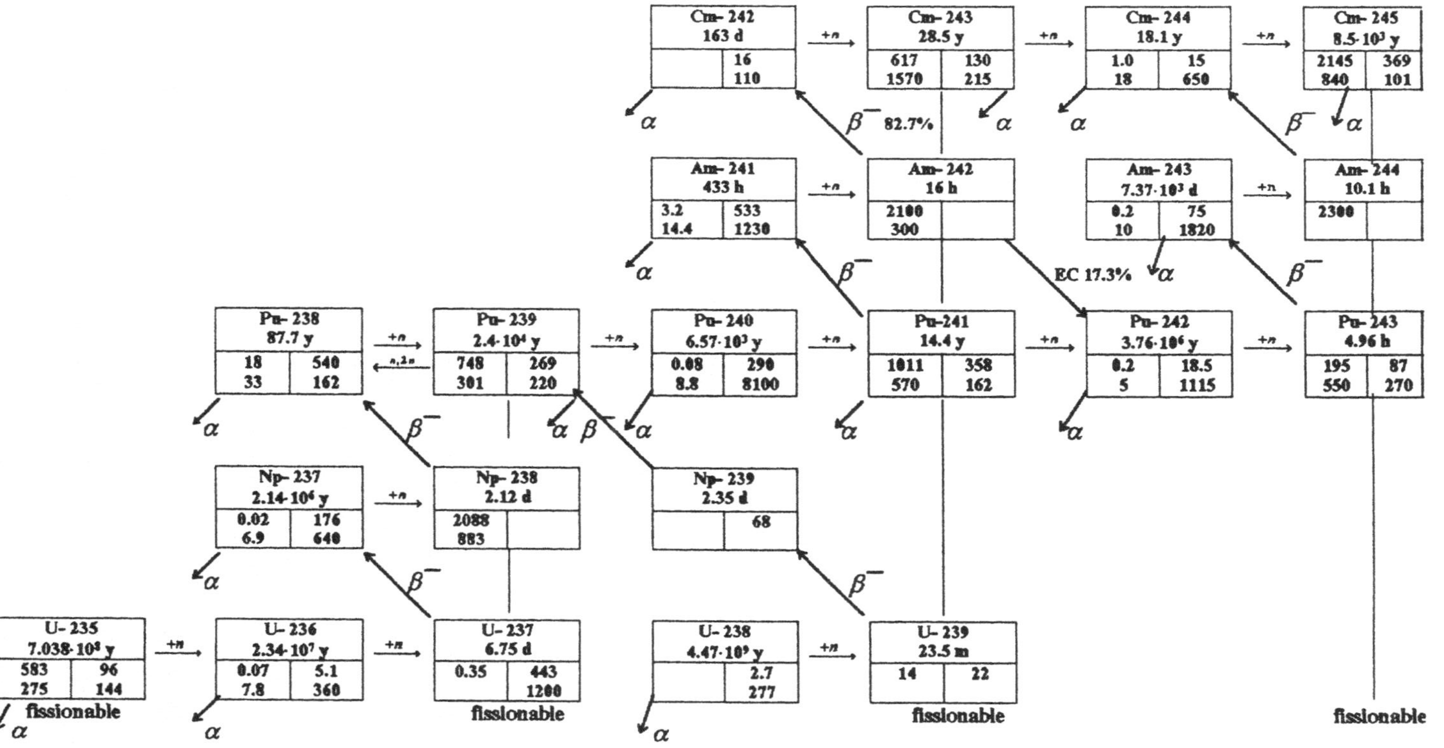

Fig 1. The scheme of isotope accumulation in thermal reactor with ^{235}U and ^{238}U

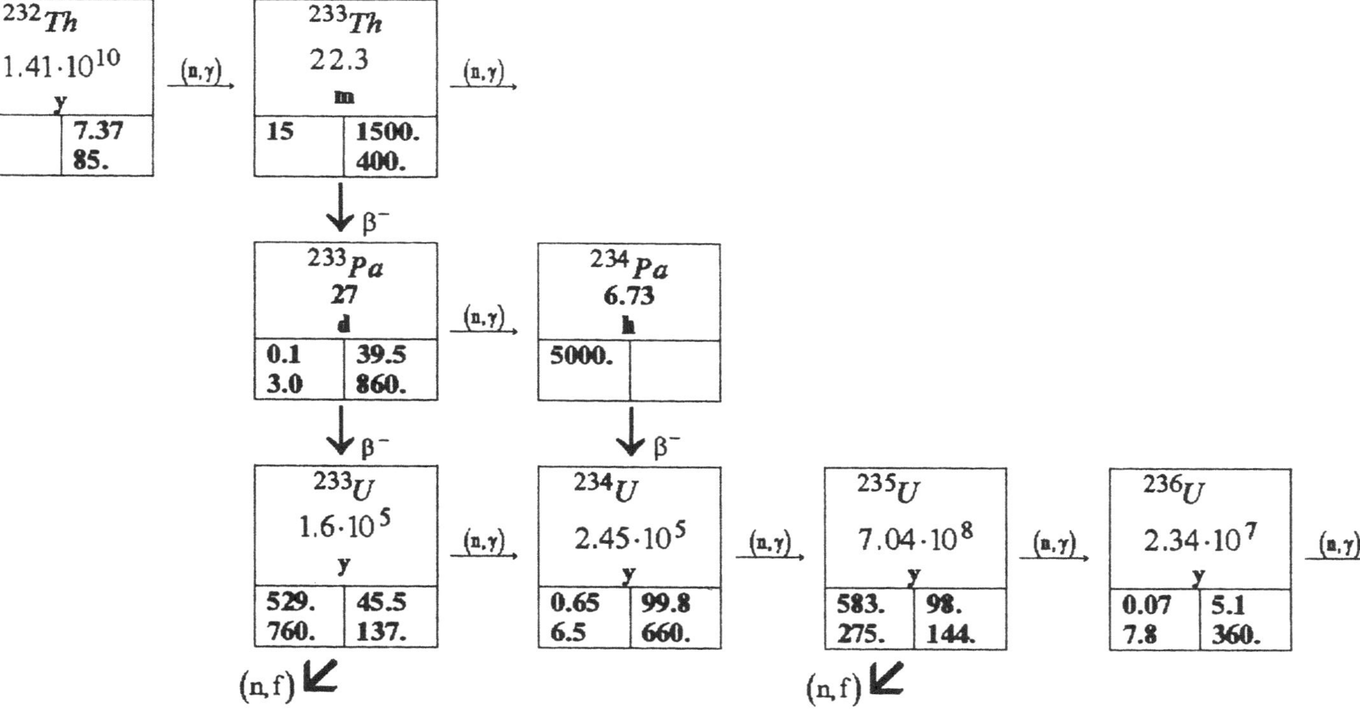

Fig. 2. The scheme of isotope accumulation in thermal reactor with Th

probabilities of its natural decay and burn-out in reactor's neutron flow F. Total burn-out cross-section σ_c^i splitting between fission σ_f^i and radiation capture σ_γ^i processes determines the ratio between probabilities of i-isotope's burn-out by fission reaction (falling out of the further actinides transformation) and i-isotope's burn-out by transition to another isobaric chain A+1. Involvement of values of thermal group cross-section and resonant integrals in resonance area according to the distribution χ of neutron flow density Φ determines the dependence of i-isotope generation and burn-out upon neutrons' energetical distribution in a reactor.

Visual comparison of scheme 1 and scheme 2 shows that generation, decay and burn-out of actinides if the reproducing isotope ^{238}U is replaced by ^{232}Th differ considerably.

The burn-out rate of ^{232}Th in the (n, γ)- process with ^{233}U generation goes ~2.7 times quicker than ^{238}U burn-out with ^{239}Pu generation in thermal group, but in resonance area this ratio changes qualitatively. ^{233}U burn-out rate with respect to ^{239}Pu is ~2 times lower in the thermal region. The equilibrium ^{233}U content in $^{235}U_\alpha\ Th_{1-\alpha}$ being approximately 5 times higher than equilibrium ^{239}Pu content in ordinary $^{235}U_\alpha\ ^{238}U_{1-\alpha}$ fuel. This circumstance provides the following advantages of using Th instead of ^{238}U:

-possibility of increasing energy generation during campaign time;

-5 times increase in economical reason for irradiated fuel recycling.

The ratio between neutron radiative capture and fission cross-sections in thermal area, that equal to 0.086 and 0.36 for ^{233}U and ^{239}Pu , and the ratio between resonance integrals, equal to 0.18 and 0.73, respectively, emphasize the fact that further plutonium isotopes generation rate compared to its burn-out by fission reaction is ~ 4 times higher than that of corresponding uranium isotopes compared to fission rate of ^{233}U. Comparison of generation, decay and burn-out processes of further isotopes shows similar difference, and, respectively, points out the difference in isotopes content, generated on reproducing isotopes ^{238}U and ^{232}Th. According to actinides generation parasitic neutron capture takes place.

Replacing the main fissionable isotope ^{235}U with ^{233}U is also accompanied by difference in actinides generation and parasitic neutron capture. Neutron flow density in the thermal group being the same, the rate of ^{234}U generation on ^{233}U is lower than the rate of ^{236}U generation on ^{235}U, and their ratio is:

$$\frac{\sigma_{n\gamma}(233)\cdot\left(\sigma_{n\gamma}(235)+\sigma_f(235)\right)}{\sigma_{n\gamma}(235)\cdot\left(\sigma_{n\gamma}(233)+\sigma_{nf}(233)\right)} \approx 0.55$$

In the resonance area this ratio equals to ~ 0.44. Besides, in the resonance area ^{233}U fission is ~ 2.8 times more probable than of ^{235}U. On the ^{235}U initial isotope in the course of consequent neutron capture three isotopes are generated one after another (^{236}U, ^{237}Np, ^{238}Pu), that are practically non-fissionable by thermal neutrons while on the ^{233}U initial isotope ^{234}U is followed by thermal-neutron-fissionable isotope ^{235}U.

Thus, ^{236}U and heavier actinides generation has more than an order decrease on ^{233}U initial isotope compared to ^{235}U.

It is necessary to emphasize that in the $^{235}U_{\alpha}\ ^{238}U_{1-\alpha}$ fuel intense generation of thermal-neutron-non-fissionable actinides (neptunium, plutonium, americium, curium isotopes) goes by two chains (^{235}U, ^{238}U), while in the $^{233}U_{\alpha}\ Th_{1-\alpha}$ - fuel actinides generation takes place in one chain (^{233}U). During this, long-living uranium isotopes are accumulated (^{234}U, ^{235}U, ^{236}U) on ^{233}U and , therefore, as mentioned above, in smaller quantities than on ^{235}U and ^{238}U initial isotopes. During implementation of the closed fuel cycle all isotopes of uranium are extracted from ENF (expend nuclear fuel) and returned to reactor as fuel. Thus, the remaining part of expende nuclear fuel on $^{233}U_{\alpha}\ Th_{1-\alpha}$ is practically free from usually generated in reactor long-living actinide activity, connected with isotopes of neptunium, plutonium, americium, curium.

Analysis of cross-section fission and neutrons radiative capture in thermal point (thermal group) values and values of resonance integrals of these processes in resonance energy area simply shows the preference of using the $^{233}U_{\alpha}\ Th_{1-\alpha}$ - fuel in a reactor with the most soft neutron spectrum. It is in the thermal point (group) that neutron radiative capture cross-section value on key isotopes ^{233}U, ^{235}U and ^{239}Pu are considerably below the corresponding value of the resonance integral in resonance area. The ratio between neutron radiative capture cross-section and fission cross-section for both ^{233}U and ^{235}U is also below corresponding resonance integrals.

2. Isotopic kinetics of recycled uranium fuel in the closed uranium-thorium cycle.

The questions of isotope kinetics of recycled uranium fuel in the closed uranium-thorium cycle have been considered in our previous work /2/.

One of the results obtained is the evaluation of the equilibrium uranium isotopes content in recycled fuel and the time within which their content reaches the equilibrium values. The following evaluated results have been obtained.

The equilibrium ^{233}U to Th value ratio is

$$\frac{N\left(^{233}U\right)}{N_o(Th)} \approx \begin{cases} 0.013 & \text{if } \chi=0 \\ 0.014 & \text{if } \chi=0.1 \\ 0.016 & \text{if the data of /3/ are used.} \end{cases}$$

Thus, after adding burned-out thorium and returning of uranium mixture from the previous campaign to the next campaign ^{233}U isotope equilibrium state is established, characterized by the value of 13-16 kg per one ton of thorium. The value of χ characterizes the hardness of neutrons in the system. Together with the cross-section evaluated by us, neutron cross-section from the work /3/ were used.

Reaching this equilibrium value if neutron flux $F \approx 5 \cdot 10^{13}\ n/sm^2 \cdot sec$ takes place within the irradiation time

$$t >> \begin{cases} 2.2 \text{ years if } \chi = 0 \\ 1.7 \text{ years if } \chi = 0.1 \\ 2.7 \text{ years if the data of /3/ are used.} \end{cases}$$

The ratio of the equilibrium value of ^{234}U to Th equals to

$$\frac{N(^{234}U)}{N(Th)} \approx \begin{cases} 0.006 \text{ if } \chi = 0 \\ 0.005 \text{ if } \chi = 0.1 \\ 0.008 \text{ if data from /3/ are used.} \end{cases}$$

This equilibrium state is reached within the time span of

$$t >> \begin{cases} 12 \text{ years if } \chi = 0 \\ 6 \text{ years if } \chi = 0.1 \\ 13 \text{ years if the data of /3/ are used.} \end{cases}$$

The ratio between the equilibrium values of ^{235}U and Th is evaluated by

$$\frac{N(^{235}U)}{N(Th)} \approx \begin{cases} 0.001 \text{ if } \chi = 0 \\ 0.0015 \text{ if } \chi = 0.1 \\ 0.0015 \text{ if dada from /3/ are used.} \end{cases}$$

This equilibrium state is reached practically within the same time span as reaching the equilibrium state by ^{234}U.

The ratio between equilibrium values of ^{236}U and Th is determined by

$$\frac{N(^{236}U)}{N(Th)} \approx \begin{cases} 0.018 \text{ if } \chi = 0 \\ 0.0034 \text{ if } \chi = 0.1 \\ 0.0083 \text{ if data from /3/ are used.} \end{cases}$$

This equilibrium state is reached within the span of

$$t >> \begin{cases} 230 \text{ years if } \chi = 0 \\ 22 \text{ years if } \chi = 0.1 \\ 83 \text{ years if the data of /3/ are used.} \end{cases}$$

Large discrepancies in the ^{236}U equilibrium value itself and the of reaching it are determined by difference in the values of ^{236}U neutron capture cross-section in the thermal group, radiation integral in resonance area and in neutron spectrum in fuel.

The rate of deterioration of fuel characteristics of uranium-thorium fuel happens because of ^{236}U accumulation. However, unlike processes in ordinary $^{235}U_{\alpha}$ $^{238}U_{1-\alpha}$ fuel, ^{236}U accumulation is a very slow process.

Because of that and the difference of three mass units between ^{236}U and the main fissionable isotope ^{233}U, recycled uranium fuel refinement by centrifuge seems efficient as soon as ^{236}U is accumulated in order to exclude generation of isotopes like ^{237}Np and ^{238}Pu.

The values obtained of equilibrium uranium isotopes content in recycled fuel allow to estimate the value of efficient neutron multiplication coefficient at the stationary stage in infinite medium approximation using the following relation

$$K_{\infty} = \frac{\sum_j \bar{\nu}_i \cdot \left(\bar{\sigma}_f^i + \chi \cdot f_f^i \cdot I_f^i\right) \cdot c_i}{\sum_j \left(\bar{\sigma}_f^j + \bar{\sigma}_\gamma^j + \chi \cdot \left(f_f^j \cdot I_f^j + f_\gamma^j \cdot I_\gamma^j\right)\right) \cdot c_j} \approx \begin{cases} 1.16 \text{ if } \chi = 0 \\ 1.07 \text{ if } \chi = 0.1 \\ 1.14 \text{ if the data of /3/} \\ \quad \text{are used.} \end{cases}$$

The error in the value obtained is about several units per cent. This means that at the stationary stage uranium-thorium power industry on thermal neutrons can work in the closed uranium-thorium cycle, self-supplying active uranium isotopes (^{233}U, ^{235}U), or with their minimum adding. Thus, an opportunity appears to burn all thorium in thermal neutron reactors, which would give a two-order-increase to fuel resource.

In terms of solving fissionable materials non-proliferation problem in uranium thorium power industry, let's emphasize the following.

In uranium-thorium fuel plutonium generation is practically absent. Thus, no considerable amounts of plutonium will be generated by nuclear power industry. In the uranium-thorium fuel there is no excess (in terms of nuclear power industry) of active uranium isotopes (^{233}U and ^{235}U). Reproduction coefficients of active isotopes K_R and effective neutron multiplication coefficients K_{ef} will be around unity and will be precisely determined by real reactor parameters. As soon as necessary according to uranium-thorium power industry operation, the isotopes generated must be removed from irradiated fuel and fed to the reactor as fuel for the next campaign. Thus, no active isotope accumulation will occur either in the form of irradiated nuclear fuel, or in other forms. Besides, active uranium isotopes (^{233}U, ^{235}U) are diluted by under-barrier-fissionable isotope ^{234}U ($c(^{234}U) \approx 0.5c\ (^{233}U)$) to a considerable. Thus in equilibrium state natural denaturalization takes place. In other words, both major problems of modern nuclear power industry are solved in a preventive way:

- ecological danger of actinides accumulation in irradiated fuel storage;
- proliferation of generated fissionable materials, that are not used as fuel.

Taking this into account, the solution of fissionable materials (uranium isotopes mixture) recording and monitoring problems simplifies qualitatively.

Another important factor, simplifying the solution of this problem, analysis of which is beyond the frame of this work, is ^{232}U generation on Th parallel to ^{233}U generation. In this process equilibrium content of ^{232}U in ^{233}U being on the level of $\sim 5 \cdot 10^{-3}$ certainly complicates irradiated fuel treatment till the moment of its returning to the reactor, but makes it practically impossible to use such uranium for its illegal accumulation, manufacturing and storage of explosive devices because of hard γ – quanta emission by ^{208}Pb generated in the radioactive chain of ^{232}U.

3. Utilization of weapon plutonium and uranium and power-plant plutonium in thermal-neutron reactors with thorium fuel.

One of the reasons that constrained implementation of uranium-thorium cycle together with or instead of uranium-plutonium cycle, was deficiency of thermal-neutron-fissionable isotopes ^{235}U or ^{239}Pu in the years of nuclear armament race. At present time the situation has changed absolutely:

- rather large excess of weapon plutonium (~ 100 ton) and uranium (~ 1000 ton) are available;
- tremendous quantity of power-plant plutonium (~ 1000 ton) has been accumulated in irradiated nuclear fuel storages.

At present time the necessity of utilization of these materials and prevention of continuing their accumulation is deemed to be unquestionable. Omitting other ways of solving this problem let's concentrate on considering their burning in thermal-neutron reactors with thorium.

In the course of transferring thermal-neutron reactors onto thorium fuel, highly enriched (weapon) uranium and plutonium, as well as power-plant plutonium can and must be used as initial fissionable materials. Their burn-out by fission reaction will be accompanied by ^{233}U and ^{235}U generation on thorium until ^{233}U and ^{235}U contents reach their equilibrium values. For making it possible to utilize efficiently by fission process not only well fissionable isotopes (^{235}U, ^{239}Pu, ^{241}Pu), but all the other actinides (neptunium, americium, curium, protactinium isotopes) it is advisable to create reactor operation conditions that provide minimum parasitic neutron capture.

Conceptually it can be represented as follows. All neutron-absorbing materials are removed from the reactor, or minimum quantity of them has remained. Let the system's reactivity loss be made up by fissionable materials adding. Adding rate value, on the one hand, determines feasibility of the approach, and on the other hand it characterizes the rate and depth of weapon materials or power-plant plutonium utilization.

Formally, ensuring the critical state maintaining condition (constant value of $K_\infty > 1$) by adding fissionable materials requires introducing a term, describing adding, to the system of joint equations of isotopic kinetics. In this case the system of equations is transformed as follows:

$$\frac{d\tilde{N}_i(t)}{dt}=\frac{dN_i(t)}{dt}+\gamma\,(t)\cdot q_i$$

Where $\frac{dN_i(t)}{dt}$ — is determined in a standard way by nuclei transformation during the process of their interaction with neutrons, in α and β- decays; $\gamma\,(t)-$ is the rate of substance input from an external source: q_i- is the fraction of the $i-$ th isotope in the substance of the external source, normalized using the condition $\sum_k q_k=1$.

In the computer program a calculation variant has been implemented in the above described two-group approximation.

For thermal reactors of RMBK-1000 type the magnitude of K_∞ falls within the range of $1.05<K_\infty<1.15$, the energy distribution of the flow density being characterized by the range of neutron hardness coefficient value $\chi\le 0.1$. Let thorium be used as the main reproducing isotope, and the following isotopes or isotopic mixtures be used as fissionable isotopes by which thorium is enriched:

- highly enriched (weapon) uranium;
- highly enriched (weapon) plutonium;
- power-plant plutonium;
- power-plant uranium from the uranium-thorium fuel.

Here by highly enriched uranium and plutonium we mean ^{235}U and ^{239}Pu; by power-plant plutonium we mean the isotopic mixture of VVER-400 with the following composition: ^{238}Pu-0.8%; ^{239}Pu-63%; ^{240}Pu-22.7%; ^{241}Pu-9.2%; ^{242}Pu-4.3%; by power-plant uranium from uranium-thorium fuel we mean the following mixture of uranium isotopes: ^{232}U-0.3%; ^{233}U-54.2%; ^{234}U- 23%; ^{235}U-5.8%; ^{236}U-16.7%. In the table 1 the estimated values of initial thorium fuel enrichment are represented if $K_\infty=1.05$, that corresponds to $K_{ef}=1$, depending on the type of fissionable isotopes or their mixtures and on neutrons' energy distribution.

Table 1.

The dependence between fuel enrichment (per cent) by fissionable isotopes (or isotopic mixture) and neutron spectrum hardness if $K_\infty=1.05$.

χ	^{235}U	^{239}Pu	Pu power-plant	^{233}U power-plant	^{233}U
0	1.23	0.278	0.41	1.91	1.08
0.05	1.46	0.336	0.51	2.17	1.18
0.1	1.69	0.399	0.63	2.42	1.26

Because of ^{239}Pu fission cross-section uncertainty in the thermal group due to the presence of resonance at $\varepsilon \approx 0.3$ ev, the estimated result of thorium enrichment by ^{239}Pu is characterized by the largest error. Possible K_∞ deviations from 1.05, inherent to real reactors, will influence the value of enrichment but will not change the qualitative pattern.

The following features are characteristic for the table 1. Reaching the critical state (K_∞ =1.05) is ensured by comparatively low enrichment by fissionable isotopes or their mixtures. Neutron spectrum hardening corresponding to hardness coefficient changing in the range of $0 \leq \chi \leq 0.1$ is accompanied by relatively high (~ 30%) increase of enrichment. This latter circumstance favors to ensuring nuclear safety in emergency situations, connected with increase of neutron spectrum hardness in the case of coolant(water) loss. It can also be used for maintaining the system's reactivity by softening its neutron spectrum.

Setting the reactor on a constant power and maintaining the critical state during its operation will require induction of additional quantity of fissionable isotopes. The results of adding rate and isotopic kinetics calculation for different composition of initial fuel and adding materials for a reactor with thermalized neutron spectrum ($\chi = 0$) are represented in the fig 3- fig 8.

In all the pictures reactor's characteristics are shown:

- neutron multiplication factor in an infinite medium K_∞ corresponding to the critical state of a real system;
- neutron hardness coefficient value χ.

The initial fuel composition [kg/ton] and adding composition [%] are given. In the first window of the left column the values of reproductivity coefficient of thermal-neutron-fissionable uranium isotopes ^{233}U and ^{235}U are shown depending on campaign time. In the first window of the right column specific values are shown (per a ton of initial fuel) of fissionable isotope or isotopic mixture adding fed up to the given moment of reactor operation for keeping the reactor in the critical state. In all the other windows specific content [kg/ton] is shown of certain isotopes depending on time. The data of all the figures correspond to the same energy emittance power, corresponding to the depth of heavy nuclei burn-out by fission reaction, that equals to -8.7 kg/ton year.

Using thorium as a reproducing element ensures high reproduction of the ^{233}U fissionable isotope in all the figures. However the delay in ^{233}U formation, stipulated by the half life of ^{233}Pa, that equals to $T_{1/2}(^{233}Pa)$=27 days, determines the time-dependence reproduction ranging from small values in the first days of reactor operation to K_r=0.5 for t=30 days and $K_r \rightarrow 1$ for t>100 days. Some difference in the behavior of K_r depending on use of initial fissionable isotopes and adding fissionable isotopes is stipulated by the difference in the rate of their burn-out in the processes of fission and radiative capture of neutrons and the difference in the magnitude of fission neutrons effective number. Thus, for instance, for χ=0 the value of the sum $\tilde{\sigma}_{nf}+\tilde{\sigma}_{n\gamma}$ for ^{233}U is the minimum value, and effective fission neutron number is the maximum value with respect to ^{235}U and ^{239}Pu and relatively the value of K_r is the maximum value (fig.3) at any t if ^{233}U is used as initial fuel and for adding.

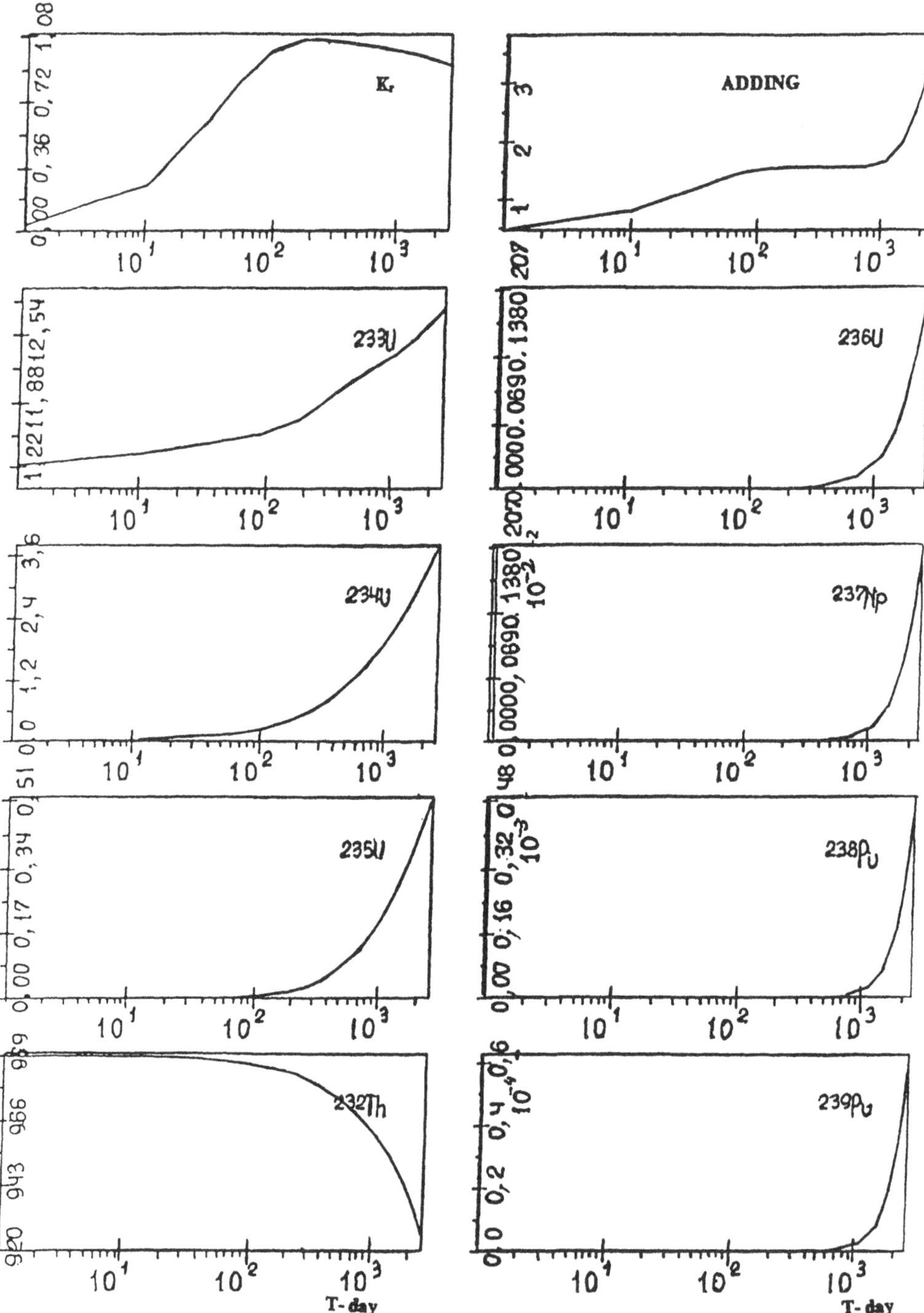

Fig.3. Reactor's characteristics: K_∞=1.05, χ=0. Initial composition of fuel, kg/ton: ^{233}U-10.8, ^{232}Th-989.2. Adding composition, %: ^{233}U-100.

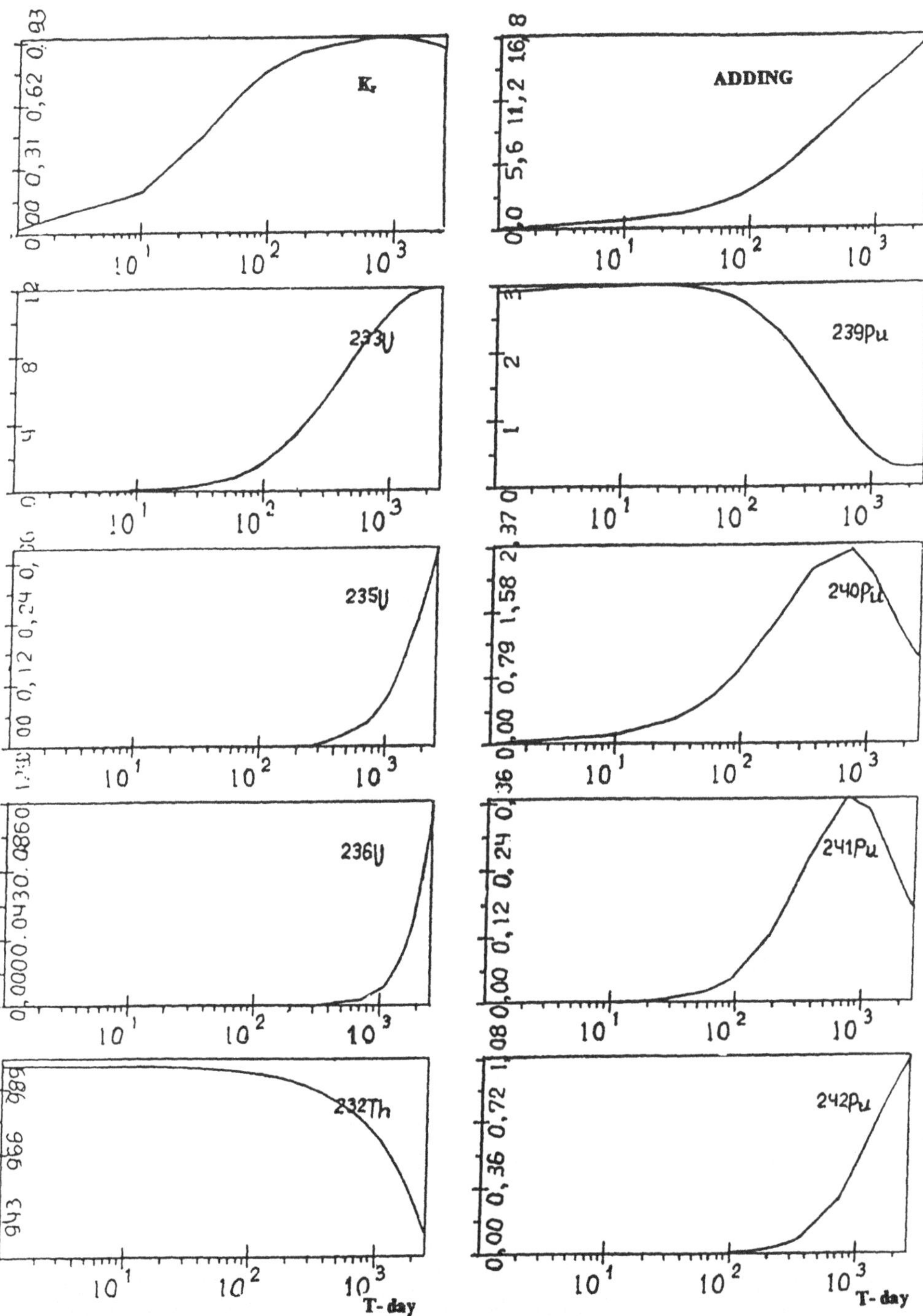

Fig.4. Reactor's characteristics: K_∞=1.05, χ=0. Initial composition of fuel, kg/ton: ^{239}Pu-2.78, ^{232}Th-997.2. Adding composition, %: ^{239}Pu-100.

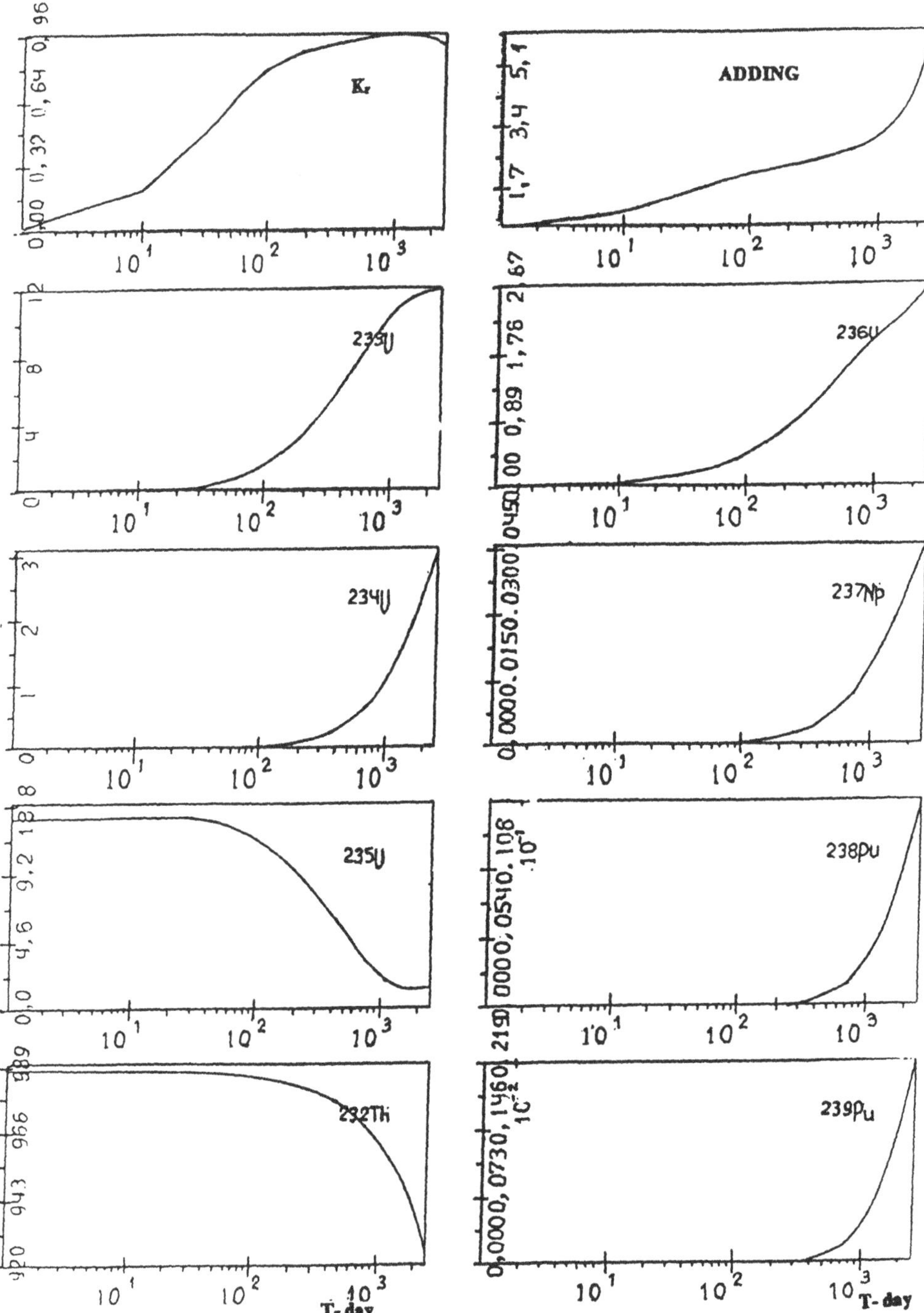

Fig.5. Reactor's characteristics: K_∞=1.05, χ=0. Initial composition of fuel, kg/ton: ^{233}U-12.3, ^{232}Th-987.7. Adding composition, %: ^{233}U-100.

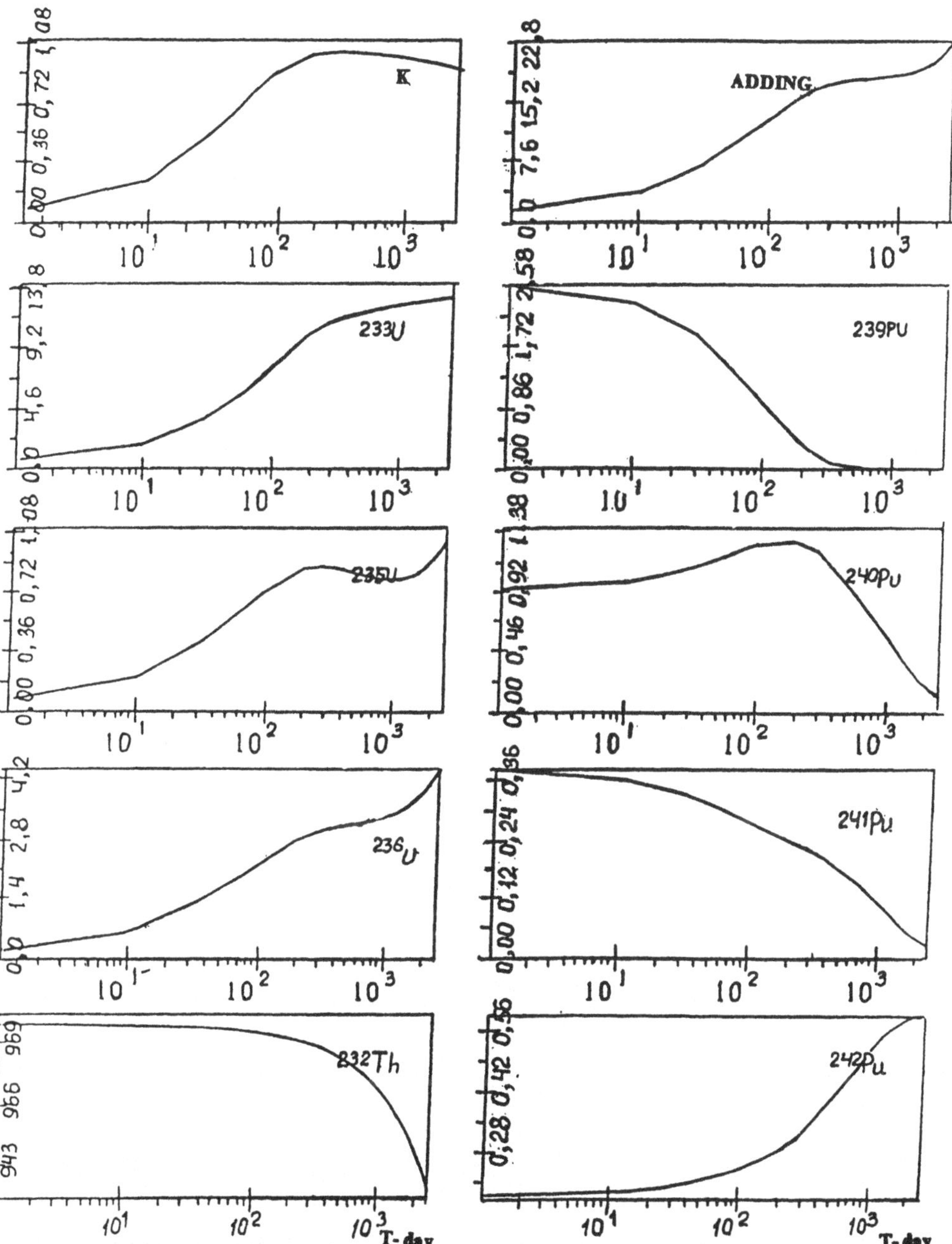

Fig.6. Reactor's characteristics: K_∞=1.05, χ=0. Initial composition of fuel, kg/ton:^{242}Pu-0.177, ^{241}Pu-0.376, ^{240}Pu-0.933, ^{239}Pu-2.59, ^{238}Pu-0.00329, ^{232}Th-996. Adding composition, %: ^{233}U-54.5, ^{234}U-23, ^{235}U-5.8, ^{236}U-16.7.

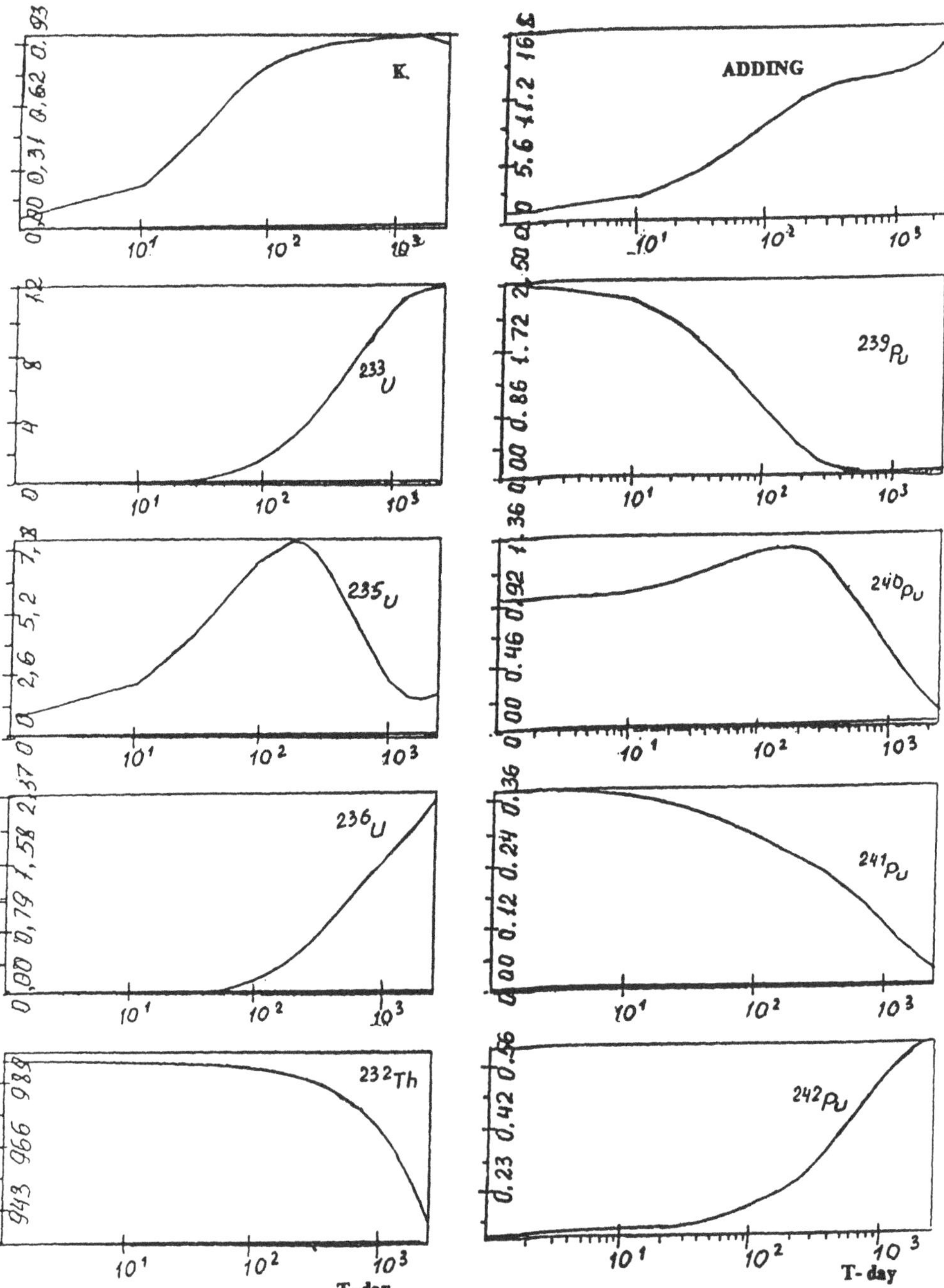

Fig.7. Reactor's characteristics: K_∞=1.05, χ=0. Initial composition of fuel, kg/ton: ^{242}Pu-0.177, ^{241}Pu-0.376, ^{240}Pu-0.933, ^{239}Pu-2.59, ^{238}Pu-0.00329, ^{232}Th-996. Adding composition, %:, ^{235}U-100.

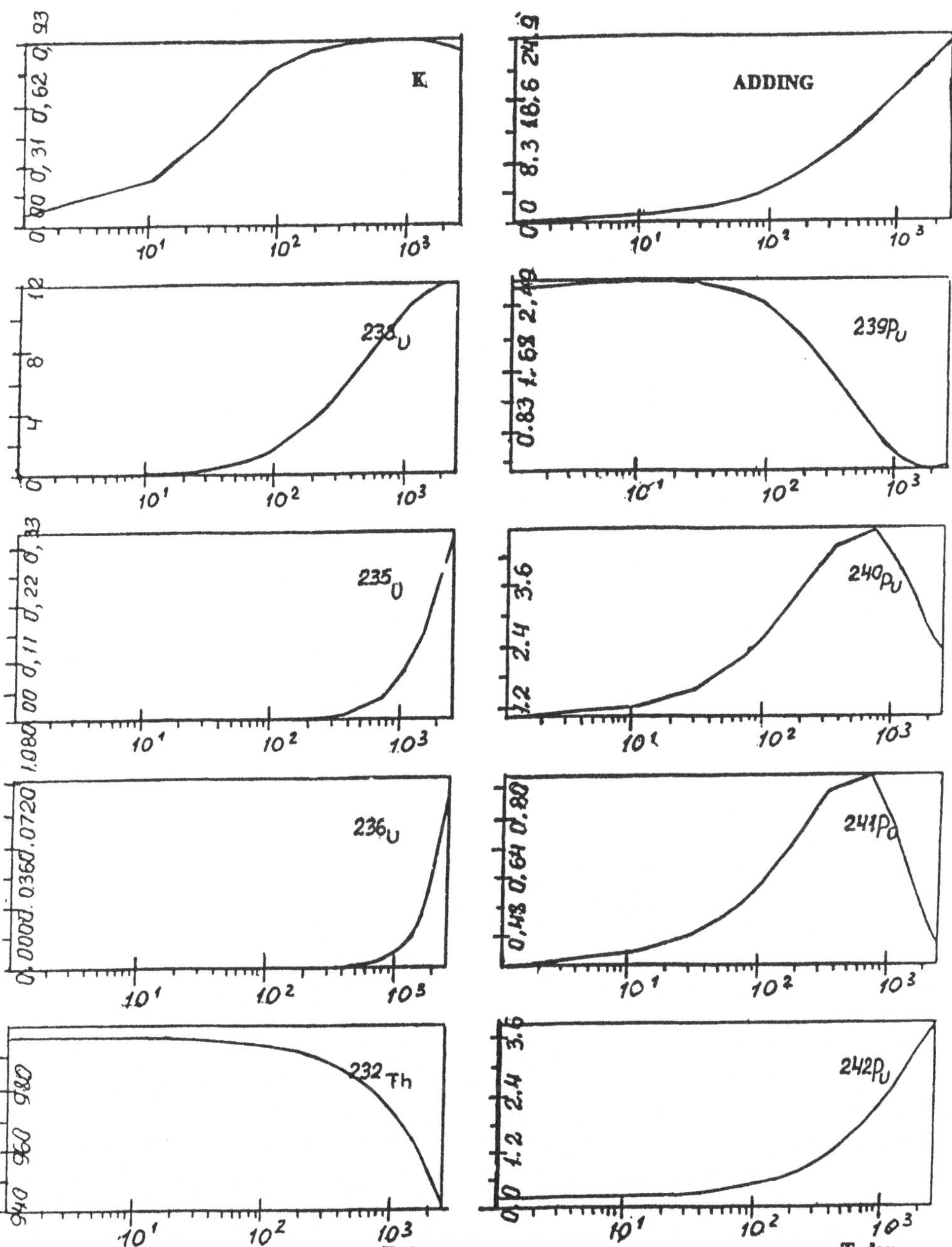

Fig.8. Reactor's characteristics: K_∞=1.05, χ=0. Initial composition of fuel, kg/ton:^{242}Pu-0.177, ^{241}Pu-0.376, ^{240}Pu-0.933, ^{239}Pu-2.59, ^{238}Pu-0.00329, ^{232}Th-996. Adding composition, %:, ^{238}Pu-0.8, ^{239}Pu-63, ^{240}Pu-22.7, ^{241}Pu-9.2, ^{242}Pu-4.3

The adding magnitude $\Delta m(t)$ with different fissionable isotopes of fuel with various composition differ considerably and change qualitatively with time t. For instance, the use of ^{239}Pu (fig.4), having high values of fission and radiative capture cross-sections in thermal group, is accompanied by low concentration of its use in initial fuel and low adding values at the initial moment of time $1 \le T \le 10$ days, then adding rate increases sharply due to its rapid burn-out. Adding (fig.3) the isotope ^{233}U to initial ^{233}U and Th fuel is characterized relatively to the case of plutonium by considerably larger rate (3 times) in the first day. However, by t=20 days the rates become equal, and at large $100 \le T \le 700$ days adding ^{233}U to the ^{233}U, Th-fuel is not required at all because of its high reproduction $K_R > 1$. At the further increase of time $t > 700$ days the adding ^{233}U needs to be recommenced because of decrease of its reproduction due to thorium burn-out and generation of actinides and fission products. The adding to ^{235}U, Th initial fuel of the isotope ^{235}U (fig.5) is characterized by a higher rate at any moment of time t compared to adding the ^{233}U isotope to ^{233}U, Th-fuel. The use of plutonium isotopic mixture (power-plant plutonium) and uranium isotopic mixture (future power-plant uranium) with the ^{239}Pu and ^{233}U as main fissionable isotopes, respectively in the initial fuel and for adding (fig.6-fig.8) is characterized by higher adding rate compared to the variants of using these fissionable isotopes.

Feeding power-plant plutonium with thorium as initial fuel and adding power-plant uranium of future uranium-thorium fuel (fig.6) is seen as a way for burning the accumulated power-plant plutonium after transferring a thermal-neutron reactor on uranium-thorium fuel cycle. At the stage of transition from uranium-plutonium fuel to uranium-thorium one there is a possibility of using power-plant plutonium with thorium as initial fuel and weapon uranium as adding (fig. 7). The quickest way of burning power-plant plutonium is to use it in initial fuel and adding (fig. 8). Comparison of fig.6-fig.8 allows to see the laws and features of these variants of burning power-plant plutonium. Changing fissionable isotopes used for adding during a campaign looks reasonable, which is supposed to be considered later.

In all the above described calculations core refill was supposed with fissionable isotopes or isotopic mixtures without diluting them with reproducing isotope. Practical implementation of refill like that can be performed by adding isotopes dilution with neutron-interaction-inert materials or water solutions. In this case the dilution level will be determined by permissible values of energy emittance and heat-take-off. Another way of practical implementation of core refill with fissionable isotopes is the variant of mobile core with different thorium enrichment by fissionable isotopes, that was used in Shippingport /5/. In this case lower ^{233}U reproductivity coefficient drop will take place, stipulated by thorium burn-out.

The refill level and rate required look tolerable even in the most difficult variant of adding power-plant plutonium (fig.8). It is essential to note that the total refill during the campaign time T_C=7 years amounts to 24.9 kg of power-plant plutonium per a ton of initial fuel specific content of ^{239}Pu isotope per a ton of fuel, however, permanently reducing from the initially introduced value of 2.59 kg/ton to 0.3 kg/ton. Specific content of plutonium isotope ^{240}Pu and ^{241}Pu reaches its maximum value after $T_C > 2$ years and decreases during further irradiation. The specific content of ^{242}Pu grows permanently with the campaign time T_C increase within tolerable limits.

Increase of neutrons' hardness in a reactor is accompanied by deterioration of uranium fissionable isotopes reproduction on thorium, necessity to increase initial fuel enrichment and adding rate in all the variants. An increase of fissionable isotopes specific content takes place at any moment of the campaign. Fissionable isotopes expenditure (initial feed plus adding) grows per unit of energy generation. All this is stipulated by increase of radiative capture cross-section to fission cross-section ratio with the increase of neutrons' hardness for all thermal-neutron-fissionable isotopes ^{233}U, ^{235}U, ^{239}Pu, ^{241}Pu.

Thus, the evaluation made shows that using thorium as reproducing isotope instead of ^{238}U is accompanied by qualitative increase of fissionable isotopes reproductivity coefficient, which allows reactor operation in the least reactivity resource mode. Adding level and rate required for that look feasible. Fuel burning efficiency increase with neutron spectrum softening.

A specific feature of uranium-thorium fuel cycle is the absence of plutonium, americium, curium isotopes generation, that takes place permanently during the use of ^{238}U reproducing isotope in the ordinary uranium-thorium fuel. Due to this fact, highly enriched (weapon) plutonium or power-plant plutonium, used in the initial fuel or for adding, first burn out on initial fuel fissionable isotope neutrons , and then on fission neutrons, reproduced on thorium, of ^{233}U and ^{235}U isotopes. Depending on time-integral neutron flux, by which plutonium (weapon or power-plant) introduced into the core is irradiated, burn-out depth is different. In this irradiation mode ^{241}Pu, generated on ^{239}Pu and ^{240}Pu, is used optimally, for in this case $\beta-$ decay rate of ^{241}Pu and ^{241}Am is negligible compared to its burn-out in the fission reaction.

The above considered weapon and power-plant plutonium burn-out in thermal-neutron reactors in thorium fuel at the stage of transition to uranium-thorium cycle is seen as an efficient transmutation for two reasons:

- burn-out by fission reaction-3/4 of each of ^{239}Pu and ^{241}Pu nuclei;
- transition of the overwhelming part of non-splitted plutonium isotopes into

^{242}Pu, characterized by the largest half life $T_{1/2}$ (^{242}Pu) = $3.8 \cdot 10^5$ years.

One should note that ^{242}Pu is α-disintegrated to ^{238}U, half life of which, $T_{1/2}$ (^{238}U) = $4.5 \cdot 10^9$ years, is particularly large, and which was used as plutonium isotopes reproducing element.

The most efficient way of power-plant plutonium transmutation is the to load it with thorium as initial fuel and to use weapon uranium as adding (fig. 7) or future power-plant uranium (fig. 6) diluted with thorium. In this case power-plant plutonium will be subject to irradiation by largest neutron flux during one campaign. In the fig. 6 and fig. 7 burn-out is illustrated of induced into initial fuel 2.59 kg - ^{239}Pu; 0.93 kg - ^{240}Pu; 0.38 kg - ^{241}Pu; 0.180 kg - ^{242}Pu versus campaign time in a fully thermalized spectrum. After the campaign of T_C=3 years this composition changes to the following: 1.62 g - ^{239}Pu; 500 g - ^{240}Pu; 97 g - ^{241}Pu and 530 g - ^{242}Pu, and after the campaign of T_C=7 years it will be 2.8 g - ^{239}Pu; 100 g - ^{240}Pu; 20 g - ^{241}Pu and 600 g - ^{242}Pu. In the neutronic spectrum with hardness coefficient χ=0.05 if 3.23 kg - ^{239}Pu; 1.16 kg - ^{240}Pu; 0.47 kg - ^{241}Pu and 0.22 kg - ^{242}Pu are fed, their content after the campaign with T_C=3 years will be 4 g - ^{239}Pu; 450 g - ^{240}Pu; 160 g - ^{241}Pu and 560 g - ^{242}Pu, and after the campaign time of T_C=7 years it will be 13 g - ^{239}Pu; 74 g - ^{240}Pu; 24 g - ^{241}Pu and 450 g - ^{242}Pu. In the neutronic spectrum with hardness coefficient

χ=0.1 after the initial feed of 3.96 kg - ^{239}Pu; 1.43 kg - ^{240}Pu; 0.578 kg - ^{241}Pu; 0.27 kg - ^{242}Pu their content after campaign of T_C=3 years is 12 g - ^{239}Pu; 460 g - ^{240}Pu; 25 g - ^{241}Pu and 610 g - ^{242}Pu, and after the campaign T_C=7 years it is 25 g ^{239}Pu; 74 g - ^{240}Pu; 74 g - ^{241}Pu and 400 g - ^{242}Pu. Thus one can see that after the campaign time T_C=3 years the quantity of ^{239}Pu decreases by more than 100 times, quantity of ^{240}Pu decreases by approximately 20 times, quantity of ^{241}Pu decreases by approximately an order, and the quantity of ^{242}Pu increases by approximately 2÷3 times. Such deep burn-out of well-fissionable isotopes ^{239}Pu and ^{241}Pu with ^{242}Pu generation makes power-plant plutonium absolutely unfit for manufacturing explosive devices, qualitatively simplifying the problems of treating it. Thus, in thermal-neutron reactor (of RBMK-1000 type) during a campaign about 1000 kg of power-plant plutonium can undergo deep processing.

In the fig.8 isotopic kinetics has been illustrated in the case of using power-plant plutonium both as initial fuel and for adding. Such scenario of burning power-plant plutonium is characterized by involvement of its considerably larger quantity during the campaign time, but smaller burn-out depth of fissionable isotopes ^{239}Pu and ^{241}Pu. In this case together with the initial feed of 1-1.2 ton of power-plant plutonium, 4.4÷4.8 ton of the same plutonium will be used as adding during a 3-year-long campaign and about 7 ton during a 7-year-long campaign.

The situation with weapon plutonium use both in initial fuel and for adding is similar, calculated results for which are illustrated in the fig.4.

The fig.5 illustrates behavior of K_R, adding and accumulation of the main isotopes versus campaign time, highly enriched weapon uranium being used in initial fuel and for adding. It is this variant that we have suggested before /4/ as weapon uranium conversion for transferring thermal-neutron reactor on the closed uranium-thorium cycle. Its efficiency is enhanced during adding realization:

- low enrichment of initial fuel;
- high reproductivity coefficient of ^{233}U;
- low adding level and rate;
- small outcome of non-uranium actinides (^{237}Np, ^{238}Pu, ^{239}Pu).

Fig.3 shows the results of calculation for the variant with the use of ^{233}U in initial fuel and for adding. One can see that use of ^{233}U instead of ^{235}U improves all the parameters. The use of closed uranium-thorium cycle of power-plant uranium instead of ^{233}U in a real situation both in initial feed and for adding will not change the essence of the dependencies shown in the fig.3.

Comparison of all the variants fig. 3- fig.8 considered shows that in reactor systems with thorium and the most soft neutronic spectra the most attractive way is to use ^{233}U both for initial feed and adding, which is the main variant in the closed uranium-thorium cycle. The use of highly enriched (weapon) uranium is efficient in transferring thermal reactors on uranium-thorium fuel. For deep transmutation of power-plant and maybe weapon plutonium they should be used for initial feed together with thorium. Transmutation efficiency increasing resources are: increasing the time of campaign and added fuel composition change during the campaign. Both trends are supposed to be considered in future.

Conclusion.

At the end let's emphasize the following points.

1. The disarmament process and creation of large excessive amounts of weapon fissionable materials (100 ton of plutonium and 1000 ton of uranium) aggravated the problem of fissionable materials and nuclear technologies non-proliferation.

2. Accumulation of weapon materials and particularly power-plant plutonium, inherent to uranium-plutonium fuel cycle, stipulates poor prospects of fissionable materials recording and monitoring as a way of solving their non-proliferation problem.

3. The absence of uranium fissionable isotopes (^{233}U, ^{235}U) surplus in the closed uranium-thorium cycle, their natural denaturalization by uranium isotopes (^{234}U, ^{236}U), as well as unfitness of uranium mixture for manufacturing nuclear weapons due to high radiological danger of disintegration of ^{232}U, generated on thorium, ensure technological banning for use of fuel materials for manufacturing explosive devices.

4. Fissionable materials accumulated (weapon and power-plant) are naturally and efficiently utilized in the uranium-thorium fuel cycle.

References

1. Holdren D, Echirn D. Bdnits R. and others. Management and disposition excess weapons plutonium. National Academy Press. Washington, 1994.

2. V.E.Marshalkin, V.M Povyshev, Yu.A.Trurnev Isotopic kinetics of recycled uranium fuel in the closed uranium-thorium cycle. VANT Theoretical and Applied Physics series 1995, issue 3/3, pp 87-94.

3. Venneri. ATW Non-Aqueous System. Joint Russian-U.S.Seminar on Transmutation. Moscow, Russia November 10-13, 1992.

4. Andryushin I.A., Anisin V.I., Marshalkin V.E., Povyshev V.M., Trutnev Yu.A., Tchernyshev A.K. Comparison of U-Pu and U-Th fuel cycles for thermal reactors. Russian-American Nuclear Power Safety Workshop. Albuquerque. June 14-18, 1993.

5. Freeman L.B., Beadoin B.R., Frederickson R. et al. Physics Experiments and Lifetime Performance of the Light Water Breeder Reactor//Nucl. Sc.Eng.1989 Vol. 102 P 341-364.

Destruction of Plutonium and Other Nuclear Waste Materials Using the Accelerator-Driven Transmutation of Waste Concept

F. Venneri, Los Alamos National Laboratory, USA

Each large nuclear power plant produces about 300 kilograms of actinides and about 120 kilograms of long-lived fission product wastes per year, with major constituents in terms of potentially dangerous radiation. The plutonium is a basic end-product of the reactor cycle using uranium fuel, and is considered especially troublesome in terms of its potential damage to humans either directly or by clandestine conversion to nuclear weapons.

Neutron bombardement can be used to convert these waste nuclei to benign or short-lived forms. The challenge is to generate the proper form of neutrons in a workable system. A reactor is efficient for making uranium fission but leads to waste accumulation with no neutrons left over to use for waste modification. A key, therefore, is to introduce more neutrons from another source. This can be done by using a modern particle accelerator.

The process and system, now called Accelerator-Driven Transmutation Technology (ADTT), is presently unique in having the potential for burning all types of nuclear waste, and for providing an energy supply bridge for millennia, to a time when practical control of fusion may be possible. The ADTT plant configuration proposed by the Los Alamos National Laboratory is shown in Figure 1.

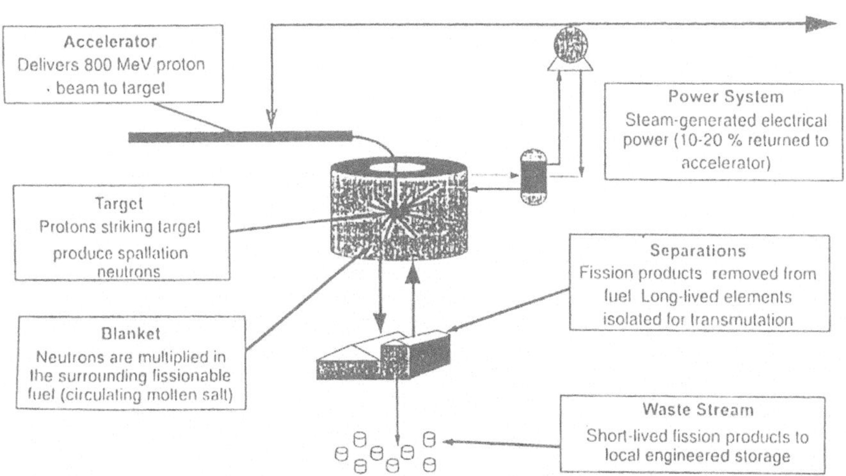

Fig. 1: Components of an ADTT system. ADTT is a closed nuclear system

E. R. Merz and C. E. Walter (eds.), Advanced Nuclear Systems Consuming Excess Plutonium, 259–263.
© 1997 *Kluwer Academic Publishers.*

More details of the target/blanket transmutation system are shown in Figure 2.

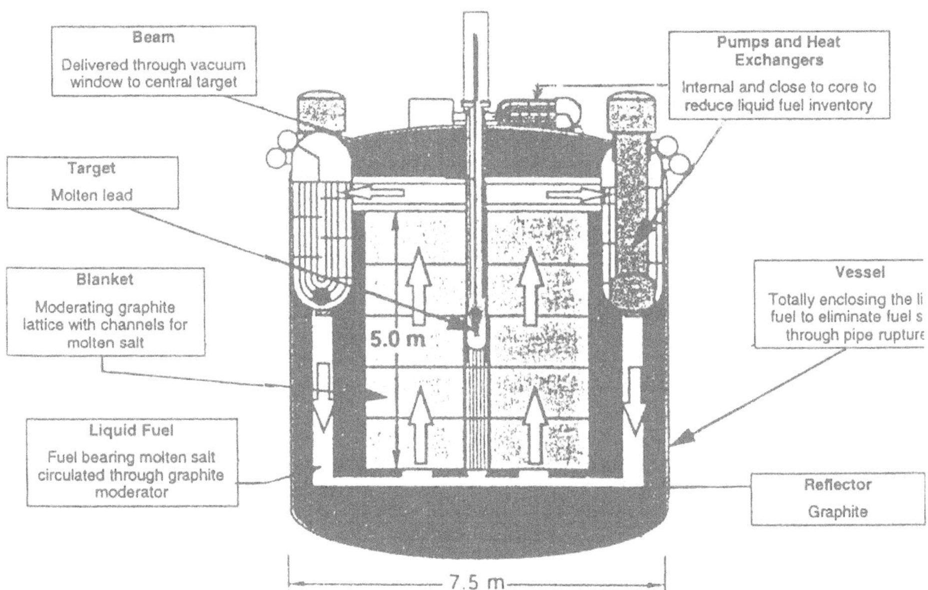

Fig. 2: The prototype ADTT molten-salt blanket. This blanket can be used for all ADTT applications. Graphite-moderated, with the waste and/or fuel dissolved in molten salt, the ADTT blanket improves on the 1960s Oak Ridge molten salt reactor concept, also incorporating many improvements introduced by newer liquid-metal reactor "pool" designs. Typical power is 500-1000 MWt.

The ability of modern accelerators to deliver large amounts of beam current for conversion into neutrons means that it is not at all necessary to maintain a critical chain reaction in the burner. Instead, one can use each linac-generated neutron to produce a multiplied number of neutrons (about 20 from the combined target spallation process and fissions of the waste or fuel nuclei) that results in a fast waste burning rate but is kept far short of the level where the chain reaction would be self-sustaining (Typically a factor of 20, or an effective criticality (k_{eff}) of 0.95).

ADTT gives advantages in the way the burn rate is achieved. The accelerator drive affords a large neutron supply even with a subcritical system. High-energy neutrons would be best to break apart the actinide wastes into fission products, and some projects are concentrating on actinide burnup using an accelerator-driven system. However, fast neutrons are ineffective for burning the long-lived fission products. Bombardment by thermal neutrons is best for burning fission products, and is also quite adequate for burning the actinide wastes. By tailoring the neutron energy toward thermal, ADTT gains large factors in the probability that the neutrons will produce the desired transmutation of both actinides and fission products.

During the burning, one will continuously process part of the molten salt solution to separate and remove fission products. There are three main types of fission products, and one would use a different physical separation process for each. Volatile fission products and noble gases have low boiling points and tend to come out of the molten salt very easily. Helium gas is bubbled through the solution to collect, or sparge, away these by-products in a fast process with a removal cycle of about one minute. Noble and semi-noble metal by-products have low solubility in the molten salt, and can be plated out in passive extraction cells, on a time scale measured in hours. The remainder of the fission products, including the rare earths, have good solubility in the salt and will tend to build up in concentration.

An actinide separation goal of 1000 : 1 to 10,000 : 1 is sought. The actinides and long-lived fission products remain in the main molten salt stream and are returned to the furnace. The stable and short-lived fission products are separated and further concentrated to reduce the volume, and sent to managed storage facilities. It is emphasized that the ADTT back-end processing only requires fission product separation from the fuel, not actinide isolation and extraction.

Using the Thorium Cycle for Excess Plutonium Destruction

The present nuclear reactor power plants use uranium fuel to produce plutonium as a major waste. The cycle is not closed, i.e., there is no method at present to destroy the waste. Figure 3 shows the world inventory of plutonium, now about 1,000 metric tonnes, of which about 90% is from commercial reactors and about 10% from weapons.

The curves show the rate at which plutonium will be accumulated using the present light-water-reactor (LWR) system, at a constant power demand. With a deeper burn of the fuel pins, the slope can be reduced somewhat. LWRs with reprocessing would bring the plutonium inventory down at first but eventually it would rise again. Using liquid metal reactors, the accumulation can be brought to a slow accumulation rate or an asymototic level of some 1,500 tonnes. With an ADTT waste-burnup and thorium fuel cycle, the steady-state world plutonium inventory could be reduced to about 200 tonnes. If the ADTT system were to be eventually phased out for fusion, it would first burn the wastes from all plants.

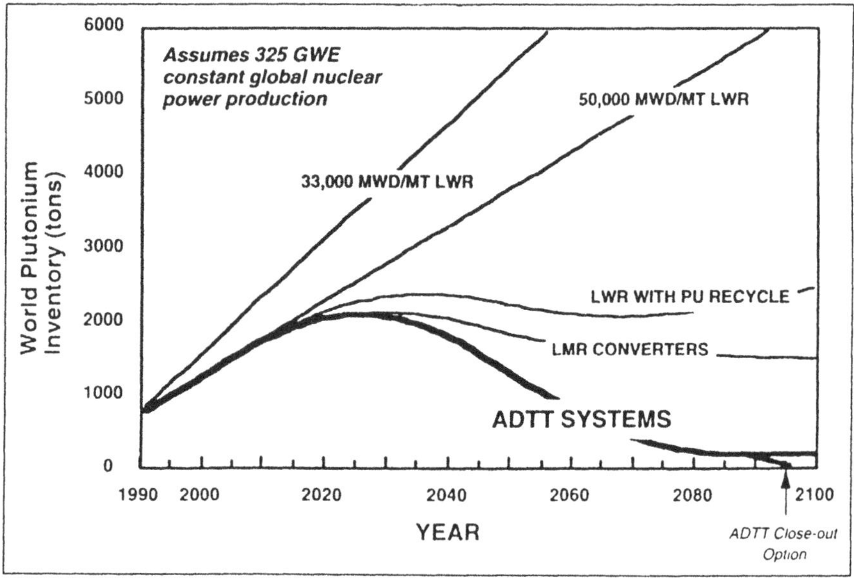

Fig. 3: The impact of ADTT on global plutonium inventory

ADTT Impact on Repository Storage

Waste that is destroyed has no risk and does not need to be stored. The accumulation of plutonium from the LWR U-Pu cycle would require a large new deep-underground repository to be opened about once every ten years. If ADTT would destroy long-lived wastes, the need for repositories would eventually become small or non-existent. The potential of ADTT for waste elimination indicates that temporary repository storage should place the wastes in recoverable form, so they can be recovered later and burned.

One can use the existing regulations for managed, near-surface-level storage of radioactive wastes to establish the burnup goals for ADTT. Waste termed Class C is the most difficult waste as defined by present regulations. It must be stabilized in a material like concrete and buried, but need not be more than five meters underground, and after 500 years, must be completely safe to humans. Figure 4 shows that a separation factor of 10^{-4} for actinide wastes would leave remnants considerably below the Class C regulations. Figure 5 shows that separation factors of only 10-300 are required for the long-lived fission products, which are reasonable goals for the methods outlined above. If these goals can be met, there would be no requirement to place remnant waste in deep geologic storage.

sotope	Annual Production (Atoms/Year)	Half life (Years)	Decay Rate (Curies)	Reduction by Separation[c] (Curies)	Concentration After Stabilization[d] (Nanocuries/gram)	Class C Decay Rate Limit[a] (Nanocuries/gram)
^{38}Pu	1.13×10^{25}	88	7.5×10^4	7.5	68	20,000[b]
^{39}Pu	41.6×10^{25}	24,100	1.0×10^4	1.0	9.1	100
^{40}Pu	19.2×10^{25}	6,560	1.7×10^4	1.7	15.	100
^{41}Pu	6.4×10^{25}	14.4	2.7×10^6	270.0	2454.5	3500
^{42}Pu	3.9×10^{25}	375,000	6.1×10^1	.006	0.05	100
^{37}Np	3.7×10^{25}	2,140,000	1.1×10^1	.001	0.009	100
^{41}Am	4.1×10^{25}	433	5.5×10^4	5.5	50.	3500[b]
^{43}Am	0.73×10^{25}	7,370	5.9×10^2	.059	0.54	100

a. Used decay limits from 10 CFR 61.55
b. Used decay limit for parent from 10 CFR 61.55
c. For a separation factor of 10,000
d. Stabilization with 50 m^3 of concrete

Fig. 4: Separation factor of 10^{-4} for actinide wastes would leave remnants below Class C regulations

Isotope	Annual Production (Atoms/Year) $x10^{25}$	Half life (Years)	Decay Rate (Curies)	Separation factor and decay rate (Curies)	Concentration After Stabilization[a] (Curies/m^3)	Class C Decay Rate Limit (Curies/m^3)
^{79}Se	0.13	65,000	11.9	10 / 1.2	0.024	0.2
^{90}Sr	9.0	29.1	1,860,000	10 / 190,000	3800	7000
^{93}Zr	15	1,500,000	60	1 / 60	1.2	2.0
^{99}Tc	15	213,000	426	100 / 4.26	0.85	3.0
^{107}Pd	4.1	6,500,000	3.8	1 / 3.8	0.076	2.0
^{126}Sn	0.46	100,000	27	10 / 2.7	0.054	0.2
^{129}I	2.7	15,700,000	1.0	10 / 0.1	0.002	0.08
^{135}Cs	4.2	2,300,000	8.4	100 / 0.084	0.0017	0.08
^{137}Cs	14	30.2	2,800,000	100 / 28,000	560	4600
^{151}Sm	0.16	90	11,000	300 / 37	0.74	2.0
^{85}Kr	1.0	10.7	560,000	1 / 560,000		

a. Stabilized with 50 m^3 of concrete

Fig. 5: Separation factors of 10-300 leave fission product remnants below Class C standards

ENVIRONMENTAL & SAFETY PROBLEMS IN PU UTILIZATION & POWER GENERATION

L.A.Bolshov, R.V.Arutyunyan, V.P.Kiselev,
B.P.Maksimenko, V.N.Nosatov, O.A.Pavlovsky
(IBRAE RAS, Moscow, Russia)

1. As to the isotopic composition, the available plutonium can be divided into two kinds: weapon plutonium (93% Pu-239) and reactor plutonium (obtained in conventional reactors) which is a mixture of different isotopes.

In view of the nuclear disarmement underway, the question arises on further fate of weapon plutonium excess, a considerable portion of which must be eliminated by the year 2000. Any solution of the problem must meet the requirements of nuclear weapon non-proliferation, assurance of population health and safety, and environmental protection. The above means that of top priority is conversion of weapon plutonium into such forms that would reliably preclude its use for military purposes. One of such forms is spent nuclear fuel. There exist two technical ways to achieve the so-called "spent fuel standard".

2. The alternative known as the choice in favor of reactors is based on conversion of weapon plutonium into mixed oxide fuel (MOX fuel) to be burnt in thermal reactors. Here, plutonium is a valuable kind of energy resources.

The MOX fuel obtained by mixing reactor plutonium with natural or enriched uranium has been used in Western Europe for a long time [1]. Over 300 tons of MOX fuel have been produced, which allowed to utilize over 15 tons of the reactor plutoium accumulated in light water reactors. By now, 30 PWRs and BWRs have been licensed for utilization of MOX fuel and 15 reactors are being constantly loaded with such fuel.

Some experience in utilization of plutonium and MOX fuel has been accumulated in Russia as well. Two cores made of weapon plutonium have been investigated at a BR-10 experimental reactor. A large number of control rods fabricated of a mixture of uranium oxide and plutonium oxide (for a few isotopic compositions of weapon plutonium) have been employed at a BOR-60 testing reactor. Reactor tests of MOX fuel (350 kg of weapon plutonium) have been performed at a BN-350 fast breeder. By now, over 2,000 fuel rods have been made and they are now under testing at BN-350 and BN-600 reactors. 2,300 kg of plutonium as the initial charge and 1,600 kg to support its operation for a year are supposed to be utilized in the BN-800 reactor being developed now. The program of recycling the MOX fuel intended for fast reactors is now at the stage of consideration.

Among the efforts already accomplished, the following should be mentioned:

- the RT-1 plant for reprocessing of irradiated fuel from VVER-440 and BN reactors is currently operating;
- about 50% of the work on construction of a plant for production of fuel rods for the BN-800 reactor ("MAYAK" Association) is completed;
- an experimental plant for fabrication of mixed uranium-plutonium fuel for BN reactors has been put into operation;
- an experimental plant for production of plutonium fuel, using the vibration technology, has been put into operation (Scientific Research Institute of Atomic Reactors, NIIAR).

The principal advantages of MOX fuel are as follows:

- from the engineering viewpoint, the utilization is well elaborated and readily realizable;
- short (3 to 5 years) terms required to construct plants for MOX fuel production;
- a high control and safety level is ensured by the fuel production technology itself;
- weapon plutonium is converted into reactor plutonium already protected (i.e., it is of no value from the military viewpoint);
- spent MOX fuel can be reprocessed similarly to conventional fuel unloaded from a light water reactor, which presents no additional organizational problems).

As applied to new conditions here, in Russia, it is expedient to consider three directions of plutonium utilization in reactors:

- conventional BN-800 with a breeding factor of 1. The core consists of PuO_2+UO_2 mixture, screened by UO_2;

E. R. Merz and C. E. Walter (eds.), Advanced Nuclear Systems Consuming Excess Plutonium, 265–273.
© 1997 *Kluwer Academic Publishers.*

- BN-800 with a new core on the basis of ceramic fuel with dissolved Zr or Th, e.g., PuO_2 screened by metal Th. This version has been evolved in order to utilize Pu combined with reprocessed uranium-233 in the future;
- VVER-1000 or VVER-500 with a ceramic fuel composition prepared from a mixture of plutonium oxide and natural-uranium oxide in zirconium (PuO_2 + UO_2 + Zr).

3. Another alternative, namely, the choice of in favor of vitrification, implies mixing of plutonium with high-level waste, followed by vitrification and burial of the mixture. Under this approach, plutonium is simply considered “garbage”, which is unjustified wastefulness from the economic standpoint.

Today, plutonium vitrification and storage is only a concept. It should be supported by some experience of industrial application at the same safety level as that of MOX fuel handling.

Apparently, the expenses for vitrification and storage will depend primarily on the technical condition of plutonium in the final product (glass), dimensions of blocks containing vitrified plutonium, and the permissible ratio of plutonium and high-level waste. At the same time, the content of plutonium in a glass block will be limited by such factors as solubility, criticality, concentration, protection against dispersion.

4. The reactor technology of weapon plutonium conversion can be realized in a shorter time as compared to the vitrification-based one. Additionally, is is more efficient from the viewpoint of preventing reuse of plutonium for military purposes. It is due to the fact that the isotopic composition of plutonium changes and diminishes in the case of fuel burn-up as distinct from the vitrification-based technology.

According to the data of the *Uranium Institute*, the total amount of weapon plutonium in the former USSR is 105 to 130 tons. Specialists of NIKIET have estimated the energy potential of 120 tons of plutonium at 1,000 to 8,000 TW-hr of electric power, depending on the power production technology employed (open or closed cycle thermal reactors with a breeding factor of 1.00-1.05). In Tables 1 and 2, some estimates concerning the possibility of plutonium utilization in Russian reactors are presented. It can be readily seen that the most efficient plutonium “incinerators” are fast reactors, followed by VVER-1000 on mixed fuel.

Table 1. Annual Balance of General Plutonium Isotopes. Open Fuel Cycle, kg/[GW(el)·year]

	(A) VVER-1000 UO_2	(B) VVER-1000 PuO_2+UO_2+Zr	(C) BN-800 PuO_2+UO_2	(D) BN-800 PuO_2+Th
Input:				
^{239}Pu	–	764.2	1910	2421
^{240}Pu	–	49.4	124	94
^{241}Pu	–	4.8	12	7.5
^{242}Pu	–	0.3	0.7	0.3
Pu	–	818	2047	2522
Output:				
^{239}Pu	124	130	1798	1690
^{240}Pu	55.2	144.2	262	186
^{241}Pu	30.9	73.2	22.8	12.5
^{242}Pu	12	37.2	2	0.8
Pu	222	384	2085	1889+770kg ^{223}U
Output – Input:				
^{239}Pu	124	-634	-112	-731
^{240}Pu	55.2	95	139	92
^{241}Pu	30.9	68.4	10.8	5
^{242}Pu	12	36.8	1.3	0.5
Pu	222	-434	39	-633+770kg ^{223}U

Table 2. The Possibility of Utilization of Plutonium Ii Russian Reactors
(Data Presented by Physics and Energetics Institute from the Report
"Engineering Analysis of Production of Uranium-Plutonium Fuel from Weapon Plutonium
and its Possible Utilization in Nuclear Energetics",
Minatom of Russia - Siemens - GRS. 1995)

Reactor	Loading Pu, kg/yr	Yield Pu, kg/yr	Balance, kg/yr
1 VVER-1000	0	223	+223
2 VVER-1000	254	308	+54
3 VVER-1000	364	395	+31
4 BN-600	1141	1053	-88
5 BN-800	1637	1508	-129

The above suggests that utilization of plutonium as nuclear fuel will seemingly become the principal direction of weapon plutonium conversion. In this connection, some research has been carried out at IBRAE RAS into estimation of possible radiological consequences of a major NPP accident with enriched Pu-239 fuel core melting. The research included:

- model calculation of the possible activity of Pu-239 release into atmosphere in the case of an accident at a VVER-1000 type reactor using enriched Pu-239 fuel;
- calculation of the density of territory contamination by plutonium isotopes as a result of such an accident;
- calculation of exposure levels to personnel and population as a result of the Pu-239 release.

As the information base, the following data were employed:

- factual data on levels of contamination of environmental bodies by plutonium isotopes after the Chernobyl accident;
- experimental data on the content of plutonium isotopes in organisms of people in the areas adjacent to the Chernobyl NPP and the plants for plutonium reprocessing at the "MAYAK" Association in the Chelyabinsk Region;
- up-to-date dosimetric estimates on impact of plutonium isotopes upon human organism.

5. To calculate the effects of major accidents at NPPs with water-water reactors, a modified "MELCOR" computer code is applied at IBRAE RAS, which takes into account the design peculiarities of domestic reactor facilities [2]. In Fig.1, the structural layout of a VVER-1000 reactor, which is employed in calculations using the "MELCOR" code, is displayed.

When developing the basic scenario of a severe accident at a VVER-1000 reactor with enriched Pu-239 fuel, the sequence of events leading to the most grave consequences was chosen. A rupture of the main circulation pipeline (within its hot portion with an equivalent diameter of 850 mm) in one of four loops, with a two-way leakage of the coolant, was taken as the primary event. Additionally, the availability of "hidden" defects in plant equipment was postulated, which caused a failure of safety systems, namely, full loss of electric power supply from both internal and external sources and, hence, a failure of the ECCS's active part, as well as a failure of the sprinkler system of vapor condensation in the containment.

It was assumed in the calculation that the fuel charge as of the time of the accident was 75,000 kg of UO_2 and 1,500 kg of PuO_2. The fission products were assumed to have been released under the containment and then into the environment through the designed seepage in the containment. The sequence of principal events of the accident, evolved on the basis of our estimates, is presented in Table 3.

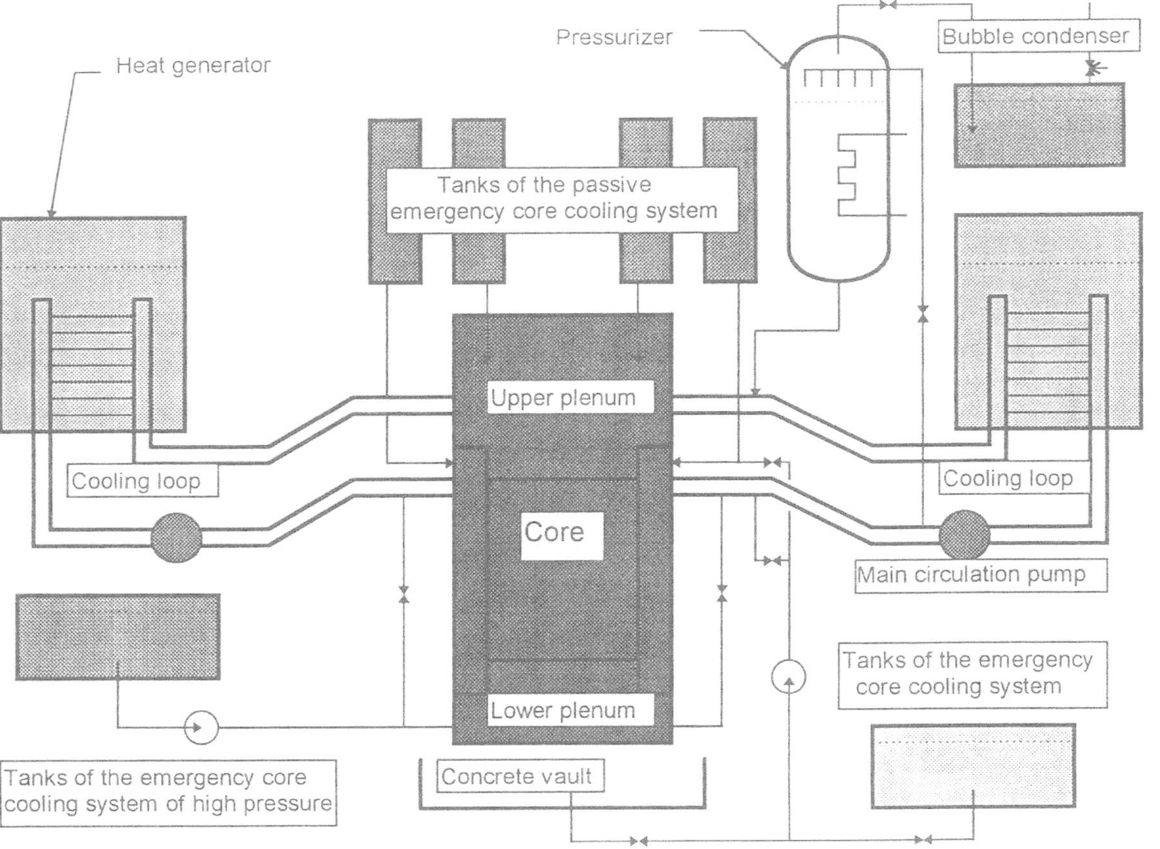

Fig.1. VVER-1000 nodalisation for MELCOR calculation

Table 3. Sequence of principal events of the accident process

Event description	Time from calculation start, sec
Initial event — failure of the main circulation pipeline	10.0
Response of the emergency protection system; start of MCP rundown	12.0
Connection of ECCS hydroreservoirs; delivery of boron solution to the reactor	15.0
Start of reactor core's drying and heating-up	17.0
Start of vapor-zirconium reaction with hydrogen release in the core	20.0
Seal failure in fuel elements and start of fission products release into atmosphere	26.0
ECCS hydroreservoirs are empty	174.0
Start of fuel-element cladding melting	347.0
Full evaporation of water in the reactor and start of heating of core debris at the bottom	1300.0
Melting through the reactor vessel bottom and fallout of debris to the reactor shaft	1660.0
Full evaporation of water in the reactor shaft	25100.0

Upon a failure in one of loops of the main circulation pipeline, an intense outflow of the coolant (first the water and then, after the level drops below the edge of the emergency pipeline, the vapor) into the reactor shaft starts. The process is attended by a sharp drop in pressure and level of the coolant in the first circuit and as early as in 2 seconds from the accident start the emergency protection system comes into action. The reactor is shut down and the rundown of the MCP is initiated. Further lowering of the pressure must result in connection of high-pressure ECCS and delivery of boron solution into the reactor. However, because of the availability of the afore-mentioned "hidden" defects the connection does not take place and the passive-type ECCS hydroreservoirs are connected after the pressure drops down to 5.9 MPa. Operation of these hydroreservoirs reduces only slightly the rate of level decrease in the reactor and dessication of the core starts as early as at the 7th second. The water evaporates in the lower, flooded, portion of the core and, in vapor form, comes into the reactor shaft and then to the spaces adjacent to the shaft. Out of more than 220 tons of water in the first circuit, about 125 tons pass to the shaft and the rest in vapor form extends over other rooms of the reactor building.

The power removed from the dried portion of the core is insufficient for efficient cooling of the fuel elements and the temperature of upper portions of the elements reaches the lower threshold of the vapor-zirconium reaction (about 1,000 K) at the 10th second after the accident starts. Liberation of hydrogen begins, which spreads over all rooms of the plant. The reaction is attended by an intense heat release, which serves additionally to further heating of the core. In 16 seconds after the accident starts, when the temperature of fuel-element claddings reaches about 1,300 K, the elements undergo depressurization and gaseous fission products leave the inter-element gap for the atmosphere. Later, releases of volatile fission products from the solid-state matrix become more and more intense, which decreases essentially the share of residual energy release in fuel. By the 337th second from the accident beginning, the temperature of fuel elements reaches the zirconium melting point and core destruction occurs. At first, the cells of the second zone (with the highest level of energy density) and a few seconds later the central zone is destroyed. As this takes place, unmelted fuel pellets also fall out to the reactor vessel bottom. The water still remaining there boils up sharply and cools the debris fallen out. By that time, the amount of water in the first circuit is sufficient for flooding the debris at the reactor bottom. However, evaporation of the water becomes more intense and in about 1,300 sec after the accident start the water is completely absent from the reactor vessel. The debris of two central zones heat the bottom and in 1,660 sec the bottom is melted through. The melt penetrates the reactor shaft, and both intense evaporation of water in the shaft and disintegration of the concrete under a high temperature start. In the reactor vessel, further destruction of core remnants occurs, and by the end of calculation less than 5% of its mass is still unmelted. Water in the reactor shaft is completely evaporated in 25,000 sec, i.e., within about 7 hours.

As an illustration to thermohydraulic processes taking place at the given accident scenario, in Fig.2 the time dependences of the pressure inside the VVER-1000 containment and the mass of UO_2 released into the environment are displayed.

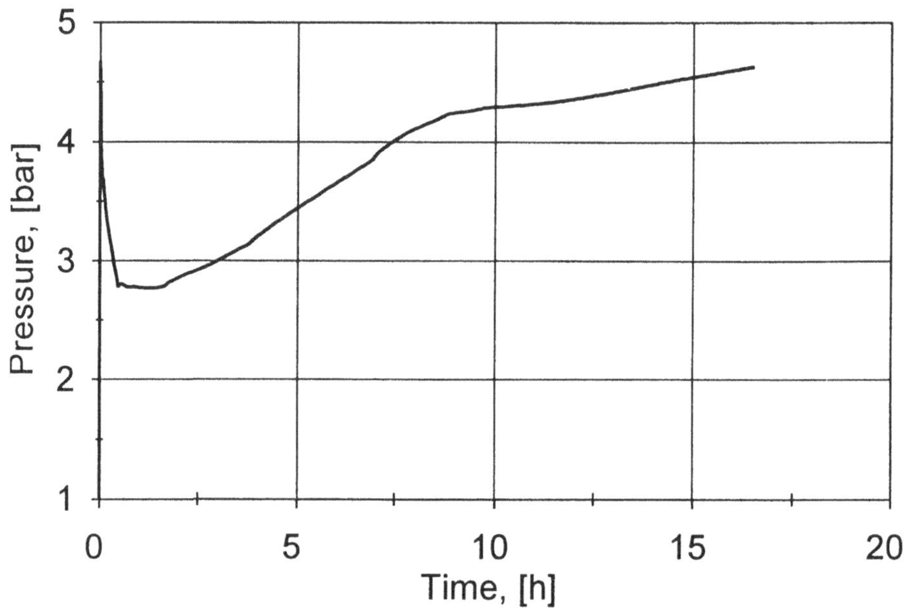
Pressure in VVER-1000 containment
in severe accident with core melting
Pressure, [bar]
Time, [h]
5
4
3
2
1
0
5
10
15
20

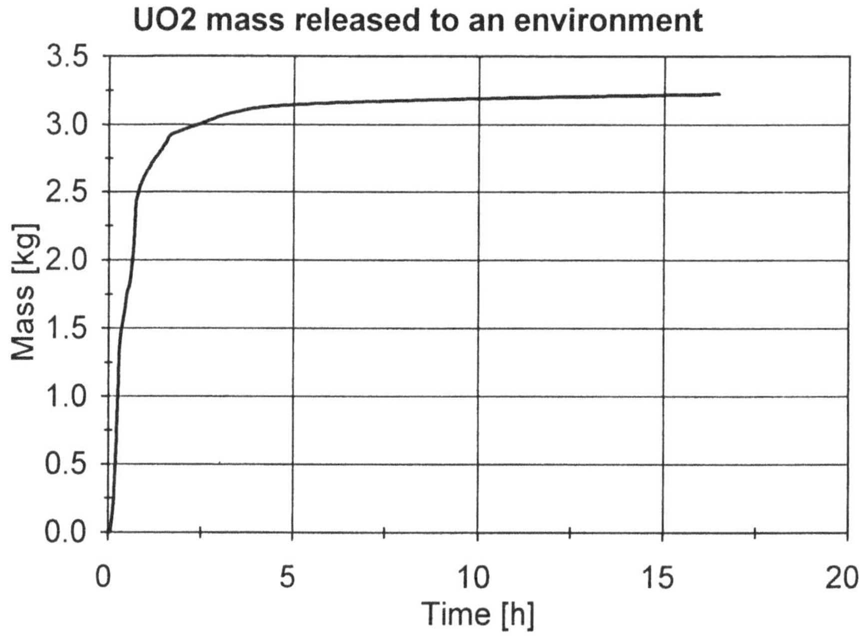
UO2 mass released to an environment
Mass [kg]
Time [h]
3.5
3.0
2.5
2.0
1.5
1.0
0.5
0.0
0
5
10
15
20

Fig. 2.

6. Using the obtained parameters of the radioactivity release, the trajectory of radioactive-cloud dispersion has been calculated and a prediction of the probable radiation situation in the near field around the plant made.

At IBRAE RAS, such calculations are executed on the basis of both conventional Gaussian models of release dispersion, which have proved their reliability for a range of a few tens of kilometers from the release source ("TRACE" computer code), and more sophisticated Lagrangian models ("NOSTRADAMUS" computer code) allowing calculations for distances of up to several hundreds of kilometers from the plant [2]. The afore-mentioned computer codes make possible to predict concentrations of radionuclides in the surface air layer, densities of territory contamination after the released-cloud passage, gamma radiation dose rates in open air, and possible doses of external and internal exposure to people. The results of such calculations are of considerable current use in Russia in the course of preparation and pursuance of business games simulating major radiation accidents when estimating the expediency of various measures to protect population (concealment of people, population evacuation, iodine preventive treatment, etc.) in areas of NPP siting, as well as when analyzing the necessary and sufficient forces and means [2-4].

The calculations performed have demonstrated that, at normal weather conditions and the accident scenario described above, the density of Pu-239 contamination at a distance of 5 km from the NPP may amount to 1.3 mBq/m^2, which is about 60 times higher than the average global levels of soil contamination as a result of accomplished nuclear weapon tests (according to estimates of the UN SCEAR [5]).

7. To pass from territory contamination levels to population exposure ones, an analysis was performed of actual and calculated data on levels of the exposure doses, related to plutonium isotopes, to population in the territories contaminated as a result of the Chernobyl accident (Ukraine, Belarus, Russia). The highest levels of contamination by Pu-239 and Pu-240 were observed in the nearest zone around the ChNPP. For instance, in an essential portion of the Polessky radiation-ecological reservation, the surface activity in Pu-239 and Pu-240 exceeds 3.7 kBq/m^2 and reaches even 5,000 kBq/m^2 at some spots. Beyond the exclusion zone, primarily in the territory of the Kiev Region, the contamination by plutonium isotopes was in excess of 0.4 kBq/m^2 over an area of more than 5,000 km^2. In the most contaminated Regions of Russia, relatively low levels of contamination by Pu-239 and Pu-240 up to 0.7 kBq/m^2 were registered in south-west districts of the Bryansk Region and up to 0.3 kBq/m^2 in the Tula and Kaluga Regions. The activity of alpha radiating radionuclides in bottom sediments of the rivers flowing through the contaminated areas is lower than that in the soil layer of catchment areas, though higher than that in river waters (Pu-238 — from 0.05 to 9.0 Bq/kg, Pu-239, Pu-240 — from 0.1 to 28 Bq/kg, Am-241 — from 0.07 to 16 Bq/kg). Migration of the radionuclides to underground waters has not yet led to stable contamination of the waters over large areas. Penetration of radionuclides to underground waters is most intense at the sites of temporary localization of radioactive waste where plutonium concentrations of up to 0.5 Bq/l are currently observed.

An analysis of actual data on the content of plutonium isotopes in organisms of people living in contaminated territories suggests [6] that at densities of teritory contamination by Pu-239 and Pu-240 long-lived isotopes in the Gomel Region of about 3.7 kBq/m^2 , the presence of Pu-239 and Pu-240 in amounts of up to 0.27 +/- 0.15 Bq was detected in tissue samples taken from 126 adults from 17 different areas of the Region in 1990-1991. The rate of plutonium isotopes deposition in human organisms was 0.006 Bq/year.

If one uses the ICRP relationships given in [7] to calculate the effective yearly dose of exposure to humans by plutonium isotopes, the maximum effective yearly dose for Gomel Region residents may reach 68 μSv/year by the year 2050 while now (10 years after the accident at the ChNPP) it is close to 10 μSv/year. If we take the period of 1986-1992, the effective people exposure dose accumulated for the afore-mentioned period may have amounted to 27 μSv.

This dose value calculated from biopsy data can be compared to the estimates made on the basis of measurements of the plutonium isotopes content in air and locally produced foodstuffs. According to the estimates of effective dose of internal exposure by plutonium radionuclides given in [8] for residents of the controlled territory of the Bryansk Region of Russia, Belarus, and Ukraine, this dose component for the period of 1986-1992 is about 25 mSv in average, which practically coincides with the estimate obtained on the basis of the ICRP relationships.

Thus, the results of actual and calculated data on doses of population exposure by plutonium isotopes in the territories contaminated as a result of the Chernobyl accident show that 10 years after the accident the ratio between the density of territory contamination by Pu-239 and Pu-240 (in mBq/m^2) and the yearly effective dose of exposure to people (in μSv/year) may be 2.7 and in 70 years, 19.0. The integral value of 70-year effective dose normalized to 1 kBq/m^2 of density of soil contamination by Pu-239, Pu-240 is 0.66 mSv in this case.

In Table 4, some data are presented on correlation of the radiation burden upon population due to plutonium isotopes and other radionuclides due to a number of real accidents (the Chernobyl accident, that at the "MAYAK" Association in the Chelyabinsk Region) and the accident scenario considered above.

Table 4. Comparison of Radiation Burden upon the Population due to Plutonium Isotopes and other Radionuclides

Place and source of exposure	Measured content of plutonium 239, 240 in a human body, Bq	Effective commitment doze due to plutonium isotopes, mSv	Effective doze of external and internal human exposure, mSv
Pripyat', Chernobyl Accident	100	13	100-200
Gomel', Chernobyl Accident	0.27 ± 0.15	2 .4	200
Chelyabinsk-65, MAYAK PA releases	3.75 ± 1.45	3.3	78 (by 1992)
5 km along the axis, accident at VVER-1000 with effluent of 12 Ci (0.044 TBq) Pu-239		0.22 (inhalation due to cloud release) 0.016 (inhalation due to secondary dust production) 0.0022 (peroral intake)	80000 (in 70 years) 25000 (in a year) 1000 (in 3 days)
5 km along the axis, accident at VVER-1000 with effluent of 6 Ci (0.22 TBq) Pu-239		1.1 (inhalation due to cloud release) 0.08 (inhalation due to secondary dust production) 0.011 (peroral intake)	80000 (in 70 years) 25000 (in a year) 1000 (in 3 days)
5 km along the axis, accident at VVER-1000 with effluent of 600 Ci (0.22 PBq) Pu-239		110 (inhalation due to cloud release) 8 (inhalation due to secondary dust production) 1.1 (peroral intake)	80000 (in 70 years) 25000 (in a year) 1000 (in 3 days)

On the basis of the foregoing, we may affirm that, if after the accident at an NPP using enriched Pu-239 fuel the density of territory contamination at a distance of 5 km from the plant does not exceed 1.3 kBq/m2, the average effective accumulated doses of exposure to population due to Pu-239 releases will not exceed 0.85 mSv, i.e., 1.2% of natural radiation background for the same period of time, even 70 years after the accident. For professionals working in zones of intense dust formation, such doses may be 5 to 10 times higher, but nonetheless they will present no danger to population health. So, the use of enriched Pu-239 fuel at an NPP with VVER-1000 reactors will not result in aggravation of effects of an accident, even with core melting, for population in the vicinity of the plant.

REFERENCES

1. Goldschmidt P., Verbeek P. "The Deposition of Plutonium from Dismantled Warheads: a West European Utility View", Nuclear Europe Worldscan, N 5/6, 1994, p.49.

2. "Polyarnye Zori-95" Command and Headquarters Exercise. Apatity, Murmansk Region, May 29 - June 2 1995, final report, IBRAE RAS, Moscow, 1996.

3. Arutyunyan R.V., Linge I.I., Pavlovsky O.A. et al. "Franco-Russian Role-Play on Decision Making in the event of Radiological Contamination of Large Areas of Land", Portsmouth-94 Proceedings of the IRPA Regional Congress, Nuclear Technology Publishing, 1994, pp. 329-332.

4. "Kalinin NPP-94" Command and Headquarters Exercise. Moscow, November 22-24, 1994, final report, NSI RAS (IBRAE), 1995.

5. United Nations. Sources and Effects of Ionizing Radiation. United Nation Scientific Committee on the Effects of Atomic Radiation, 1993, Report to the General Assembly, with Scientific Annexes. United Nations sales publication E.94. IX.2. United Nations, New York, 1993.

6. Popov V.I., Kochetkov O.A., Molokanov O.A. et al. "Formation of Internal Exposure Doses to Chernobyl NPP Personnel and Missioned Workers in 1986-1987". Sov. J. of Medicinal Radiology", 1991, pp. 33-41.

7. ICRP-60, 1990 Recommendation of the International Commission on Radiological Protection. Annals of the ICRP, vv.1-3, Pergamon Press Oxford, 1991.

8. Ivanova N.P., Shvidko N.S., Ershov E.B., Balonov M.I. "Population Doses in Russia from Plutonium Fallout following the Chernobyl Accident". Rad. Prot. Dosimetry, v.58, N4, 1995, pp.255-260.

Accelerator-Driven Transmutation System with Liquid Lead Cooling - Basic Conditions, Layout Goals and Results -

Peter-W. Phlippen

Keywords/Abstract: transmutation / Plutonium / subcritical nuclear system / safety requirements / power generation / liquid lead / liquid nuclear fuel / liquid metal

An effective transmutation of Plutonium and minor actinides can be achieved by subcritical neutronical systems. Criteria like low pressure, low actinide inventory, low beam power combined with maximum power rating, extended safety levels and easy accessibility of fuel for cleanup are aimed on. This paper discusses some general requirements for transmutation systems and describes a thermal system, operated with liquid lead as coolant and fuel carrier as well as for the spallation target.

Graphite is chosen as moderator. Low actinide oxide concentrations of 1.2 wt-% dispersed in liquid lead are sufficient to operate the system at a multiplication ratio of 0.95. A radial constant power rating is obtained by increasing the moderation ratio over the radius resulting in a power peaking below 1.5 with a blanket outer radius of 2.5 m. The annual Pu consumption is about 480 kg.

INTRODUCTION

The today's proposed strategies to dispose of high level waste (HLW) in final repositories do not satisfy at all the critics of this disposal form. This is mainly due to the extremely long time of hundred thousands of years, in which the nuclides in question must be kept away from the environment. This restriction imposes high requirements to the geological ages hosting final repositories. Hazardous nuclides are transuranium isotopes (TRU) (Np, Pu, Am, Cm) consisting mainly of Pu and due to its mobility in the underground, the long-lived fission products (LLFP) Tc-99 and I-129. Adding Partitioning and Transmutation (P&T) steps to the fuel-cycle allows for the introduction of disposal strategies offering the possibility to reduce the long-term radiological impact of nuclear waste to a few thousands of years. For this, engineered barriers, for example made of steel [1-3], lasting up to one thousand years are sufficient to guarantee the exclusion. Consequently different disposal requirements arise.

Another view rises up from the Pu-generation in spent LWR fuels. Today, its isotopic composition must be regarded as weapon grade and thus spent LWR fuels represent a permanent proliferation risk. Even the Pu-bounding in highly active spent fuel is no ultimate safe exclusion for that task. Additionally, the world needs to care about ex-weapon Pu, which represents already a relatively easy to handle status. It seems not acceptable simply to dispose it off after vitrification or after one MOX-burning cycle in LWR. There is a strong need to destroy this material; for disposal reasons, the destruction should include the minor actinides (MA) (Np, Am, Cm) too.

REQUIREMENTS

In general, TRU consuming installations must fulfill high level safety requirements. Kugeler et. al. [4] describe a so called „catastrophe free“ nuclear technology which is characterized by a deterministic exclusion of

1. early deceases,
2. delayed deceases,

E. R. Merz and C. E. Walter (eds.), Advanced Nuclear Systems Consuming Excess Plutonium, 275–282.
© 1997 *Kluwer Academic Publishers.*

3. contamination of the environment as well as
4. evacuation or resettlement

after severe accidents by system layout and construction. These requirements coincide with a deterministic exclusion of accidents with levels above four of the IAEA INES-scale. Thus, the consequences of accidents are to be restricted to the nuclear facility itself by layout!

There are some basic nuclear mechanisms which may help to fulfill these requirements in a technical installation. Reactivity initiated accidents can be avoided, if i.e. subcritical systems as well as low or nearly no excess reactivity in combination with large prompt negative reactivity feedback coefficients are installed. In combination with a low system pressure, these provisions also reduce driving forces for activity releases. Passive or self-acting afterheat removal systems ensure a reliability of the fission product barriers.

Some additional measures are helpful to achieve the safety goals. Avoiding an accumulation of fission products enhances not only the neutron economy but facilitates the technical requirements for the afterheat removal. Reduced TRU inventories limit the possible releases. Thus avoiding re-creation of burnt TRU by breeding processes in U is an adequate measure for further reduction of the initial TRU content. Additionally, this enhances the TRU burning capacity.

Due to their special safety and performance features, subcritical systems driven by accelerators (ADS) represent promising devices to transmute the aforementioned nuclides while fulfilling the extended safety requirements.

SYSTEM DESCRIPTION

A subcritical device based on liquid fuel operating with thermal neutrons is proposed. Such a device offers the advantage of minimizing the actinide inventory in the system and offers the possibility of destroying LLFP in especially adapted incineration zones. ***Figure 1*** presents the general layout. It is driven by external neutrons produced by spallation reactions of high energy protons with a liquid lead target. These neutrons are moderated by graphite in the blanket surrounding the target and introduce the desired incineration-reactions in common with fission neutrons. The target is designed as a cylindrical liquid lead column in a vertical position [5]. For the neutronic layout we choose a target-radius of 40 cm liquid lead due to its favorable thermo-physical and neutronic properties. The heat deposited in the target is removed by circulating the liquid material through external heat exchangers. To prevent lead-vapor entering the accelerator vacuum, a proton-beam window separates the area above the lead surface from the accelerator tube. For this purpose a rotating SiC beam window can be applied [6]. In its favorable position shown in ***Figure 1*** it

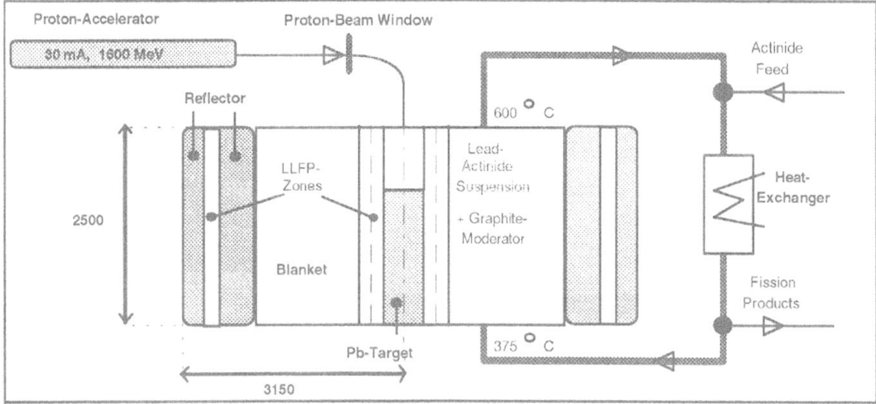

***Figure 1**: Schematic overview of the fluid fuel transmuter*

is protected from supplementary radioactive and thermal radiation by the target. A chemical separation and reprocessing facility connected to the blanket is assumed to extract fission products and to feed in actinides into the system in a semi-continuous way. The external system feed consists exclusively of LWR-type TRU waste and optionally LLFP.

The blanket of the system is subdivided into two regions: a "fission product region" and an "actinide region". As shown in ***Figure 1,*** the fission product region consists of three different zones. The first zone, directly surrounding the target, includes a two column layer containing Technetium-pins in a triangular lattice. The corresponding pitch is 2.2 cm and the coolant liquid lead fills the space between the pins. Assuming that the energy produced by (n,γ)-reactions and β^--emissions of the transmutation products is deposited locally, we get an average power density in the Tc-pins of 4.5 MW/m^3. Technetium is exposed in metallic form. For the neutronic calculations we used 1 mm thick Zr-clad with an inner diameter of 1.8 cm (**Table I**). The following reasons can be stated for including a Tc-layer in that position: First the Tc-layer protects effectively the graphite from fast spallation neutrons flowing out of the target. Otherwise and because of the strong absorption resonance at 4.5 eV, it seems to be useful to transmute Tc with epithermal neutrons and finally, Tc absorbs thermal neutrons scattered back from the blanket avoiding increased absorption losses in the target structures.

Volume Fraction [%]	N_C/N_{HM} [-]	number of tubes	zone radii [cm]
18.75 (Fp-zone)	1270	547	45.-101.
18.75	950	158	101.-110.
17.5	1070	1280	110.-169.
16.25	1190	1370	169.-218.
15	1340	1336	218.-260.
Totals		4690	

	number of LLFP-pins	Inventory [kg]
Tc (targ.)	215	1570
Tc (refl.)	179	1305
Ioidine	136	365
Totals	530	3240

Table I: Some layout data of the actinide- and LLFP-regions

To avoid neutrons to leave the system, we include a second Tc-zone within the radial reflector. In this zone Tc is mainly transmuted by well thermalized neutrons using effectively the cross section behavior of the nuclide. The structure of the Tc-layer corresponds to the one already described.

The third fission product zone contains Iodine. This zone surrounds the first Tc-zone and consists of actinide-tubes and Iodine-pins in a triangular graphite-moderated lattice with a pitch of 6.60 cm. Each Iodine-pin is surrounded effectively by four actinide-tubes. For the neutronic calculation we use 1 mm thick Zr-clad with an inner diameter of 2.0 cm. Between moderator and clad we include a 4 mm thick annulus filled with liquid lead to remove the heat from pin and graphite. The average power density in the pin is 1.3 MW/m^3 (**Table I**). To find an appropriate form to expose Iodine is much more problematic than for Technetium. The composition to be found should be thermodynamic stable (appropriate values for vapor- and dissociation pressure), compatible with clad materials (corrosion) and should have appropriate neutronic characteristics (cross section, density). Some potential candidates are Ce_3I, PbI_2 and NaI. For our calculations we use PbI_2 with an isotopic composition of 75 % I-129 and 25 % I-127. The irradiation time for all the LLFP-pins is correlated with the graphite service life, which in this case is 4 years.

The main part of the blanket consists of the graphite-moderated actinide-region and is located in between the Iodine zone and the reflector extending from 101 cm to 260 cm. The actinide-region is subdivided into four zones with different moderation ratios characterized by the fuel volume fraction VF (**Table I**). Surrounding the blanket a 55 cm thick graphite-reflector is included. Best moderation is achieved beside the reflector. This design is

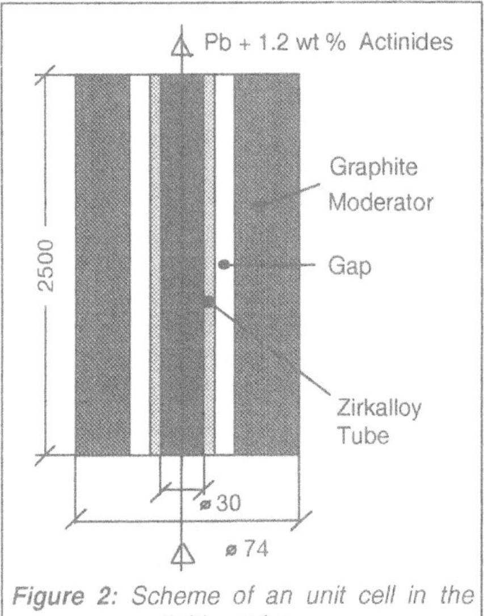

Figure 2: Scheme of an unit cell in the actinide region

necessary for obtaining a radial homogenized power density- and graphite service life distribution. In each zone tubes containing the actinides within a triangular lattice are placed. In ***Figure 2*** a characteristic unit cell of the actinide-region is shown. For the neutronic calculations we used 2 mm thick Zr-clad with an inner diameter of 3 cm. Surrounding the clad we include a 3 mm gap.

The actinides are exposed as oxide particles forming a actinide-lead suspension. Liquid lead is used due to its superior neutronic and thermo-physical properties serving as carrier material and, in a first stage, as coolant. In this design the actinide-lead suspension is pumped through the blanket and cooled in external heat-exchangers. The fuel power density and the lead velocities in the tubes are adjusted to get an outlet temperature of 600 °C.

A more sophisticated design includes an internal cooling mechanism. In this case the actinide-lead suspension is circulated at a velocity just sufficient to have turbulent flow characteristics and to hold the actinides in suspension, while filling the gap between graphite and clad with lead serving as coolant. To improve the stability of the suspension, an adjustment of the density of the actinide oxides between 80 % and 90 % of their theoretical density establishes nearly equal bulk densities of the fuel components and diminishes the sedimentation force. This pure lead is pumped through external heat-exchangers and removes the fission heat produced in the suspension and the heat deposited in the graphite by (n,γ)-reactions [7]. The most important consequence of this design variant results in a reduced out-of-pile time obtaining a smaller actinide inventory.

To operate the actinide region, we choose a batch-wise mode with variable cycle length. After each cycle a certain amount of the inventory consisting of actinides and fission products is removed from the system, denoted by the reprocessing factor, and reprocessed in a chemical facility. While the fission products are separated and conditioned in an appropriate way for final storage, the actinides are treated in order to be recycled into the system after a certain decay-time. The feed into the system at the beginning of each cycle consists of two streams: the recycled stream, which is the main part, and the external feed to compensate for burnt actinides.

SYSTEM REQUIREMENTS

The optimization of the transmuting system has been done mainly with regard to neutronic parameters. In a further step the technical feasibility has been assessed. The most important goal aimed on is to obtain maximized incineration rates for actinides and LLFP operating with accelerator currents as low as possible. Therefore we designed a system with an multiplication ratio of $k_{eff} \approx 0.95$. Consequently an optimized overall system efficiency is achieved. As constant thermal power generation during a cycle has been deemed best, a minimum decrease of the multiplication ratio from the beginning of an equilibrium cycle (BOEC) to the end of the equilibrium cycle (EOEC) due to burn-up effects was identified as a further significant goal. This implies an increase of the accelerator current during an equilibrium cycle, which should be minimized. The fuel power density in the blanket is limited by fast neutron damage to the graphite moderator. Hence, a trade-off between graphite service life and incineration rate has to be achieved. In addition, the fuel power density distribution should be homogenized to guarantee radial constant lead velocities. Further, this has positive effects on corrosion, thermal stresses and irradiation damage.

	BOEC	EOEC
k_{eff}	0.95	0.94
I [mA]	25	30
ϕ_{ave} [n/cm²·s]	$5.65 \cdot 10^{14}$	$5.75 \cdot 10^{14}$
ϕ_{Act} [n/cm²·s]	$7.20 \cdot 10^{14}$	$7.30 \cdot 10^{14}$

Table II: Main system parameters at BOEC and EOEC

RESULTS

To meet the above mentioned criteria and considering the influence of each design parameter on the neutronic behavior we chose the following operation mode for the actinide-region: actinide density in the suspension α = 1.2 wt %; reprocessing factor Rp = 50 %; recycling 100 % of the actinides and no fission products; cycle length T_b = 60 days; decay time T_d = 120 days; and average fuel power density L_f = 165 MW/m³.

The neutronic calculations have been performed with modules from the SCALE 4.0 system [8] and the HETC code [9] to obtain the external neutron source. An operative equilibrium state in the actinide region is achieved approximately after 3 years. The system operates with constant thermal power P_{th} =1360 MW yielding a mean power density in the blanket of L_b = 26.5 MW/m³. Main parameters describing the equilibrium cycle of the system are listed in **Table II**. Note the increase of the system average neutron flux ϕ_{ave}. This is due to the changing actinide composition and the fission product build-up during irradiation.

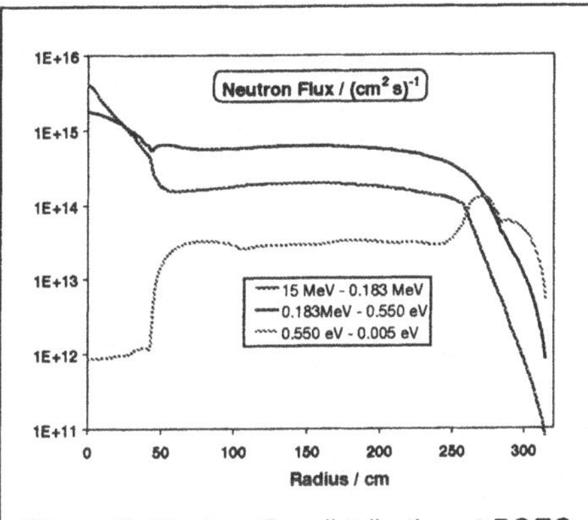

Figure 3: *Neutron flux distribution at BOEC*

Figure 3 shows the radial neutron flux behavior in a three group structure at BOEC. The radially increasing thermalization of the system yields a stabilized thermal and fast flux. Since the thermal flux determines the power distribution in the system, the required constant radial power generation is linked to a radial increasing thermal flux, as the thermal fission cross section of the main fuel isotope (Pu-239), tends to decrease with softer neutron spectra, due to its relatively high resonance at 0.3 eV. ***Figure 4*** shows the fuel and the mean power density distribution at BOEC. The power peaking factors for the calculated case are P_{max}/P_{ave} = 1.44 and P_{min}/P_{ave} = 0.35 respectively and are localized at the boundaries of the actinide region. In the bulk of the system a quit flat behavior is achieved and a homogenized power density distribution is obtained.

The service life of the blanket is determined by the accumulated fluence of graphite at the calculated temperatures. In our case the limit where the absolute swelling starts is $2.3 \cdot 10^{22}$ n/cm² for neutrons with energies E ≥ 180 keV [10]. For the given system a homogenized graphite service life distribution is achieved. The neutron flux ϕ(E) determining the graphite service life behaves extremely flat (***Figure 3***). The maximal value for the flux ϕ (E) within the actinide region (VF=17.5 %) is ϕ(E) = $1.82 \cdot 10^{14}$ n/cm²·s yielding a lifetime of about 4 years.

The actinide inventory of the system at BOEC and EOEC is shown in ***Figure 5***. The inventory at BOEC is 4170 kg and is dominated by the Plutonium isotopes. Due to the chosen operation mode an extensive Curium build-up, especially Cm-244, is avoided. Mainly Plutonium (the isotopes Pu-239 and Pu-241) is fissioned. The incineration of MA occurs predominantly by fissioning the corresponding daughter nuclides. Additionally, an appropriate decay time outside the system offers further advantages to obtain fissile isotopes. The incineration of Am-241, for instance, is attained by fissioning Am-242m, Cm-243 or Pu-239, the last isotope resulting from an α-decay of Cm-242 (162.8 d period) via Pu-238. All calculated inventories include the volume of the external piping system. A ratio of internal to total inventory of 1/4 was assumed. Regarding safety aspects we should mention the positive contribution to the net temperature coefficient of Pu-239. Fortunately, the system con-

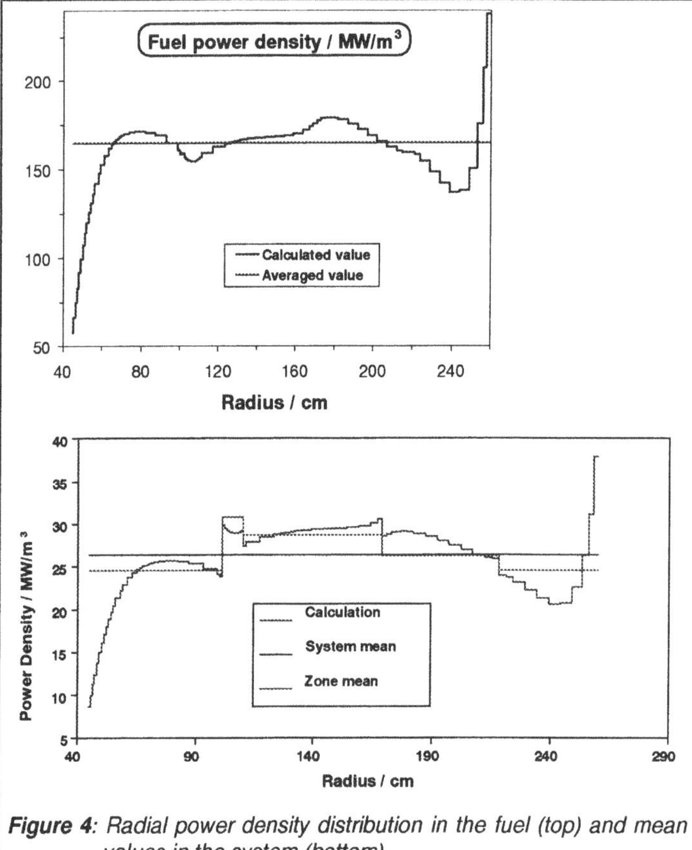

***Figure 4**: Radial power density distribution in the fuel (top) and mean values in the system (bottom)*

tains an appreciable amount of Pu-240, which due to its resonance absorption at 1.0 eV leads to a net prompt negative temperature coefficient in the order of 10^{-6}/K to 10^{-5}/K (***Figure 6***).

Table III shows the LLFP-inventory in the corresponding fission product zones at the beginning and at the end of the irradiation period. For these zones we chose, according to the graphite service life, a 4 year irradiation period. The transmutation product obtained by irradiating Iodine is mainly Xenon (the isotopes Xe-128 and Xe-130). Due to the neutron flux level in the Iodine zone and the rather small thermal absorption cross section of the mentioned Xe isotopes, negligible amounts of Cs- and Ba-isotopes are produced. Note the remarkable amounts of Te-128, which are obtained via β^+-emission of the isotope Xe-128.

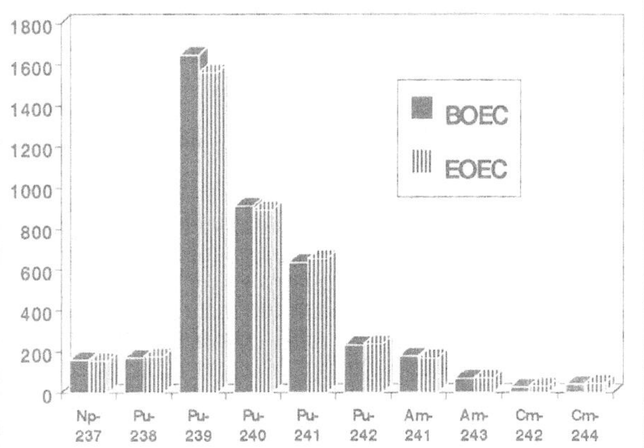

Figure 5: Actinde inventory in the system at BOEC and EOEC

Burning Technetium in different spectral zones yields a completely different proposition of transmutation products. The main reasons are the different flux levels and spectra as well as the effective Tc-absorption cross sections. The average neutron flux in the Tc-zone adjacent the target is $1.05 \cdot 10^{15}$ n/cm²·s, while the average flux within the reflector is $1.09 \cdot 10^{14}$ n/cm²·s.

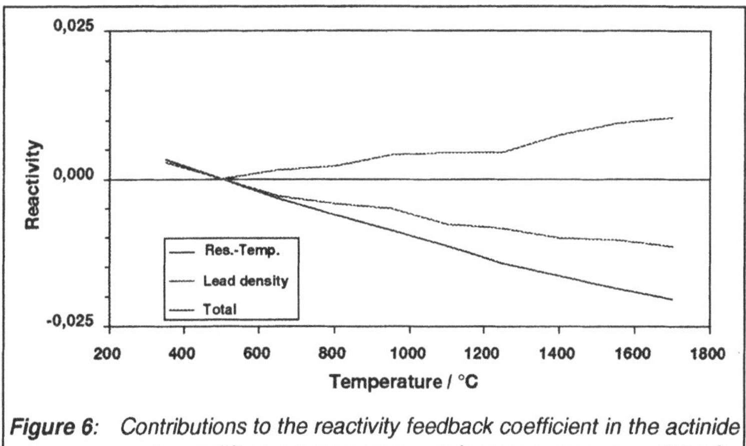

Figure 6: *Contributions to the reactivity feedback coefficient in the actinide region at different temperatures (reference temperature 500 °C)*

Consequently we obtain the following incineration rates for actinides and LLFP in the system:

- Actinides: 540 kg/a
- Technetium-99: 49 kg/a
- Iodine-129: 7.5 kg/a

In addition to the incineration, appreciable amounts of TRU-isotopes and LLFP the system produces a surplus of electric energy. Assuming an efficiency of the thermal cycle $\eta_{th} = 0.40$ and a conversion ratio to transform electric power to beam power of $\eta_{acc} = 0.50$ [11] we get an overall electric efficiency of $\eta_{tot} = 0.33$. This means that on the average 96 MW are needed to run the accelerator and app. 465 MW electric power is feed to the grid.

	Iodine-Transmutation [g]	
	BOC	EOC
Te-128	-	$7.87 \cdot 10^{2}$
I-127	$9.00 \cdot 10^{4}$	$7.66 \cdot 10^{4}$
I-128	-	$2.15 \cdot 10^{-1}$
I-129	$2.75 \cdot 10^{5}$	$2.45 \cdot 10^{5}$
I-130	-	$1.31 \cdot 10^{1}$
Xe-128	-	$1.27 \cdot 10^{4}$
Xe-129	-	$2.76 \cdot 10^{1}$
Xe-130	-	$2.98 \cdot 10^{4}$
Xe-131	-	$1.89 \cdot 10^{1}$
Xe-132	-	$4.58 \cdot 10^{1}$
Cs-133	-	$1.48 \cdot 10^{-1}$
Cs-134	-	$5.10 \cdot 10^{-2}$
Ba-134	-	$1.37 \cdot 10^{-2}$

	Technetium-Transmutation			
	Target-Zone [g]		Reflector-Zone [g]	
	BOC	EOC	BOC	EOC
Mo-96	-	$5.10 \cdot 10^{-1}$	-	$1.30 \cdot 10^{-2}$
Tc-99	$1.57 \cdot 10^{6}$	$1.44 \cdot 10^{6}$	$1.36 \cdot 10^{6}$	$1.23 \cdot 10^{6}$
Tc-99m	-	$9.74 \cdot 10^{0}$	-	$9.47 \cdot 10^{-2}$
Tc-100	-	$2.21 \cdot 10^{-2}$	-	$1.33 \cdot 10^{-2}$
Ru-99	-	$1.96 \cdot 10^{1}$	-	$1.62 \cdot 10^{1}$
Ru-100	-	$1.21 \cdot 10^{5}$	-	$7.46 \cdot 10^{4}$
Ru-101	-	$4.90 \cdot 10^{3}$	-	$1.36 \cdot 10^{3}$
Ru-102	-	$1.15 \cdot 10^{3}$	-	$2.02 \cdot 10^{1}$
Ru-103	-	$3.74 \cdot 10^{0}$	-	$6.69 \cdot 10^{-3}$
Rh-103	-	$7.98 \cdot 10^{0}$	-	$3.53 \cdot 10^{-2}$
Pd-104	-	$1.60 \cdot 10^{1}$	-	$6.69 \cdot 10^{-3}$

Table III: LLFP-inventory in a 4 year cycle at BOC and EOC

CONCLUSIONS

The discussion of the proposed subcritical accelerator-driven fluid fuel system shows that the requirement of transmuting long-lived radioactive waste effectively can be met. In addition, electric power is produced to run the accelerator and to feed into the grid.

The goals to be met are at least partially satisfied. Especially the variation of the moderation ratio radially yields the desired effect of homogenized power density- and graphite service life distribution. More over, the following additional advantages should be recalled:

- Subcriticality allows for a much more flexible system design, especially regarding safety aspects (reactivity insertion).
- A thermal system with liquid fuel offers the advantage of minimizing the actinide inventory.
- Since liquid lead acts as an inert diluent for TRU, this system avoids breeding of fresh TRU and thus operates on the highest possible consumption ratio.
- The combination of liquid lead with a graphite moderator allows the system to operate at a pressure only high enough to overcome the pressure drops.
- The semi-continuous removal of a certain amount of fission products from the facility includes the advantage of a lower relative afterheat production in the subcritical design than in a reactor thus diminishing the accident potential of the system.
- The waste streams of the proposed system contain just short-lived fission products in addition to activation and spallation products, without producing materials suitable for nuclear weapons.

Despite of this preferable behavior, a lot of questions remain to be answered by future investigations, some important ones are:

- The general technical layout features must become more detailed in order to validate a the realization in principle.
- Safety aspects including afterheat removal, evaluations of severe accidents and ultimate safety behavior need to become more detailed.
- The behavior of AcO_2-lead-suspension needs more detailed analysis on its chemical and thermo-chemical behavior.
- Fuel reprocessing needs to be more detailed.
- Other fuel options are to be investigated (carbides, nitrides dispersed or solid).
- Including breeding zones (i.e. Th) may overcome the relatively short cycles between fuel reprocessing without decreasing the multiplication ratio to an inadmissible extent.

ACKNOWLEDGEMENT The author thanks Dipl.-Phys. Pablo Lizana-Allende for his engagement in preparing this paper.

REFERENCES

[1] Smailos, E. et al., *Untersuchungen zur Eignung keramischer Behälter als Korrosionsschutz für hochradioaktive Abfallprodukte bei der Endlagerung in Steinsalzformationen,* Kernforschungszentrum Karlsruhe, KfK 4244, April 1987

[2] Schwarzkopf, W. et al., *In-Situ Corrosion Studies on Selected High-Level Waste Packaging Materials under Simulated Disposal Conditions in Rock Salt Formations,* Kernforschungszentrum Karlsruhe, KfK 4324, Jan. 1988

[3] Schwarzkopf, W. et al., *In-Situ Corrosion Studies on Cast Steel for a High-Level Waste Packaging in a Rock Salt Repository,* Materials Research Society Symp. Proc. **127** , 1989, pp. 411-418

[4] Kugeler, K.; Phlippen, P.-W., *Catastrophe free nuclear technology for the future world energy supply,* World Energy Council, 16th Congress, Tokyo, October 1995, No. 2.3.07, pp. 99-114

[5] Jansen, Ch; *Protonenfenster und Flüssigmetalltarget für eine beschleunigergetriebene Transmutationsanlage;* Phd-thesis; RWTH-Aachen; 1995

[6] Jansen, Ch; Lypsch, F.; Lizana, P.; Phlippen, P.W.; *Proton-Beam Window Design for a Transmutation Facility Operating with a Liquid Lead Target;* Int. Conf. on Accelerator-Driven Transmutations Technologies and Applications; Las Vegas; 1994

[7] Lypsch, F; *Aspekte zum Einsatz von Systemen zur Transmutation radioaktiver Stoffe -Neutronik, Technik, Sicherheitstechnik-;* Phd-thesis; RWTH-Aachen; 1995

[8] Hermann, O. W. et al., *SCALE 4.0 - A Modular Code System for Performing Standardized Computer Analysis for Licensing Evaluation* , ORNL, Radiation Shielding Information Center, 1990.

[9] Cloth,P. and Filges, D. et al., *HERMES - A Monte Carlo Program System for Beam-Materials Interaction Studies,* Research Centre Juelich, KFA-Report Jül-2203, 1988.

[10] Haag, G. et al., *Development of Graphite,* Journal of Nuclear Materials **171**, 41-48, 1990

[11] Lengeler, H.: *Nuclear Waste Transmutation using High-Intensity Proton Linear Accelerators,* Report: CERN AT/93 DI, Genf, Schweiz, 1993.

PLUTONIUM PROBLEM AND ONE OF THE SOLUTION TECHNIQUES

Yu.A. Trutnev, V.E. Marshalkin, V.M. Povyshev
RFNC-VNIIEF

It is shown that the problems of plutonium and long-lived actinide α activity are the problems of U-Pu fuel cycle rather than those of the entire nuclear energy. The conversion of modern thermal reactors to the closed U-Th fuel cycle basically simplifies the solution of the most important problems in the modern nuclear energy.

The ecological problem of handling long-lived actinides of exposed nuclear fuel including plutonium isotopes is one of the main difficulties delaying the evolution of nuclear energy.

Recently this problem became more acute because of nuclear disarmament and emergence of excessive (~100 t [1]) high-enriched (weapon-grade) plutonium. The ecological aspect was complemented by the risk of potential proliferation of this material and uncontrolled manufacturing of explosive devices. The urgency is due to the following circumstances:

- large scale amount of free plutonium allowing to manufacture ~10^4 explosive devices;
- well validated technical and technological objectives relating to the manufacturing of plutonium explosive devices.

In addition, the plutonium problem has a social-economical aspect. One cannot forget that the creation of nuclear weapons and nuclear energy have taken from mankind enormous material, social and other costs. In this view the zero trivial solution seems to be inappropriate relative to the generation that created nuclear weapon and nuclear energy:

- elimination of nuclear weapons;
- closing the nuclear energy.

The fission of heavy nuclei with the simultaneous release of enormous specific energy and the amount of neutrons sufficient for chain reaction is a unique physical phenomenon whose practical implementation - as a nuclear energy source - has taken surprisingly short period. The uranium fission process was discovered in 1939, the first atomic bomb was shot in 1945, the first atomic power plant became operative in 1954. By this time the solution is found for the complex of science and engineering, technological problems, obvious progress is made in the adoption of nuclear energy source.

We believe that the solution of plutonium problem should be sought for through the analysis of causal relations that formed the up-to-date situation.

Without considering multiple aspects, note the following. In natural conditions there exists only one isotope (^{235}U) whose fission by the neutrons is super-threshold and is accomplished by the entire spectrum of the fission neutrons. Its content is 0.72% in natural uranium with the basic component being represented by ^{238}U isotope reaching 92.275%. The fission neutrons of ^{235}U in the radiation capture of ^{238}U form ^{239}U that because of short half life rapidly decays to neptunium and then to the ^{239}Pu isotope. ^{239}Pu isotope is characterized by a relatively long half life ($T_{1/2}$≈24000 years), high average number of spontaneous fission neutrons $\bar{\nu}(\varepsilon) \approx 2.87+0.13\cdot\varepsilon$, a relative large neutron fission cross section through the entire energy range and is still the most preferable isotope for the design of explosive devices. The solution of problem for sufficient enrichment of uranium mixture by ^{235}U isotope allowed to put into operation the first reactors for ^{239}Pu isotope generation to develop nuclear weapons. The solution of problem for radiochemical separation of plutonium and uranium from the exposed fuel meant the solution of the problem for the closed U-Pu fuel cycle. The efforts to improve this fuel cycle and reactors naturally led to the emergence, evolution and further development of peaceful nuclear energy.

Thus we can say that nuclear energy is a side product of the industrial generation of weapon-grade plutonium and both science and engineering solutions were achieved through the evolution of U-Pu fuel cycle. The special-purpose generation of weapon-grade plutonium in special reactors naturally converted to plutonium generation in power reactors. Therefore the plutonium problem is not the problem of nuclear energy but that of U-Pu cycle and its zero solution seems to be the rejection of U-Pu fuel cycle.

During the past four decades the efforts of multiple specialists were made to develop the U-Th fuel cycle. The researchers reasonably [2,3] indicated several advantages of using U-Th cycle in nuclear energy as compared to intensively developed till today U-Pu fuel cycle. Particularly, the experimental

E. R. Merz and C. E. Walter (eds.), Advanced Nuclear Systems Consuming Excess Plutonium, 283–285.
© 1997 *Kluwer Academic Publishers.*

proof was obtained [4] for the feasibility of self-reproduction of active ^{233}U isotope from the raw ^{232}Th isotope in thermal reactor.

Currently there is a unique situation allowing to believe that the next evolution phase of nuclear energy would be the conversion of modern thermal reactors to the closed U-Th fuel cycle.

This fuel cycle lacks the plutonium generation. Available from the dismantling, high-enriched (weapon-grade) plutonium and uranium together with the accumulated power plutonium would be most efficiently used in the transition phase to self-sustained closed U-Th fuel cycle. The neutrons produced by fission of ^{239}Pu, ^{235}U and ^{241}Pu isotopes by thermal neutrons when replacing the reproducing ^{238}U isotope by ^{232}Th would provide the conversion of these active isotopes to ^{233}U fissioned by thermal neutrons. Thus in the closed U-Th cycle plutonium is not generated and the accumulated plutonium will be burnt in the reactor conversion phase to this fuel cycle. The orientation to thermal reactors in the next evolution phase of the nuclear energy would allow to take the full advantage of the acquired science and technology achievements, validated technologies and experience in the operation of the most advanced reactors. This seems to be particularly important in view of the approaching end of resources of the operational reactors and their replacement.

The closed U-Th cycle has undoubtful advantages over the U-Pu cycle in the solution of the most important problems in the modern nuclear energy:

- handling the actinides of exposed fuel and high-active waste;
- nonproliferation of fissile materials and nuclear technologies;
- reasonable use of high-enriched (weapon-grade) uranium and plutonium;
- transmutation of generated long-lived actinides and power plutonium;
- increase in the efficiency of using nuclear fuel resources;
- increase in the operational safety of nuclear reactors.

It is known that ^{233}U isotope generated from the breeding ^{232}Th isotope is characterized with the lowest probability of random neutron capture reaching ~8% at the thermal point (group) and ~15% in the resonance region. ^{234}U isotope produced by the reaction $^{233}U(n,\gamma)^{234}U$ is a well breeding isotope of the active ^{235}U isotope. Thus it is seen that before the trivially fissioned ^{235}U isotope there appears ^{233}U isotope that is most efficiently fissioned by thermal and resonance neutrons. In this view the generation probability of ^{236}U and subsequent isotopes is

$$({}^{233}U \to {}^{236}U) \approx \begin{cases} 1.1\% & \textbf{by thermal neutrons} \\ 5.1\% & \textbf{by resonance neutrons} \end{cases}$$

It is easily seen that the spurious generation of long-lived actinide ^{236}U per fission in Th cycle is less as compared to plutonium by more than ~13 times if thermal neutrons are used and by ~7 times if resonance neutrons are used.

It can be shown that the recycling of U-Th fuel would result in that uranium isotopes will reach the equilibrium states /5/. The degradation rate of the fuel properties of the mixture would result only from the accumulation of ^{236}U. However unlike the usual $^{235}U_{\alpha}\,^{238}U_{1-\alpha}$ fuel the accumulation of ^{236}U is an order of magnitude slower process. Because of this fact and the difference by three mass units between ^{236}U and the main fissile ^{233}U isotope it seems to be efficient to introduce centrifuge purification of recycled uranium from ^{236}U isotope as it accumulates in order to avoid the degradation of fuel properties of recycled uranium fuel and the generation of isotopes such as ^{237}Np, ^{238}Pu. In the energy where the burning of Th is followed by the generation of ^{236}U there will be the conservation of natural equilibrium of long-lived alpha activity.

Currently it is recognized that the basic barrier against the proliferation of nuclear weapons is the restriction of access to fissile materials. The basic solution in limiting the access to the fissile material to use them in weapons is, first, the lack of these materials in storage facilities and, second, unacceptability for using the components of nuclear fuel for the manufacturing of explosive devices. Both requirements are met by the thermal reactor operating in closed U-Th cycle.

The multiplication factor of active isotopes is close to unity and will be exactly determined by actual reactor parameters. And given the need determined by the functioning of U-Th energy the generated uranium isotopes will be removed from the exposed fuel and loaded in reactor as a fuel for the next campaign. Thus there will be no accumulation of active isotopes at the storage facility in the form of exposed fuel and other forms.

Another important factor facilitating the solution of the nonproliferation problem for fissile materials is the generation of ^{232}U isotope from ^{232}Th in parallel with the generation of ^{233}U. And the equilibrium content of ^{232}U in the uranium mixture of course complicates the handling the exposed fuel till its return to the reactor while making actually impossible the use of such uranium for manufacturing and storage of explosive devices because of the time increase of intensity of hard γ quanta from ^{208}Pb formed in the radioactive series of ^{232}U. In addition, the energy release from the decay of all isotopes of ^{232}U radioactive series is ~42 MeV, the equilibrium is reached in ~10.6 years. This heat release undoubtfully complicates the use of such uranium for the manufacturing of explosive devices.

The U-Th cycle naturally simplifies the accounting and control of uranium

Finally, the active isotopes (^{233}U and ^{235}U) are relatively diluted underthreshold fissionable isotopes (^{234}U and ^{236}U). The further denaturation of the equilibrium uranium mixture is possibly by adding ^{238}U isotope.

The production of high enriched (weapon-grade) uranium and plutonium is associated with high cost. Their use as the fuel for power reactors operating at 2-4% enrichment by active isotopes is not efficient. The use of these materials to start the U-Th fuel cycle with self-breeding of active isotopes (or with a slight feeding) to provide the burning of the entire amount of Th is of course the most reasonable conversion to nuclear energy.

The fission of ^{233}U is followed by the release of the highest effective number of fission neutrons that can be used not only to reproduce ^{233}U from ^{232}Th but also to transmute the accumulated actinides and power plutonium. Given the replacement of high breeding ^{238}U isotope by ^{232}Th and the lack of generation of plutonium and other actinides except for ^{236}U would allow the most efficient transmutation of accumulated power plutonium and other actinides in U-Th cycle. The deep burning of weapon-grade and power plutonium will be followed by the formation of plutonium mixture dominated by ^{242}Pu. This will make this plutonium unsuitable for the manufacturing of explosive devices and will minimize the plutonium α activity given the decay of ^{242}Pu and ^{238}U.

In thermal reactors about 60% of energy release is achieved through the fission of natural ^{235}U isotope and about 40% through the fission of nuclei of reproduced plutonium. The closed U-Th cycle will burn natural ^{232}Th with the fission efficiency for the generated uranium isotopes (^{233}U and ^{235}U) at fuel level 95-98%. This basically increases the resource of nuclear fuel of thermal reactors.

The physical, neutron physics, radiation and chemical properties of thorium and uranium generated from thorium in U-Th cycle more suitable for reactors relative to the start mixture of isotopes generated in U-Th cycle, the U-Th mixture opens additional possibilities for increasing the operational safety of reactors. Because of high breeding of active uranium isotopes in U-Th cycle there is no need for usually used redundancy of reactivity. As this redundancy decreases, the nuclear reactor safety increases. Maintaining the amount of active isotopes at the same level is followed by the stabilization of neutron flux during the campaign. Higher burnout together with the increase of the content of active U isotopes in U-Th relative to the content of plutonium in U-Pu cycle naturally increases the cost efficiency of reprocessing of exposed nuclear fuel. The progressive nature of cross-sections in the thermal region and higher cross-section of the neutron capture for Th as compared to ^{238}U facilitate the reactor control. Larger portion of independent outcome of ^{135}X by ~5 times in the fission of ^{233}U relative to ^{235}U, accordingly, to its burnout reduce the xenon pollution of U-Th reactor. The Th fuel is characterized by a higher radiation resistance which permits to increase the burnout depth per campaign [3]. The melting temperature of thorium and thorium dioxide is higher than that of uranium and uranium dioxide. Thorium is technologically more suitable than uranium.

REFERENCES

1. Holdren D., Ehirn D., Bdnitz R. et al. Control and disposal of excessive plutonium from weapons. National Academy Publishers, Washington, 1994.

2. Rahn F., Adamantiades A.G., Kenton J.E., Brann Ch. A guide to Nuclear Power Technology. John Wiley and Sons New York, 1984. Translation edited by Academician V.A.Legasov. Reference manual for nuclear energy technology. M., Energoatomizdat, 1989.

3. Murogov V.M., Troyanov M.F., Shmelev A.N. Using thorium in reactors. M., Energoatomizdat, 1983.

4. Freeman L.B., Beandoin B.R., Frederickson R.A., et al., Physics Experiments and Lifetime Performance of the Light Water Breeder Reactor. N.S.E., v. 102, p. 341-364, 1989.

5. Marshalkin V.E., Povyshev V.M., Trutnev Yu.A. Isotopic kinetics of recycled uranium fuel in the closed U-Th cycle. Submitted to VANT, 1995.

THE UTILITARIAN APPROACH TO THE DISPOSITION OF PLUTONIUM

CARL E. WALTER
Pleasanton, California, USA

1. Introduction

The U.S. National Academy of Sciences (NAS) report [1] on the management and disposition of weapon-grade plutonium makes a number of valid observations and recommendations. Among the observations are that: (1) weapon plutonium presents a "clear and present danger" and (2) there is little advantage to annihilating weapon plutonium in view of the large amount of plutonium contained in spent fuel resulting from operation of today's nuclear power plants. The latter observation led to a two-part recommendation that begins with conversion of weapon plutonium into a form that is similar to that of plutonium in spent fuel. NAS refers to this step as meeting the "spent-fuel standard" and considers it to be a sufficient first step in disposition of excess weapon plutonium. The second part of the recommendation is that attention must subsequently be given to the proliferation risks associated with the much greater, and growing, amount of plutonium contained in the world's nuclear spent fuel. This risk arises from the fact that technology for separating plutonium from spent fuel is well known and radiation barriers decay with time. We direct our attention here to the second step, but it also resolves the first step.

NAS takes the position [1] that disposition of weapon plutonium is of such paramount concern, that no other conflicting priorities should be appended in its implementation, for to do so would extend the period of clear and present danger. The U.S. Department of Energy (DOE) considers that use of an advanced liquid metal reactor (ALMR) system for disposition of weapon plutonium would be costly and not timely [2]. However, quantitative comparisons of time and cost with other disposition options in the U.S. and in Russia were not published. Nevertheless, DOE has dismissed the ALMR as a prime candidate for disposition of weapon plutonium, and in addition has suspended the successful development effort on the ALMR/Fuel Recycle System that had been ongoing. The latter action is based on a narrowly held opinion that recycling plutonium is counter to nonproliferation objectives [3]. In view of advances in reactor design and pyro-metallurgical processes, as well as the sizable inventory of light water reactor (LWR) spent fuel, there is little basis to support such an opinion today.

The decision by DOE to dismiss the ALMR as a viable option for disposition of weapon plutonium was made without regard to long-term energy resource conservation or nuclear spent fuel management. We believe that this decision process is flawed. It is incumbent on each nation to consider the overall consequences of any action that affects weapon proliferation, safety, the environment, energy resources, and the economy. In consideration of these factors, an all-encompassing effort that begins with disposition of weapon plutonium in ALMRs and ends with a viable nuclear power system that has the

E. R. Merz and C. E. Walter (eds.), Advanced Nuclear Systems Consuming Excess Plutonium, 287–296.
© 1997 *Kluwer Academic Publishers.*

capability to meet future national needs will be found to be both cost-effective and timely. There is synergism between (1) development and demonstration of an ALMR/Fuel Recycle System, (2) the first step of disposition of excess weapon plutonium, and (3) the second step that resolves the nonproliferation and environmental issues of LWR spent fuel. The advantages of this synergism should be recognized and the appropriate actions should be implemented. This is the utilitarian approach to the disposition of plutonium that could be used for weapons.

2. Energy Needs

World-wide energy needs over the next half-century (and beyond) will increase significantly. In particular, world electricity generation is projected to increase from about 12 petawatt-hours (PWh) per year (PWh= 10^{12} kWh) in 1995 to about 25 PWh/y in 2025 and about 40 PWh/y in 2050 [4]. Variations in projected energy needs depend on assumptions made in the projections relative to population, energy efficiency, and emission controls. These approximate nominal values represent an average annual growth of electricity generation of 2%.

Although the U.S. is already a "developed" nation, its electricity demand is also expected to increase– from just under 3 PWh/y in 1995 to 4.1 PWh/y in 2010 [5]. This increase in demand represents an annual growth rate of 2.1%. Unless the increased amount of electricity is produced in an environmentally benign way, the quality of life will tend to deteriorate. It does not seem reasonable to expect that photo voltaic, wind, and hydro power can provide all the world needs for electricity during the next century. Even strong, informed, proponents of renewable energy [4] agree on this point. Energy conservation can also be pursued, but extreme conservation would degrade our quality of life. Nuclear power must represent a substantial fraction of the electricity generated if the deleterious environmental effects of coal-fired electricity are to be avoided.

At present, emissions of the "greenhouse" gases (carbon dioxide and nitrogen oxide) from transportation sources and fossil-fueled electric utilities are comparable, although transportation produces far more carbon monoxide, nitrous oxides, and methane emissions [6]. Transportation emissions could be significantly reduced with increased reliance on electric-powered transportation, provided that the electricity is not generated by fossil-fueled plants. Such a change in transportation would further increase the demand for electricity. In the U.S., for example, depending on the efficiency and extent of electric transportation, an increase of 50% in electricity generation over currently predicted amounts could result [7]. Deployment of an improved nuclear power system thus merits a high-priority effort in the U.S. as well as in other countries.

3. Energy Resources

The energy potential in present and expected spent fuel from U.S. LWRs is appreciable. Fission of the plutonium contained in spent fuel from LWRs, would produce ~6 PWh, or about one and a half times the total electric energy estimated to be needed in the U.S. in 2010. A similar amount of energy could also be realized from the contained uranium-235. In a recent report [8], NAS recognizes the value of the energy resource in LWR spent fuel. NAS concludes that access to the spent fuel should be maintained in the presently presumed geologic repository for a reasonable period of time (~100 years) to avoid foreclosing alternative fuel strategies that may be in the national interest. In view of this conclusion, interim placement of the spent fuel in a repository designed for perpetuity seems illogical. Instead, a facility for monitored retrievable storage is called for. It seems imprudent to waste an energy resource by burying it in a geologic repository. Not only would its retrieval be costly, but the plutonium would continue to be a potential source of nuclear weapon material well into the distant future.

More than a thousand times greater energy potential, can be realized from the LWR spent fuel through utilization of the ALMR/Fuel Recycle System and depleted uranium. For comparison on the same basis, we calculated the energy potential of the ALMR/Fuel Recycle System, LWR/Once-Through Fuel Cycle, and indigenous (U.S.) fossil fuel resources. These values, based on current international and DOE energy resource information [9,10] are given in Fig. 1. In the U.S., using ALMRs vs. LWRs would result in ~345 times more energy sufficiency. In comparison, the large U.S. coal resource, generally considered to be vast, contains only about one-tenth the energy that could be provided by ALMRs with fuel recycle. Gas and oil resources can provide, respectively, only 5.3 and 4.2 times more energy compared to LWRs. In assessing the ALMR/Fuel Recycle System energy potential, we utilize depleted uranium as make-up "fuel". The amount of uranium available for LWR/Once-Through Fuel Cycle is assumed to be the reasonably assured resource in the U.S. at less than $130/kg [10]. The amount of depleted uranium in the U.S. includes the existing stockpile [11] and the additional tails expected to result from enrichment of uranium used to fuel existing

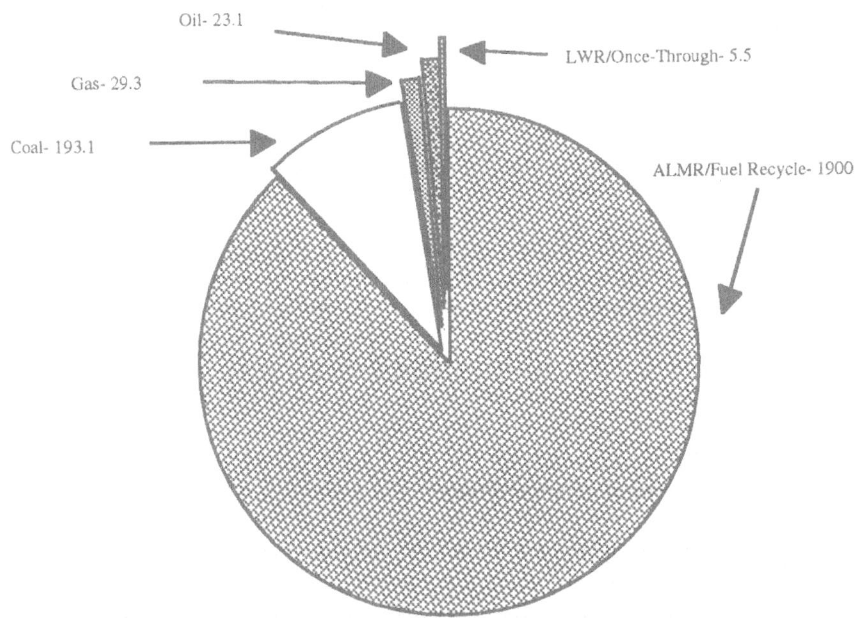

***Figure 1*. Energy (TWy) in indigenous U.S. resources. The life span of these resources can be inferred by considering that primary energy use in the U.S. in 1994 was 2.85 TWy [7].**

LWRs operated over their 40-y design life. Beneficial use of depleted uranium is the utilitarian approach to conservation of energy resources.

4. LWR Spent Fuel Disposal

At the end of 1995, the cumulative discharge of spent fuel from U.S. LWRs was ~32,300 t and will increase to 84,500 t by 2030 in the case of no new reactor orders [12]. Disposal of spent fuel from nuclear reactors has been discussed for years, but progress on this issue has been barely perceptible. Even in 1975 the pace of progress was characterized as "glacial speed". Now, 21 years later, approval and availability of the first geologic repository is still under investigation in the U.S. for LWR/Once-Through Fuel Cycle spent fuel. But even rapid progress is not likely to yield a satisfactory solution. The approach being taken suffers from a fatal flaw: no one can be held accountable for a malfunction of a mined geologic repository for even a minuscule fraction of the time (over a million years) that the warranty must be valid. From a nonproliferation viewpoint, there is no guarantee that the repository will not be breached, covertly or overtly, by a future society.

The historical background of geologic disposal, strongly influenced by political factors, has resulted in ever changing waste forms and geologic repository performance requirements. The proposed time of isolation has increased from 600 years to 10 million years during the 40-year time of the geologic repository "program". Now, performance requirements are leaning toward assuring that water immediately adjacent to the repository remain potable forever. Also, there is no general agreement within the program as to what the metrics of performance should be: e.g., dose, release, or risk.

Faced with this situation, an alternate, responsible, approach must be devised and implemented for proliferation-prone spent fuel. Some improvement is achieved by reprocessing spent fuel and recycling uranium and plutonium in mixed-oxide fuel in LWRs because the amount of spent fuel is reduced by a factor of three or four. The latter approach is being followed or planned in some countries, e.g. France, Russia, Japan. But even this reduction is insufficient to render the geologic repository unattractive as a source of plutonium for use in weapons. The optimum solution to prevent proliferation is to eliminate all but unavoidable dilute amounts of the actinides in the waste. Orders-of-magnitude reductions, as can be achieved by using ALMR/Fuel Recycle Systems, are required. This is the utilitarian approach to management of nuclear spent fuel.

5. Ethics of Geologic Disposal of Spent Fuel

In the past, unfortunately, technology has often been developed without regard for its ethical consequences. Lately, a closer tie is being made between technology use and its environmental and societal consequences. The Radioactive Waste Management Committee of the Organization for Economic Cooperation and Development, Nuclear Energy Agency (OECD/NEA) recently published its collective opinions on the ethical aspects of geologic disposal [13]. Although nonproliferation was not an explicit consideration in the formulation of these opinions, they appear to be strictly applicable to nonproliferation as well. Some of the collective opinions of this body are:

(1) The distinction must be made between radioactive waste that can and cannot be recycled.
(2) Generation of radioactive waste shall be minimized.
(3) Future populations must not be committed to continued expenditure of resources to provide future population protection.
(4) Interdependencies among all steps in radioactive waste generation and management must be appropriately addressed.

These collective opinions on ethics appear to be violated by geologic disposal of LWR spent fuel. Placing spent fuel that could be recycled in a repository clearly is contrary to opinions (1) and (2). Concern about future diversion of materials with high plutonium content will require continued surveillance and expenditure of resources. Thus, using the LWR/Once-Through Fuel Cycle violates opinion (3). Lack of consideration of the interdependency between national energy resources and once-through fuel cycles violates opinion (4). The opinions of this international committee are entirely consistent with a utilitarian approach. Conversely, the approach being pursued by DOE with respect to disposition of excess weapon plutonium, disposal of LWR spent fuel, and stewardship of energy resources is non-utilitarian.

6. Maturation of Nuclear Power in the U.S.

Currently, 109 LWRs with an installed capacity of 100 GW are in operation in the U.S. [14]. The potential power deficit (62 GW by 2025) [15] in the nuclear-electric power sector as LWRs reach their design life and are retired can be avoided by replacing retired LWRs with ALMRs. Beyond replacement of LWRs, additional ALMRs will be required in view of the predictable growth in electricity demand, the adverse environmental consequences of fossil-fueled plants, and the dim prospects for cost-competitive, ubiquitous, renewable energy electric plants. This is an appropriate transition to an advanced nuclear power sector that is safe, environmentally sound, and proliferation resistant.

ALMRs for a 100-GW capacity could be fueled with all of the plutonium and some of the uranium reclaimed from the LWR spent fuel. Calculations [15] indicate that plutonium from this source must be augmented by an additional supply of about 375 t of plutonium or highly enriched uranium (HEU). The additional material can be derived in several ways. The easiest resolution, however, is to use the 38.2 t of excess weapon plutonium [16] and some of the more than sufficient surplus of weapon HEU.

The relatively small amount of excess weapon plutonium takes on a disproportionate importance because it allows direct fabrication of fuel elements for the initial inventories of ALMR cores without the need for processing LWR spent fuel. Until the U.S. public becomes more comfortable with processing spent fuel, it would be advantageous to use excess weapon plutonium in the initial ALMR fuel. This comfort level will be reached by an appropriate public information campaign that describes the advanced pyro-metallurgical and electrorefining techniques that would be used, as opposed to the formerly only available aqueous reprocessing method. In addition, the public must become convinced of the need for continued reliance on nuclear energy. Use of excess weapon plutonium in this manner would also be consistent with the disposition approach preferred in Russia, and could lead to earlier action in both countries.

7. ALMR/Fuel Recycle System Characteristics

The possibility of breeding fissile material is greatly enhanced in a fast (unmoderated) reactor such as the ALMR. The conversion ratio in a fast reactor refers to the ratio of the plutonium produced in uranium to the amount of plutonium fissioned. This ratio is adjustable by core design [17]. The reactor can be tailored to make use of excess neutrons in various ways. Initially, the potential for breeding more fuel than was consumed was considered the most significant advantage of the fast reactor. Early power-sector scenarios envisioned a mix of LWRs and fast reactors operating at a conversion ratio greater than one. Spent fuel from both types of reactors would be reprocessed and incorporated into fresh fuel for the fast reactor and a number of LWRs. Alternatively, the reactor can be operated at a conversion ratio of 1.0, with no net production of fissile material. Exclusive use of ALMR/Fuel Recycle Systems has numerous advantages.

Three sources of fertile material for ALMR fuel may be considered: recycled uranium from LWR spent fuel, depleted uranium in enrichment tails, and newly mined natural uranium. The amounts of these resources are shown in Fig. 2. As can be seen from Fig. 2, less than 15% of the uranium in the accumulated LWR spent fuel would be needed for the initial inventories and makeup for 100-y operation of a 100-GW ALMR power sector. The remainder could be used for fueling the retiring complement of LWRs in U.S. or foreign reactors. The depleted-uranium stockpiles (enrichment tails) contain several times the mass of uranium in the spent fuel expected to be discharged from LWRs and would extend the viability of ALMRs for thousands of years. Expedient recycling of uranium for use in LWRs, would minimize further mining of uranium. Eventually, mining and enrichment of uranium would not be required.

ALMRs have a number of advantages over LWRs: (1) actinides are fissioned or transmuted and subsequently fissioned in the fast neutron spectrum, (2) long-lived fission products can be transmuted to shorter half-life isotopes in thermalized regions, (3) the equation of state of sodium permits higher reactor coolant outlet temperature at low ambient pressure, thus enhancing thermal efficiency and safety, (4) cores are typically ~1/3 as high, thus facilitating remote fuel fabrication, (5) use of metallic fuel elements (vs. LWR oxide fuel) with improved thermal conductivity reduces fuel material temperature and also simplifies fuel fabrication, and (6) in the steady state, with a conversion ratio of 1.0, the reactor consumes most of its own long-lived waste and requires only addition of uranium-238 as a fertile source of plutonium. In addition, initial safety reviews conclude that the modular ALMR design is passively safe and recent detailed cost analyses indicate that the cost of power from improved ALMR designs can be the same or lower than from LWRs [18].

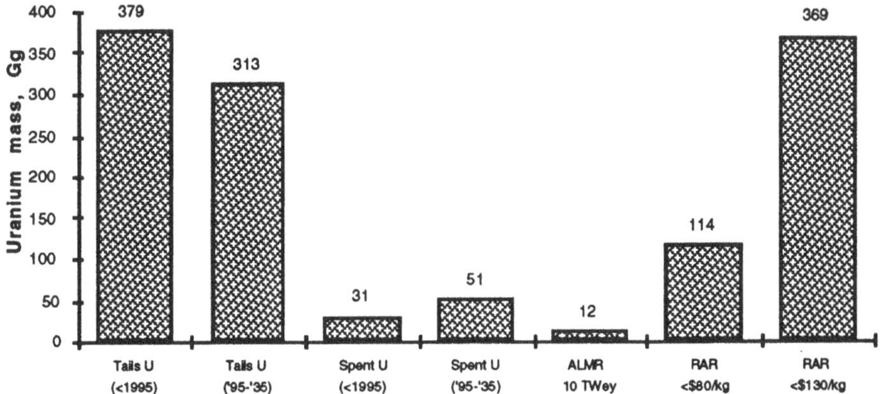

Figure 2. **Amounts of uranium contained in enrichment tails, LWR spent fuel, and reasonably assured resources (RAR) in the ground [10]. The uranium amount required for a 100-GWe ALMR sector for 100 years is shown for comparison.**

Increased demand for electricity would be accommodated by appropriate adjustment of the conversion ratio in all or selected ALMRs. The additional plutonium produced in this manner would be aggregated to provide first-core inventories for new ALMRs as needed.

8. Nonproliferation Aspects of ALMRs

Central to the issue of disposition of excess weapon plutonium, is determination of what constitutes disposition. The objective of disposition is clear– to prevent the reuse of plutonium in a nuclear explosive. Annihilation of excess weapon plutonium, as pointed out by NAS [1], does not solve the proliferation issue. The relatively much larger stockpile of plutonium isotopes contained in LWR spent fuel, when separated from the spent fuel, is not much more difficult to use for a weapon than the isotopes present in weapon plutonium. Therefore, to avoid proliferation of nuclear weapons, one must consider both excess weapon plutonium and the plutonium present in commercial spent fuel– worldwide!

There is a solution for managing plutonium in a power generating system. Accountability of all the plutonium in collocated reactor and fuel facilities would be maintained under International Atomic Energy Agency (IAEA) safeguards. At equilibrium, the incoming fuel-feed material to the ALMRs would be depleted or natural uranium and the outgoing material (waste) would consist of encapsulated short-half-life fission products with negligible actinide impurities. Generally, weapon-capable material would not be transported into or out of the power plant. This technology is well represented by the ALMR/Fuel Recycle System as discussed in several papers presented at the Global '95 conference [19].

The ALMR/Fuel Recycle System is based on pyro-metallurgical processing utilizing solvent electrorefining. Considerable process improvements have been made recently. These improvements not only produce a better product, but the cost of processing is greatly reduced, such that the resulting cost of power from the ALMR is competitive as stated above. Of particular importance to this discussion is the inherent proliferation resistance of the process.

Fuel would be recycled in facilities adjacent to the ALMRs, thus avoiding the need for public transportation. Initially, the recycling equipment would be used to partition LWR spent fuel into three products. One product would be a metal containing impure plutonium, all the minor actinides, and rare earth fission products. This product would be loaded into adjacent ALMRs within a short interim storage/process time. The second product would contain very pure uranium (< 1% uranium-235) that could be used directly for fabricating ALMR fresh fuel or combined with HEU for use in LWRs with remaining life. The third product, consisting of short-half-life fission products and no actinides, would be in a waste form suitable for geologic disposal. Long-half-life fission products would be recycled in the ALMR.

When ALMR fuel is being recycled, only two products result. One is a metallic mixture of uranium, plutonium, all the minor actinides, and some of the rare-earth fission products. This mixture would also contain zirconium, as the metallic fuel developed for use in the ALMR is typically 70% U, 20% Pu, and 10% Zr [17,18]. The second product consists of a waste form containing fission products with essentially no actinides. The waste form would be suitable for a high-level waste repository with significantly reduced performance requirements from those now considered for a spent fuel geologic repository.

The ALMR/Fuel Recycle System is extremely proliferation resistant. At no time does pure, separated, plutonium exist. The presence of the minor actinides in the plutonium makes it unusable directly for a nuclear explosive, as plutonium-238 and the minor actinides produce heat and preclude, or greatly impede, the construction of an explosive device. The rare-earth fission products remain with the plutonium and provide a significant radiation barrier for a few years. Fresh plutonium-bearing fuel would generally not be transported on public roads, as each complex of ALMRs would have fuel recycling capability. Material protection, control, and accountability would be easily implemented under stringent provisions for safeguards and security under IAEA inspection.

Relative proliferation resistance of alternative technologies can be evaluated by considering the possible scenarios. We consider the following three 100-GW electricity-generating scenarios for the U.S. in 2050: (1) retirement of all nuclear power when the existing LWRs complete their design life and replacement of the resulting power deficit by non-nuclear means, (2) long-term continuation of 100-GW generation using the LWR/Once Through Fuel Cycle, and (3) very long-term continuation of 100-GW generation using ALMR/Fuel Recycle Systems. In these scenarios, we assume that geologic repositories would each have the spent fuel capacity of Yucca Mountain. This is the site being investigated by DOE as a potential permanent geologic repository for high-level waste and 63,000 t of spent fuel.

The first scenario above would have an inventory of ~770 t of plutonium in two geologic repositories. The last LWR would have been retired in ~2035 [15], and the replacement power would be produced largely from coal-fired plants.

The second scenario would have an inventory of ~1280 t of plutonium in three geologic repositories. This inventory would grow beyond that amount at the rate of 18 t/y, so that by 2105 four repositories would be filled with 2270 t of plutonium. Every 32 years after that, a new geologic repository would need to be provided. The material in these repositories could be expected to be unattended by distant future societies. We note in passing that reasonably assured resources of uranium in the U.S. would be expended before 2050. Thus, continued use of LWRs would depend on imported uranium or the discovery of new U.S. resources.

In the third scenario, the plutonium inventory in the ALMR/Fuel Recycle Systems would remain at 1150 t under the continuous surveillance of a sophisticated international safeguard system. This is <u>the only utilitarian scenario</u>.

We assert that nonproliferation is best served by transparent management of plutonium in an on-going power producing operation. The alternatives, abandonment of spent fuel in multiple geologic repositories that are not likely to be safeguarded in the distant future, are proliferation prone. Not only is the third scenario most resistant to proliferation, it is the most benign environmentally and it is the only scenario that provides indigenous U.S. power that is sustainable for centuries.

Eventually, as a better source of electricity is found, the plutonium inventory in the ALMR/Fuel Recycle Systems would be reduced in a pre-planned, systematic, manner. This could be accomplished in various ways. Obvious solutions include operation at conversion ratios less than 1.0 and replacing uranium with a non-fertile material in ALMR fuel. Special means for annihilating the relatively small amount of remaining plutonium (~3 t) in the last ALMR module would be used.

9. Conclusions

Exclusive use of the ALMR/Fuel Recycle System provides a utilitarian approach to disposition of excess weapon plutonium and plutonium contained in spent fuel. This utilitarian approach appears to resolve the energy resource, safety, proliferation, environmental, and economics issues that are of concern. There is an orderly transition strategy for achieving a nuclear electric power sector that provides the U.S. (and the world) a sound method for meeting a growing demand for electricity. The systematic retirement of LWRs as they complete their design life and their replacement with ALMR/Fuel Recycle Systems appears to be an appropriate evolutionary advancement in electric power generation. Indeed, the utilitarian approach to disposition of plutonium becomes one of safe, secure, ethical management of plutonium.

By adopting a utilitarian approach, a proliferation-resistant nuclear power sector that observes internationally accepted principles of ethics can be maintained into the distant future. Further, this utilitarian approach can accomplish the disposition of excess weapon plutonium synergistically in a timely manner. Only a small amount of plutonium will need to be annihilated when nuclear power is no longer needed.

References

1. *Management and Disposition of Excess Weapons Plutonium*, National Academy of Sciences, Committee on International Security and Arms Control, National Academy Press, Washington, DC (1994).
2. *Storage and Disposition of Weapons-Usable Fissile Materials– Draft Programmatic Environmental Impact Statement, Volume I*, Department of Energy, DOE/EIS-0229-D, Washington, DC, February 1996.
3. *Summary Report of the Screening Process*, Department of Energy, DOE/MD-0002, Washington, DC, March 29, 1995.
4. *Renewable Energy: Sources for Fuels and Electricity*, Chapter 1, edited by T. B. Johansson, et al, Island Press, Washington, DC (1993).
5. *Electricity for the American Economy*, Edison Electric Institute (1995).
6. *Emissions of Greenhouse Gasses in the United States: 1987-1992*, Department of Energy/Energy Information Administration, DOE/EIA-0573, Washington, DC, November 1994.
7. Borg, I. Y. and Briggs, C. K., *U. S. Energy Flow –1993*, Lawrence Livermore National Laboratory, UCID-19227-93, Livermore, CA, October 1994.
8. *Nuclear Wastes– Technologies for Separations and Transmutation*, National Academy Press, Washington, DC (1996).
9. *International Energy Outlook 1995*, Department of Energy/Energy Information Administration, DOE/EIA-0484(95), Washington, DC (1995).
10. *Uranium: 1993 Resources, Production, and Demand*, Organization for Economic Cooperation and Development, Nuclear Energy Agency (1994).
11. Schwertz, N., et al, Depleted Uranium Hexafluoride: Waste or Resource?, published in *Proceedings of the International Conference on Evaluation of Emerging Nuclear Fuel Cycle Systems, Global 1995*, Versailles, France, September 11-14, 1995.
12. *World Nuclear Capacity and Fuel Cycle Requirements 1993*, Department of Energy/Energy Information Administration, DOE/EIA-0436 (93), Washington, DC, November 1994.
13. *The Environmental and Ethical Basis of Geological Disposal*, Organization for Economic Cooperation and Development, Nuclear Energy Agency (1995).
14. World List of Nuclear Power Plants, American Nuclear Society, *Nuclear News*, vol. 39 no. 3 pp. 29-44, March 1996.
15. Walter, C. E., A Strategy for an Advanced Nuclear-Electric Sector– Proliferation-Resistant, Environmentally Sound, Economical, published in *Proceedings of the International Conference on Evaluation of Emerging Nuclear Fuel Cycle Systems, Global 1995*, Versailles, France, September 11-14, 1995.
16. *Technical Summary Report for Surplus Weapons-Usable Plutonium Disposition, Department of Energy*, DOE/MD-0003, Washington, DC, July 17, 1995.
17. Magee, P. et al, Performance Analysis of the 840 MWt PRISM Reference Burner Core, published in *Proceedings of the 3rd JSME/ASME Joint International Conference on Nuclear Engineering* (p. 819), Kyoto, Japan, April 23-27, 1995.
18. Boardman, C. E. et al., "Integrating ALWR and ALMR Fuel Cycles," presented at the 4th ASME–JSME International Conference on Nuclear Engineering (ICONE-4), New Orleans, LA, March 10–14, 1996.
19. International Conference on Evaluation of Emerging Nuclear Fuel Cycle Systems, Global 1995, Versailles, France, September 11-14, 1995.

NEUTRON SPECTRUM CONCERNING EX-WEAPON PLUTONIUM DEGRADATION AND PLUTONIUM UTILIZATION IN THE FUTURE

H. SEKIMOTO
Research Laboratory for Nuclear Reactors
Tokyo Institute of Technology
O-okayama, Meguro-ku, Tokyo 152, Japan

1. Introduction

In Japan about 31% of the total electricity comes from nuclear power stations and extensive research and development are proceeded for peaceful uses of nuclear energy. Most Japanese nuclear specialists consider the incineration or degradation of ex-weapon plutonium as an urgent and important problems but not as their jobs, and almost no systematic study for this subject is performed in Japan.

In the present paper neutron spectrum concerning ex-weapon plutonium degradation and plutonium utilization in the future is discussed from scientific point of view. Figure 1 shows the total perspective of the present paper.

In Japan many LWRs are operated successfully for long periods. The amount of spent fuel has been accumulated very rapidly. The amount of plutonium in the spent fuel becomes an important problem. It is planed to be utilized soon as a MOX fuel in LWRs to decrease its stockpile.

The geological disposal is the main option of the planned high level waste final treatments, but it does not seem to obtain enough public acceptance. The research and development program on incineration of high level wastes (HLWs) is aggressively performed in Japan as the OMEGA project.

The ex-weapon plutonium incineration or degradation is a short term problem, though it seems an urgent problem. Even the LWR usage with HLW incinerator is also a short-term system, when the nuclear energy is considered as a future energy using total natural uranium (not only ^{235}U) which can be utilized for 10^4 to 10^6 years.

The transformation of weapon-grade plutonium to reactor-grade is discussed in Chapter 2. The MOX utilization in LWR and incineration of HLWs are discussed in

E. R. Merz and C. E. Walter (eds.), Advanced Nuclear Systems Consuming Excess Plutonium, 297–306.
© 1997 *Kluwer Academic Publishers.*

Chapter 3, and the future equilibrium nuclear state is discussed in Chapter 4.

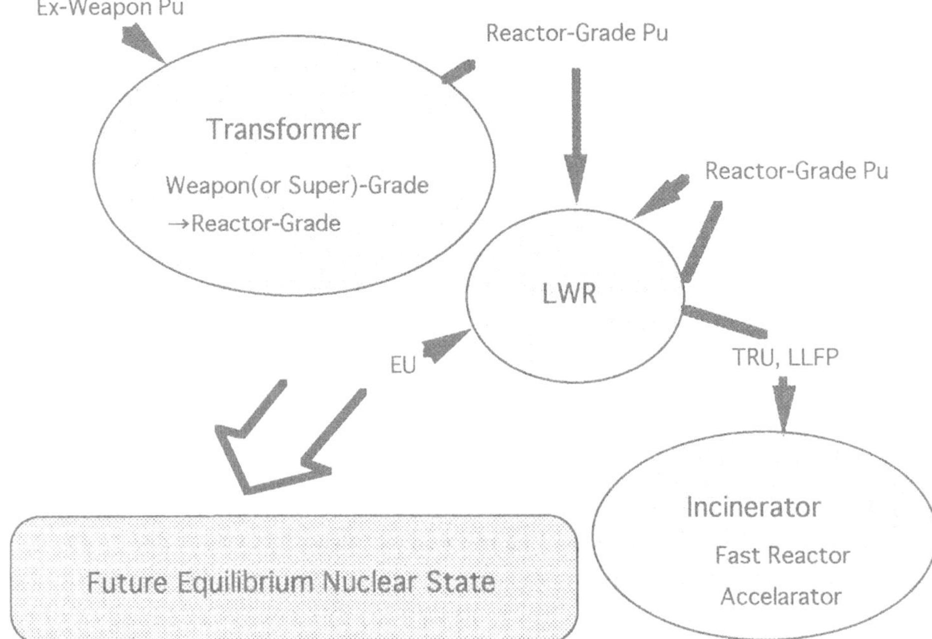

Figure 1. Total perspective of the present paper.

2. Transformation of Weapon-Grade Plutonium to Reactor-Grade

The isotopic concentrations of plutonium employed in this paper are shown in Table 1 for super-grade, weapon-grade and reactor-grade. In this paper it is premised that the super-/weapon-grade plutonium cannot be handled in conventional commercial nuclear facilities, but should be treated in special facilities. On the other hand the reactor-grade plutonium can be handled in the conventional commercial nuclear facilities. The degradation from super-/weapon-grade plutonium to reactor-grade plutonium or the complete incineration of super-/weapon-grade plutonium is required to solve the problem. The degradation seems easier and more rapid than the complete incineration. In the present paper the neutron spectrum effect on plutonium degradation is discussed. For the neutron spectrum, fast reactor spectrum (sodium-cooled oxide fuel reactor) and thermal reactor spectrum (PWR) are compared.

TABLE 1. Isotopic composition of various grade of plutonium

	Isotope				
	Pu-238	Pu-239	Pu-240	Pu-241*	Pu-242
Super-grade	0.00000	0.98000	0.02000	0.00000	0.00000
Weapons-grade	0.00012	0.93800	0.05800	0.00350	0.00022
Reactor-grade**	0.01300	0.60300	0.24300	0.09100	0.05000

* Pu-241 plus Am-241

** from PWR, 33Mwd/kg-burnup, 10yr-cooling

Figure 2. a shows the nuclide concentration change of the weapon-grade plutonium in the fast reactor spectrum along burnup. The super-grade plutonium shows the similar change. On the other hand the thermal reactor spectrum makes the change more rapid as shown in Figure 2. b. The total amount of plutonium decreases to a half of the original value with 2.6years of irradiation for the fast reactor spectrum and with 1.1years for the thermal reactor spectrum.

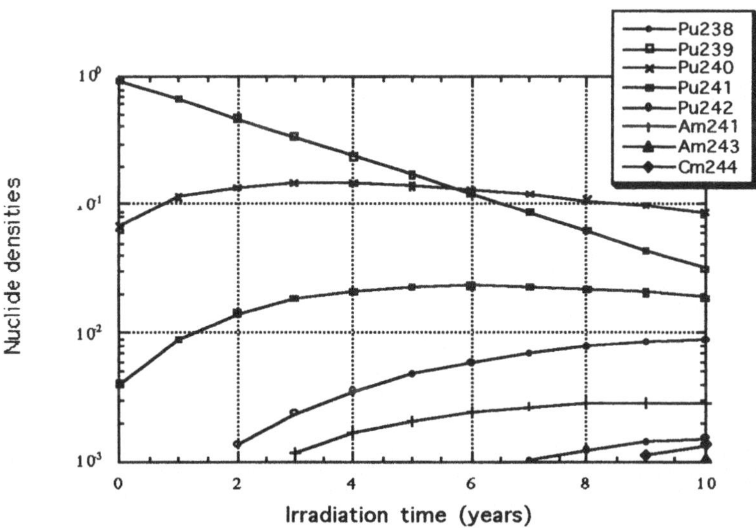

Figure 2. a. Nuclide number density change of the weapon-grade plutonium in the fast reactor spectrum along burnup

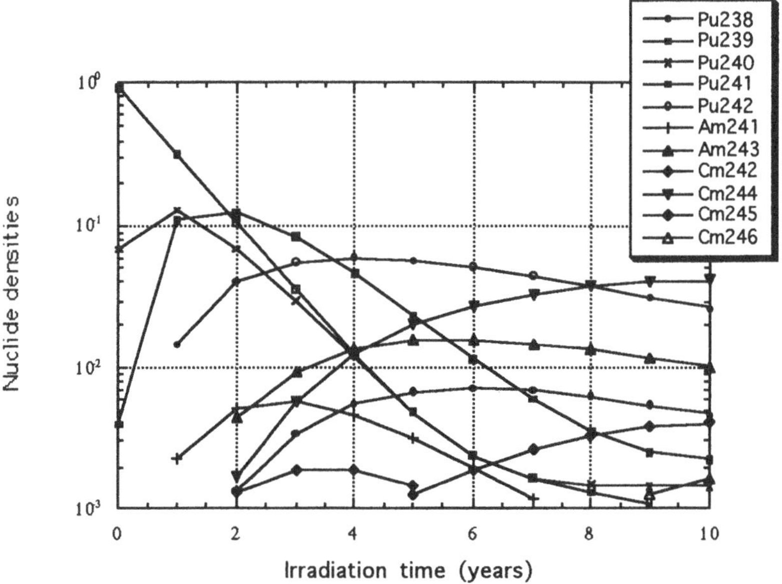

Figure 2. b. Nuclide number density change of the weapon-grade plutonium in the thermal reactor spectrum along burnup

TABLE 2. Irradiation time necessary for each characteristics of ex-weapon plutonium becoming reactor-grade

	Super-Grade		Weapon-Grade	
	Thermal	Fast	Thermal	Fast
Pu-239 isotopic abundance	1.0	4.5	0.9	3.6
η-value of total Pu	1.1	4.1	1.0	3.4
Total number of spontaneous fission neutrons after 3-yr cooling				
from actinides	0.7	2.3	0.5	1.5
from Pu	1.5	3.9	1.5	4.6
Decay heat after 3-yr cooling	1.2	5.8	1.1	5.1

The time required for each characteristics of super-/weapon-grade plutonium becoming reactor-grade is shown in Table 2. The difference between super-grade and weapon-grade is small. On the other hand the spectrum effect is considerable.

From both points of rapid decrease of total plutonium amount and rapid degradation from super-/weapon-grade to reactor-grade, the thermal reactor spectrum is superior to the fast reactor spectrum.

3. Japanese R&D Programs for Plutonium Utilization and HLW Incineration

In Japan LWRs are utilized successfully for many years, and their spent fuels have been accumulated to the level of almost full capacity of storage. In Long-Term Program for Research, Development and Utilization of Nuclear Energy in Japan issued in 1994 it is intended to start using MOX fuel in a few LWRs in the second half of the nineties and to increase the number of such reactors in a planned fashion, allowing for flexibility, to about ten by the year 2000 and ten odd by 2010. Once the super-/weapon-grade plutonium is transformed to reactor-grade, it is technically possible to be utilized in the Japanese MOX utilization scenario.

The high-level waste (HLW) disposal from LWR system is an important problem in Japan. The Atomic Energy Commission of Japan submitted in 1988 a report entitled "Long-Term Program for Research, Development on Nuclide Partitioning and Transmutation Technology". The program is called "OMEGA" which is the acronym derived from Option Making Extra Gains from Actinides and fission products. In this program JAERI has been carrying out studies on the partitioning and transmutation program in the following areas[1];

- Development of the four-group partitioning process,
- Design study of the transmutation systems;
  - Actinide burner reactor,
  - Accelerator-driven subcritical reactor,
- Development of an intense proton accelerator,
- Development manufacturing and pyrochemical reprocessing of nitride fuel,
- Basic research for supporting the development of transmutation systems.

The neutron spectra employed in these systems are fast energy spectra, since the harder neutron spectrum reduces the production of higher actinides. It is just an opposite way to the intention for the weapon-grade plutonium burning mentioned in the previous chapter.

The following system can be considered. The weapon-grade plutonium is transmuted to the reactor-grade plutonium in the thermal reactor designed, built and operated for this purpose, and the transmuted plutonium is utilized in commercial reactors. The long-life radioactive waste will be incinerated in the special incineration reactors.

4. Future Equilibrium Nuclear State

If we can utilize the whole of natural uranium, we can use fission energy for 10^{5-6} years, which may be enough for human beings. However, fission energy has its peculiar problems. The main problems peculiar to fission energy are safety, nuclear proliferation and radioactive waste. Among them safety is partly accepted even in the present time. In the future the total amount of energy production in the world will be increased and the total risk of the reactors operated in the world may be supposed to increase, but the safety technology will be also advanced. Even at the present time we have many reactor concepts much safer than the reactors presently operated. We can expect that in the future sufficiently safe reactors will be available, and the total risk will be acceptable. For nuclear proliferation, this risk will become more independent of the choice of whether we continue the utilization of nuclear reactors or stop it, since the bomb making technology will be advanced and the necessity of plutonium from the reactors will be diminished. The important item for nonproliferation will become rather the condition of society. An important part of this condition is the stability of the society, which must be insured with a good economy. Inexpensive energy is necessary to keep society in the condition that nuclear bombs will be never used. The waste problem, the last of the above three problems, will become more severe in the future than at the present time, since the long-life radioactive waste will accumulate as time passes. In this chapter a general discussion will be made on the waste problems of fission nuclear reactors in the future equilibrium state.

When a nuclear system is operated to produce constant energy in a changeless way for a long period, each nuclide density in the system converges to an equilibrium value[2]. We have named a hypothetical future society in this state the "equilibrium nuclear society". We have almost no information about it at present, though it can be considered as a goal of our present society for the above future scenario. In our model of the equilibrium nuclear society, we consider a nuclear energy center to confine the radioactive materials in it. Natural uranium and/or thorium is supplied to the center as a fuel fed to fission reactors. All of the actinides are recycled into the reactor. The end products of the heavy-isotope decay series (lead and bismuth) and the stable fission products are taken out from the center. The discharged short-life fission products are kept in the center until decaying out, and then follow the above process. The middle-life fission products (^{90}Sr, ^{137}Cs and ^{151}Sm) are stored in a repository in the center until decaying out. Whether the long-life fission products (LLFP; ^{79}Se, ^{126}Sn, ^{99}Tc, ^{93}Zr, ^{135}Cs, ^{107}Pd and ^{129}I) are to be taken out from the center or to be incinerated in the reactor will be left as an option in the present study.

The previous study[2] shows the following characteristics of the nuclide densities in the equilibrium state. The actinides stayting in the uranium cycle are heavier than those in the thorium cycle. The plutonium density is higher for the fast reactors than for the thermal reactors, but the densities of heavier actinides increase for the softer spectrum and higher flux level. To evaluate the criticality feature of neutron balance in the equilibrium state, the *h*-value quantity is defined as

$$h = \frac{\nu\Sigma_f}{\Sigma_a} = \frac{(\nu\sigma_f\,,\,\boldsymbol{n})}{(\sigma_a\,,\,\boldsymbol{n})} \quad ,$$

where (,) is an inner product and the elements of vector $\boldsymbol{n}$ are nuclide densities n_i for both heavy nuclides and fission products generated from inserted fuel in the reactor. It is similar to the η-value, but includes the fission product contribution. It is also similar to the infinite neutron multiplication factor, but does not include the contribution from coolant, structural materials and other core components. To make the reactor critical, $h > 1$ is a necessary condition for usually employed homogeneous configurations. The thermal reactors with natural uranium show *h*-values of about unity or less, so thermal reactors with natural uranium are supposed incapable of producing the equilibrium state. However, the values for fast reactors exceed unity and are generally 1.2 to 1.3. The hard-spectrum fast reactor in particular shows a higher *h*-value, and may easily attain the equilibrium state by employing proper reactor design and fuel management. For the thorium cycle, the thermal reactor shows an h-value of about 1.1 and the fast reactor shows about 1.15 for lower neutron flux levels. When we consider the parasitic neutron absorption in the reactor and want to incinerate some toxic materials, the option of a fast reactor with natural uranium seems the best. Even for this system the neutron economy remains still as a very severe problem.

The waste problem will become more severe in the future equilibrium society than the present society, since long-life radioactive waste will accumulate as time passes even in that society. Risk may be the direct index appropriate for evaluation of this problem, but it depends on the disposal method and environmental conditions, and furthermore the evaluation of the risk usually has large uncertainties and errors. On the other hand, the toxicity per unit radioactivity is given for most nuclides, and we can use it like nuclear data. The data is in several kinds of toxicity units and this introduces some ambiguities for our choice of unit, but the differences between these units are acceptable in our study. In the present study the annual limit on intake (ALI) is employed. In the reactor the toxic nuclides are produced, but at the same time the inserted fuel nuclides, which are natural toxic nuclides, are incinerated. We can evaluate easily whether the system is a toxicity increasing system or toxicity decreasing system. Some of the discharged toxic nuclides are much more soluble in water than uranium or thorium. For these nuclides the toxicity decreasing factor (produced toxicity / removed toxicity) of the system is estimated to be much lower than the risk decreasing factor (produced risk / removed risk). In these cases the toxicity should be used carefully. From the above considerations, in the present paper we discuss the possibility of equilibrium state fission nuclear systems satisfying the condition, that the toxicity of waste discharged from the nuclear energy center is less than the original fuel toxicity incinerated in the reactor while the amount of toxicity in the energy center remains constant.

A 3-GWt reactor makes ~2.8×10^{27} fissions per year, and at the same time fuel of ~1.1 t is charged. We can charge together the long-life daughters of the fuel with only negligible change of the neutron multiplication factor of the system. The toxicity of natural uranium with its daughter is ~1.3×10^{6}ALI. Annually the reactor produces fission products of ~1.1 t that is the same as the charged fuel mass. About 11% of them are LLFPs. The total toxicity of LLFPs is ~7×10^{5} ALI.[6] This value is not so sensitive to the reactor type or inserted fuel. It remains constant for almost 10^{5} years. The total toxicity of LLFPs is lower than the corresponding equilibrium toxicity of natural uranium plus its daughters, though it is higher than the corresponding pure natural uranium toxicity. The toxicity level of produced long-life fission products becomes the same as the level of incinerated toxic nuclides ~2000 years after incineration for uranium + daughter-thorium burning and ~20 years after for uranium + daughter-thorium + daughter-radium burning. These toxicity balances may suggest that the disposal of LLFPs be manageable[3].

All of the actinides are recycled into the reactor and always confined in it in the present system. However, it is impossible to confine them perfectly in the system, and a small part may leak from the system finally into the biosphere. The most likely mechanism of leakage may be contamination of the fission products to be discharged to the environment. The separation of the middle-life fission products meets a similar problem, but their toxicity will decay to the incinerated fuel toxicity level within 600 years. The problem of the separation of actinides is more important for ensuring the condition that the toxicity in the environment outside the nuclear center will never increase. In this section we discuss the upper limit on the leakage of actinides from the reprocessing process that is acceptable from the view point of the toxicity balance between production and incineration. The toxicity of actinides depends much on the feed fuel (uranium or thorium) but little on the reactor type. In this paper we discuss uranium fuel, though the toxicity of reprocessed actinides for the natural uranium system is roughly one order of magnitude higher than for the thorium system. The toxicity changes with time after leaving the reactor, that should be compared with the incinerated toxicity. The acceptable leakage rate, which enables toxicity balance outside the nuclear center, depends on the time after leakage as shown in Table 3. In this table the time lag between fuel charging and discharging is neglected.

TABLE 3. Acceptable leakage rate of actinides at reprocessing

Decay time	Incinerated fuel		
(year)	U	U+Th	U+Th+Ra
10	1.1×10^{-8}	3.0×10^{-8}	9.5×10^{-8}
100	2.6×10^{-8}	7.2×10^{-8}	3.6×10^{-7}
1000	7.4×10^{-8}	5.1×10^{-7}	1.1×10^{-6}

After 10 years cooling, the acceptable leakage rate for the pure uranium feed case is very low ($\sim 10^{-8}$) and becomes mitigated ~10 times for the uranium + thorium + radium feed case. After 100 years cooling, it will be mitigated to $\sim 4 \times 10^{-7}$ for the uranium + thorium + radium case. After 1000 years cooling, the acceptable leakage rate will be considerably reduced to $\sim 10^{-6}$ for the uranium + thorium + radium feed case. The 1000 years confinement is about the same as that required for the middle-life fission products. Though this period seems to be too long for human control, it may be a confirmed period for waste confinement by underground artificial barrier, judging from archaeological studies. Alternatively, for the future nuclear equilibrium society, the waste may be stored in the energy center under sufficient human control.

The future equilibrium state has been investigated generally for the full scope of nuclear systems of one reactor type fed with one fuel type. These studies show us the following conclusions;

- The system of fast reactors with natural uranium can give us energy for very long periods such as 10^{5-6} years.
- In the future equilibrium nuclear society, the waste problem may become more severe than the safety and proliferation problems.
- Neutron-rich fast reactors do not require any major design changes for incineration of just actinides.
- The incineration of LLFPs requires large design changes of the reactors even for neutron-rich fast reactors.
- The total toxicity of LLFPs is about the same or less than the total toxicity of feed natural uranium and its secular equilibrium daughters.

- The requirement not to increase the toxicity in the global environment (outside the nuclear center) makes little impact on the nuclear reactor design, but applies severe criteria to separation factors for discharged materials.

5. Conclusions

From both points of rapid decrease of total plutonium amount and rapid degradation from super-/weapon-grade to reactor-grade, the thermal reactor spectrum is superior to the fast reactor spectrum.

However, for efficient utilization of natural uranium to utilize whole ^{238}U, the fast spectrum is inevitable. To incinerate HLW the fast spectrum is superior to the thermal spectrum, since higher energy neutron leads to a neutron rich system and reduces the heavier components of actinides.

6. References

1. Mukaiyama, T., Kubota, M.,Takizuka, T., Ogawa, T., Mizumoto, M., and Yoshida, H. (1995) Partitioning and Transmutation Program "OMEGA" at JAERI, *Proc. of the International Conference on Evaluation of Emerging Nuclear Fuel Cycle Systems GLOBAL'95*, vol. 1, pp. 823-830, Versailles, France.
2. Sekimoto, H. and Takagi, N. (1991) Preliminary Study on Future Society in Nuclear Quasi-Equilibrium, *J. Nucl. Sci. Technol.*, **28**, 941-946.
3. Sekimoto, H., Nakamura, H., and Takagi, N., Toxicity of Radioactive Waste Discharged from Nuclear Energy Center in the Future Equilibrium State, *Ann. Nucl. Energy*, **23**, 663-668.

PROBLEMS OF RADIOCHEMICAL REPROCESSING, RELATED TO PLUTONIUM UTILIZATION

B. Zakharkin
State Scientific Center,
Acad. Bochvar All-Russian Scientific Research Institute of Inorganic Materials, VNIINM

Plutonium is an integral part of nuclear power. Synthesized from uranium in neutron flows from nuclear power facilities, plutonium is accumulated in spent fuel unloaded from reactors.

Depending on the accepted model of the fuel cycle, plutonium can remain in a "bonded" form (in a fuel composition with heavy-nuclei fission products — "open" fuel cycle) or be separated in a pure form as a result of radiochemical reprocessing of spent fuel ("closed" fuel cycle).

From the beginning of nuclear power development in Russia, plutonium was considered a highly efficient fuel component capable of replacing (in due course) to a significant extent natural uranium in what concerns fuel supply. Such a problem could be solved only within the limits of the closed fuel cycle the concept of which dictated the goals and objectives of the efforts of scientific research, designing, etc. institutes and industrial enterprises of the Ministry of Medium Machine Building and later the Ministry for Atomic Energy. Such a choice met the goals of the most efficient nuclear power development, namely, the maximum attainable use of potential energy of heavy-nuclei fission, attainment of the regime of self-provision with fuel with secondary-fuel recycling. In this context, it was quite logical to assign the top priority to plutonium utilization in fast neutron reactors.

A significant milestone of the practical implementation of closed fuel cycle was the putting of the RT-1 plant, a multipurpose radiochemical complex, into operation in 1976-1977. For nearly two decades the plant has been providing the reprocessing of spent fuel from large- and small-scale nuclear power facilities (Fig.1). Owing to some difference in the categories of the materials being reprocessed, the problem of using secondary, unburnt, uranium has been successfully solved, i.e., the fuel cycle was closed in this component.

The problem of plutonium utilization is a much more sophisticated one. At the RT-1 plant, plutonium subjected to deep purification is separated primarily as dioxide. Energy (fuel) type plutonium may feature a significant qualitative scatter: a "low-background" material (the content of Pu-238, Pu-240 and Pu-241 is under 3 mass percent) is obtained after reprocessing of fuel from BN-600, BN-350 fast reactors whereas other categories of spent fuel give a "high-background" plutonium (the content of the afore-named isotopes is over 30 mass percent).

In conformity with the general line, all energy type plutonium should be utilized in power industry via returning it to reactors of nuclear power plants. For this purpose, R&Ds have been launched since the early 1950s to develop a technology for obtaining uranium-plutonium oxide fuel and fuel elements and assemblies based on such fuel. As a result, for the time being the Ministry for Atomic Energy possesses a wide set of reliable processes for obtaining of U-Pu fuel for fast neutron reactors, which have been successfully tested on experimental industrial facilities (see the table below). It has been confirmed that high-quality fuel can be obtained under irradiation in cores of BN-600, BN-350 and BOR-60 reactors. Research into mixed fuel for thermal neutron reactors is currently at an essentially earlier phase of development as this direction has been evolved only for the last three years.

E. R. Merz and C. E. Walter (eds.), Advanced Nuclear Systems Consuming Excess Plutonium, 307–311.
© 1997 *Kluwer Academic Publishers.*

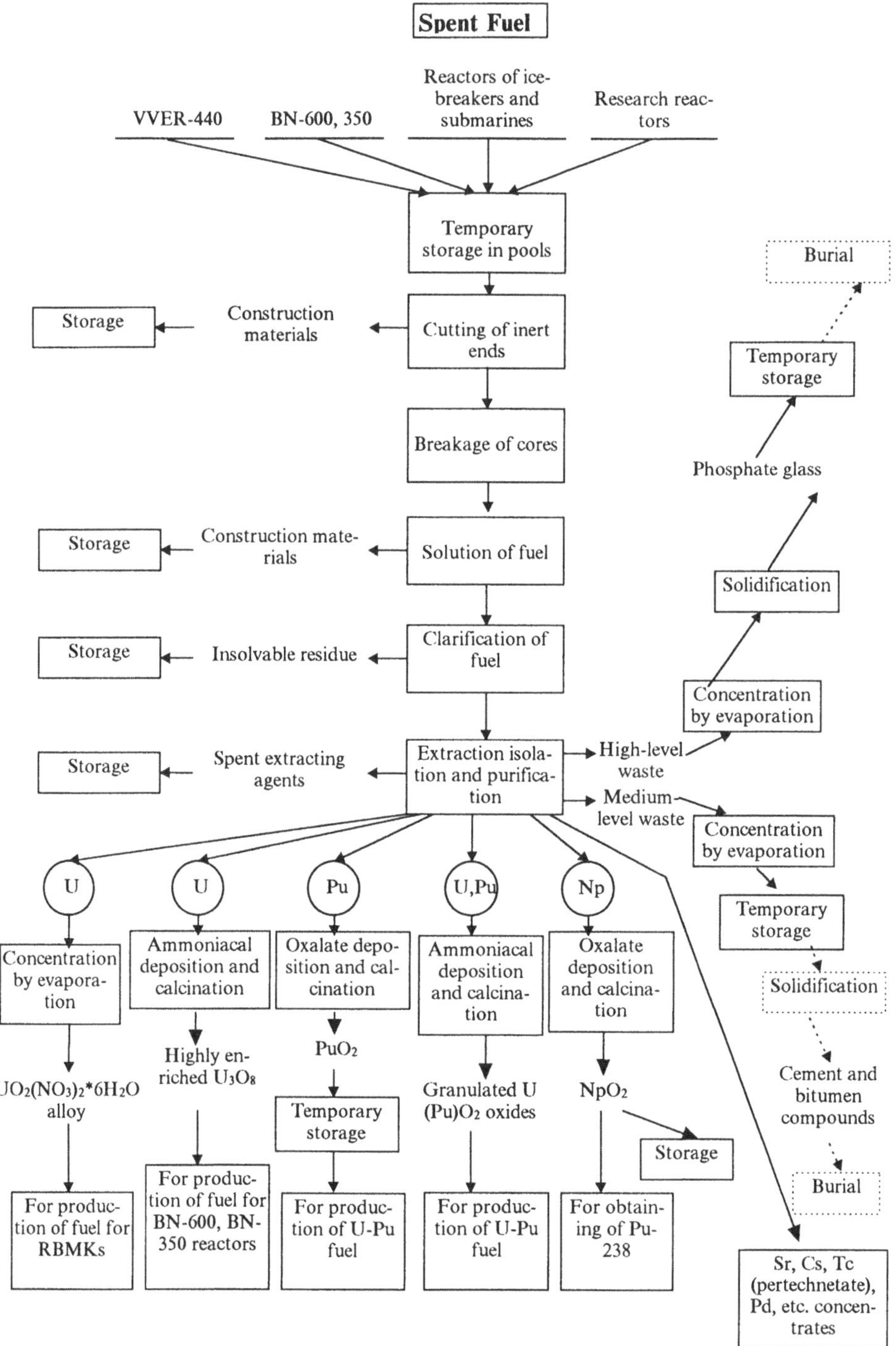

Fig.1. Functional capabilities of the RT plant. Dotted lines indicate operations capacities for which are now at the stage of construction or validation (waste burial).

Experimental and experimental industrial sites for U-Pu fuel production

Site definition	Location	Operation period	Purpose	Fuel composition and process of its obtaining	Type of Pu used	Capacity
Laboratory lines (sites)	VNIINM, Moscow	Since early 1950s until now	Obtaining of experimental fuel samples and fabrication of individual fuel elements (FEs)	Delta-phase alloys of Pu; PuO_2; $(U, Pu)O_2$ and other processes	Weapon	—
Experimental site	"MAYAK" SPA, Chelyabinsk	1960-1970s	Production of pellets and experimental FEs for research reactors	Pu alloys; PuO_2	Weapon	Total mass of used Pu — about 1 ton
Experimental complex	NIIAR, Dimitrovgrad	Since 1985	Production of U-Pu fuel, fabrication of FEs, fuel assemblies (FAs) for testing in fast reactors	$(U, Pu)O_2$; electrochemical granulation and vibrocharging of FEs	Weapon and energy	40-50 Fas per year at fuel production site (350 kg of Pu)
"Granat" experimental industrial facility	"MAYAK" SPA, Chelyabinsk ("300" complex)	Since 1988 until now	Production of U-Pu fuel for testing in fast reactors	$(U, Pu) O_2$; ammoniacal granulation of coprecipitated Pu and U compounds	Weapon quality	35 kg of Pu/year (for 10 FAs)
"Paket" experimental industrial facility	"MAYAK" SPA, Chelyabinsk ("300" complex)	Since 1980 until now	Production of U and Pu dioxides, fabrication of FEs for testing in fast reactors	$(U, Pu) O_2$; obtaining through both mechanical mixing of individual U and Pu oxides and sol-gel process and ammoniacal granulation	Weapon quality	5-6 tons of Pu/year
Experimental industrial complex for mixed-fuel production (first line)	"MAYAK" SPA, Chelyabinsk ("300" complex)	50-percent completeness of production capacities	Production of U-Pu fuel, fabrication of pellets, FEs and FAs for using in industrial type fast reactors	$(U, Pu) O_2$; obtaining through coprecipitation of U and Pu	Weapon and energy	5-15 tons of Pu/year
Experimental industrial complex for mixed-fuel production (second line)	"MAYAK" SPA, Chelyabinsk ("300" complex)	Design development (feasibility study stage)	Production of U-Pu fuel, fabrication of pellets, FEs and FAs for using in VVER type reactors	$(U, Pu) O_2$; obtaining through coprecipitation of U and Pu	Weapon and energy	

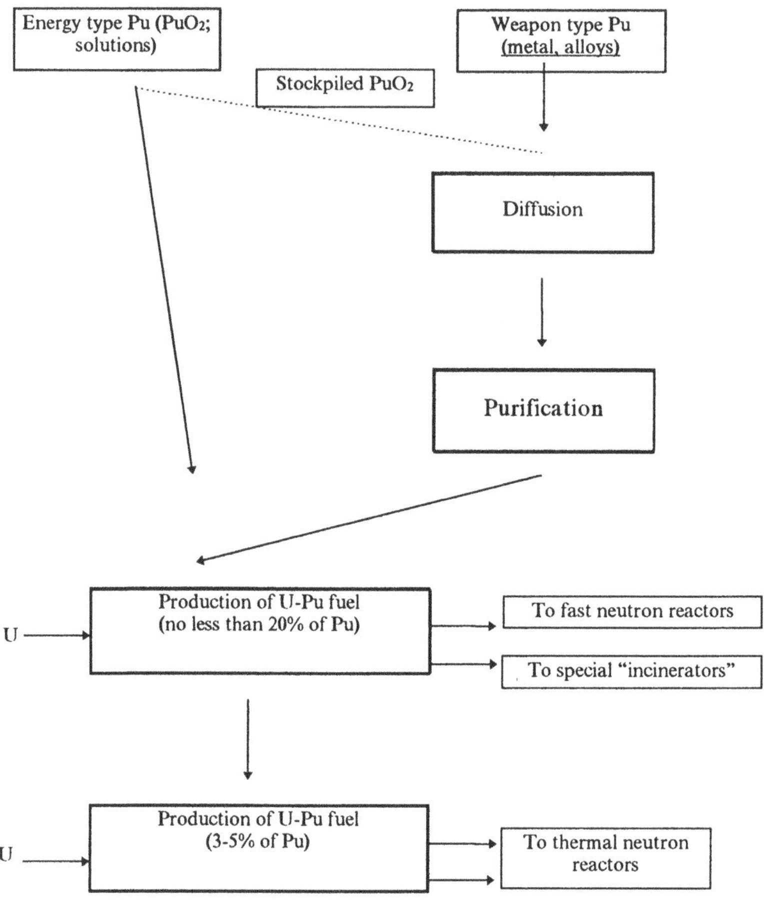

Fig.2. Schematic diagram of plutonium conversion for purposes of power industry

The first efforts aimed at practical implementation of the novel fuel direction using plutonium go back to the early 1980s. Simultaneously with the adoption of a program on development of the Ural series of BN-800 fast reactors, the construction of a special technological complex ("300") for production of U-Pu fuel with a productivity of 5-6 tons of Pu per year was initiated in the vicinity of the "MAYAK" Scientific and Production Association. However, the construction and mounting work actively started in the late 1980s was fully frozen at the stage of 50-percent completeness because of the economic crisis broken out in Russia. In the recent years, some limited-scale activities on fuel production in "Granat" and "Paket" experimental production facilities ("MAYAK" Association) were kept going on. Feasibility studies into siting of a line for mixed-fuel production for thermal reactors (VVERs) in the territory of the "300" complex or construction of an independent facility with a productivity of 1000-2000 kg of Pu per year with the capability of producing fuel assemblies for both fast and thermal reactors have been and are still being performed. The possibility of using plutonium in "Kandu" reactors is now under investigation.

The need in the afore-mentioned feasibility studies was dictated by a new problem, namely, the necessity of utilizing weapon plutonium from nuclear warheads considered excessive for defense purposes within the framework of pertinent Russian-American agreements.

Evidently, the ways and processes for implication of weapon and energy type plutonium into power industry via mixed fuel are identical, with the exception of some peculiarities due to difference in the isotopic composition and background radiation. Facilities intended for obtaining U-Pu fuel should take into account the possibilities of processing materials containing both energy and weapon type plutonium or their mixtures. In the case of weapon type material, the technology should include additionally the operation of diffusion of metal plutonium and its alloys as well as that of purification of plutonium from impurities (in particular, gallium) in conformity with the diagram given in Fig.2.

So:

- the use of plutonium in nuclear power is the only way of plutonium utilization acceptable from the technical viewpoint;
- this task should be carried out for both energy ("high-background") and weapon (excessive) types of plutonium;
- the full technological validation of specific ways and processes for plutonium conversion, irrespective of plutonium origin, into energy fuel is available;
- investments for construction of facilities to produce U-Pu fuel for fast and thermal reactors are now required.

A potential global hazard characteristic of nuclear technologies and, in particular, production of plutonium as a fissile and highly toxic material motivates both a higher responsibility of the countries possessing such technologies and the necessity of international cooperation and collaboration in this field.

French Law on Waste Management Research - General View on Partitioning Program and Current State of Partitioning Studies

J. Lefevre, CEA-DCC/DIR, Saclay, France
* With a large contribution from Michèle VIALA, CEA, Fuel Cycle Division

• Abstract

Because National Waste Management Agency (ANDRA) met any opposition in 1989 during HLW and long lived MLW disposal sites investigations, the French Government decided a temporary halt and, after a subsequent political discussion, edicted a law, December 30, 1991, to relaunch in a better public and political acceptance climate, the HLW and long lived MLW ultimate disposal management ways to choose. This law consisted in a long-term research program, until year 2006, on three research axes :

1. Partitioning and transmutation of long-lived wastes in order to minimize their long-term toxicity.

2. Assessment of the capacity of geological structures to confine radioactive wastes research performance in underground laboratories under ANDRA leadership.

3. Improvement of waste packaging techniques and prolonged surface storage.

Evaluation of this R&D program is made early until 2006 by 12 members of «Evaluation National Commission». Flowsheets with general R&D planning for these Research axes are presented.

More details are given on Partitioning and Transmutation R&D program concerning Plutonium cycle and minor actinides management.

Examples of long-lived radionuclides partitioning are given.

*

* *

In the frame of the French nuclear energy program, waste management aspects were taken in account from the beginning. In the first stage it was chiefly treatment and conditioning of low and medium level waste, then, from 1959 the vitrification of high level waste. Progressively a management structure, legally and reglementary organization, safety control systems were put in place. A coherent coordination was settled between the three large nuclear partners, EDF the national french utility, COGEMA created in 1976 as nuclear fuel cycle industrial operator, and CEA for all aspects of nuclear R&D in FRANCE. It was in 1979 that nuclear waste management entity was created, ANDRA as a national Agency but within the CEA, chiefly in charge of waste disposal problem.

At the Government level Ministry of Industry was in charge of civil nuclear program and Ministry of Defense for military nuclear problem. Nevertheless it was agreed that the waste management aspect will be treated as a common land for civil and military programs.

E. R. Merz and C. E. Walter (eds.), Advanced Nuclear Systems Consuming Excess Plutonium, 313–329.
© 1997 *Kluwer Academic Publishers.*

From 1983 to 1989 a first large program of disposal sites investigations was performed by ANDRA on 4 different host rocks, clay, granite, shell and salt. This program was stopped by a moratorium decided in April 1989 by the French Government after strong local public opposition on one of the four sites investigated.

A parliamentary was designated as a mediator to find a new approach with local representatives and public concerned to reach a better acceptance for underground laboratories sites.

It is important to underline the fact that an underground laboratory, at this stage, is devoted to the site qualification for a future effective waste disposal site.

To avoid premature opposition it was also decided to choose at least two underground laboratories, and, to select only at the end of on-site investigations (at least 8 years after) the waste underground disposal site.

The mediator, Christian BATAILLE, started by a large consultation and public hearings in order to bring to the fore different areas where local representatives and public agreed the principle of a radioactive waste disposal site. These, he found about thirty different places even with some spontaneous candidatures.

These different areas were examined by ANDRA for a first selection based chiefly on geological characteristics.

At the same time, Christian BATAILLE, proposed to the french Government a law taking in account requests or observations registered during the hearings.

The three main technical questions asked during public hearings were :

1- Is it possible to minimize global radiotoxicity of waste to dispose of ?

2- Is it possible to minimize dose rate delivered to men in the future from disposal sites ?

3- If minimization of radiotoxicity is sufficiently high, is it possible to avoid geological disposal ?

On the other part modifications were requested on procedures and organization.

First of all independancy of ANDRA which is now separated from CEA and rattached directly to three ministries : Industry, Research and Environment.

Then, a procedure of information and consultation with the creation of Local Information Committes including a large representation in each site area, creation of a National Assessment Committee described more in details further.

1. THE FRENCH LAW OF DECEMBER 30, 1991

Execpt the aspects of organizations or procedures indicated just before, the law is chiefly a R&D law with three main axis defined in rather general terms in its article 4 :

axis 1 - Research of conditions allowing the separation and transmutation of long-lived radionuclides existing in high and long-lived waste.

axis 2 - Study possibilities of disposal retrievable or not retrievable in deep geological formations, in particular by implementation of underground laboratories.

axis 3 - Study of conditioning processes and long term interim storage in surface for these wastes.

It was decided to select a research organism leader for each axis : CEA* for axis 1 and 3, and, ANDRA* for axis 2.

These organisms are in charge of research in collaboration with producers EDF* an COGEMA* and with others research entities such as CNRS*, BRGM*, Universities.

A coordination system is put in place between the two research leaders CEA and ANDRA to define a R&D Strategic plan, taken in consideration industrial aspects by a coordination with producers EDF and COGEMA.

The General R and D planning for radioactive waste management defined in 1995, in the frame of the December 30, 1991 law is represented ou **Figure 1**.

To respect the subject of this seminar, the axis 1 will be developed more in details, chiefly for the partitioning aspect because the transmutation program has been presented here by M. M. SALVATORES.

The law precised these three research programs have to be conducted with equivalent means and the same determination at least during 15 years, that is to say until the year 2006.

In the same article 4 of this law it is specified a National Assessment Committee is created to assess the three areas of research every year and produce an annual report with recommendations on the quality of researchs and orientations to follow.

This Committee is made of 12 members appointed on January 27, 1994, set up by the supervising three ministries of Industry, of Research and of Environment.

The Committee aims to be independent to ensure the quality of its assessment and the credibility of its conclusions and recommendations.

The Committee works towards the largest consensus of all its members. The Committee has set an objective of widespread communication of its assessment to Ministerial Advisors, Parliamentary Bureau, Press, Local Information Committees.

The Committee produced already two reports the first presented on June 27, 1995 and the second one on June 27, 1996.

* See glossary.

GENERAL R AND D PLANNING FOR RADIOACTIVITE WASTE MANAGEMENT

DECEMBER 1991 LAW

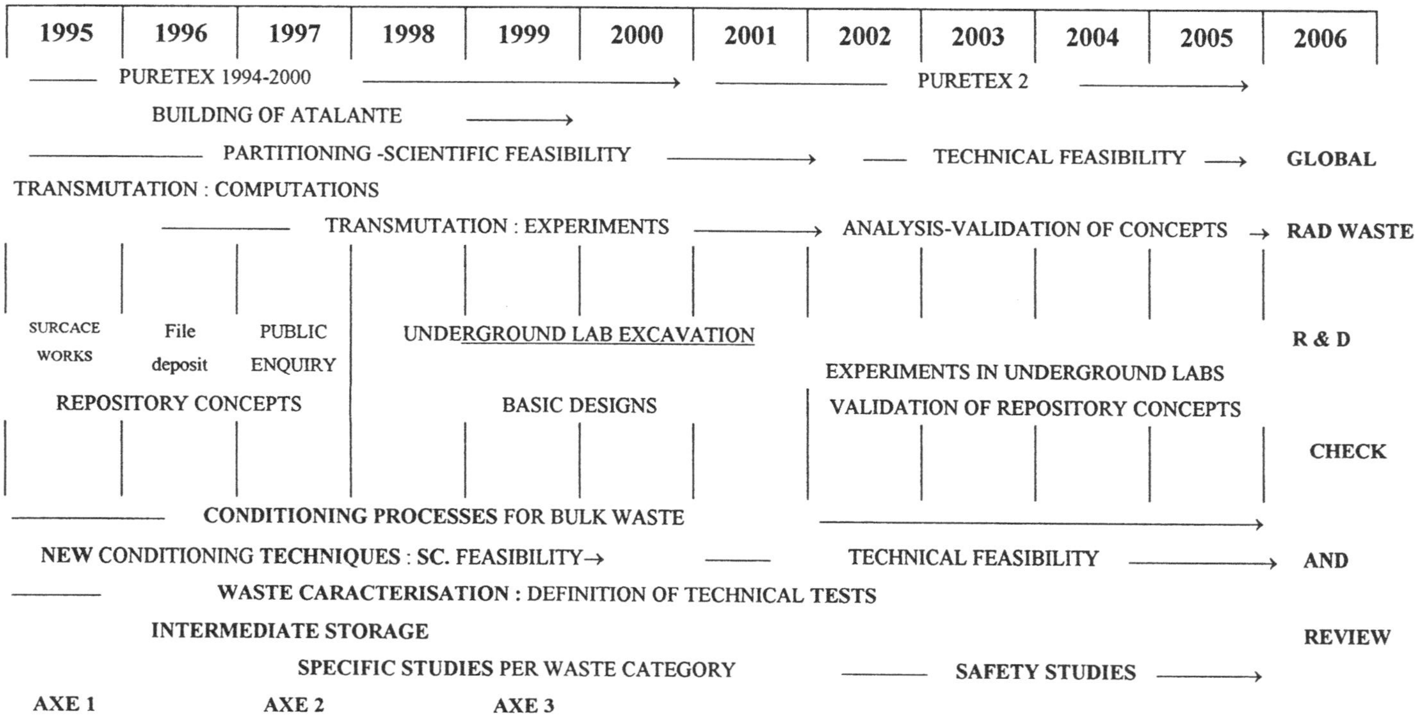

1995	1996	1997	1998	1999	2000	2001	2002	2003	2004	2005	2006

PURETEX 1994-2000 → PURETEX 2 →

BUILDING OF ATALANTE →

PARTITIONING -SCIENTIFIC FEASIBILITY → TECHNICAL FEASIBILITY → **GLOBAL**

TRANSMUTATION : COMPUTATIONS

TRANSMUTATION : EXPERIMENTS → ANALYSIS-VALIDATION OF CONCEPTS → **RAD WASTE**

SURCACE WORKS | File deposit | PUBLIC ENQUIRY | UNDERGROUND LAB EXCAVATION **R & D**

EXPERIMENTS IN UNDERGROUND LABS

REPOSITORY CONCEPTS | BASIC DESIGNS | VALIDATION OF REPOSITORY CONCEPTS

CHECK

CONDITIONING PROCESSES FOR BULK WASTE →

NEW CONDITIONING TECHNIQUES : SC. FEASIBILITY→ **TECHNICAL FEASIBILITY** → **AND**

WASTE CARACTERISATION : DEFINITION OF TECHNICAL TESTS

INTERMEDIATE STORAGE **REVIEW**

SPECIFIC STUDIES PER WASTE CATEGORY — **SAFETY STUDIES** →

AXE 1 **AXE 2** **AXE 3**

Figure 1

The final report will be presented on year 2006 with (as possible) the disposal site proposition to the Government which after Parliamentary Office for scientific and technologie decisions will submitt to the Parliament a new law to operate the Radioactive waste management and disposal site as decided

2. AXIS 1 - PARTITIONING AND TRANSMUTATION

The French law of December 1991 on radioactive waste management requires this management to comply wiht the protection of nature, environment and health, taking the rights of future generations into consideration. It explicitly indicates that solutions must be found to enable the partitioning and transmutation of the long-lived radionuclides present in these wastes.

Among these technical recommendations, the National Assessment Committee requested that thought concerning the order of precedence to be assigned to the long-lived radionuclides to be separated and transmuted be continued and that quantified goals be defined. It also requested that the studies and research be organised so as distinguish between the short and medium-term options concerning already industrial systems, or those undergoing industrialisation, and the longer-term options which are based on the overall foreseeable innovative systems. Lastly, it considered sceanario studies and their contribution in the acquisition of highly diversified data bearing on radionuclide inventories, flows of material, economics, safety, etc. These studies are necessary to assess the partitioning-transmutation systems that result in particularly complex schemes.

The planning for partitioning processess is given ou **Figure 2** and the planning for transmutation studies is given ou **Figure 3**.

As mentioned before we developed now only the partitioning program.

2.1. Partitioning : Assets and Limits

Radionuclides are found in non-reprocessed spent fuels, in type B wastes and in glass **(table 1)**.

To achieve transmutations or fissions, the radionuclides must be submitted to a sustained neutronic flux. These conditions lead to the use of ceramic-type materials able to withstand high temperatures, thermal flux and mechanical constraints. It is therefore necessary to have concentrated and purified radionuclides. (The ideal would be an isotopic purity). With the removal of the long-lived radionuclides from the wastes as a goal, the partitioning must be performed with the highest possible efficiency before proceeding to the purification, concentration and production of the ceramics.

The type B wastes represent large volumes in which the radionuclides are diluted. They must be decontaminated to a point where they can be stored on the surface. The feasibility of such operations depends on the type of waste (concrete, metals, etc.) and on the nature of the contamination (labile, fixed etc.). The performance required is beyond our present know-how. On the other hand, the quantities of radionuclides recovered would be small and easy to manage with those from the bulk of the spent fuel.

The spent fuel, when reprocessed, is quasi-totally dissolved. The valuable materials, U and Pu, are separated and purified (with a recovery yield of 99,9 %). Iodine and other fission gases (krypton,...) are discharged. The fission products and minor actinides (Np, Am and Cm) are contained in a so-called «high level effluent» (which contains 98 % of the $\beta\gamma$ activity and the α activity exclusive of plutonium). The glass is made from the high level effluent **(Figure 4)**.

The high level effluent containing most of the radioactivity is the subject of long-lived radionuclide separation studies.

In a first stage, the partitioning-transmutation applies to spent fuels only. It does not apply to type B wastes.

Extraction by solvent techniques are extremely efficient (they are the ones that enable the industrial recovery of uranium and plutonium), on condition that a good solvent is used. The latter must extract the desired element with very high affinity (recovery yield) and good selectivity (without extraction on unwanted species), allow the subsequent recovery of the element (back extraction) and enable the implementation of operations under industrial conditions without generating new waste.

Partitioning is first of all a question of molecular chemistry (discover the functionalities of the molecules of the solvent and tailor the molecule to the extraction conditions), then of chemistry and of chemical engineering (operating conditions).

For neptunium, zirconium and technetium, the research pertains to implementation, as the solvent is well-known (TBP).

For caesium, the calixarenes functionalised by etheroxide chains have been discovered. The development of a molecule for the solvent and process implementation remain to be achieved.

For the co-extraction of actinides and lanthanides, a molecule has just been developed (DMDBTDMA) and the Diamex process is to be defined (chemistry and implementation).

For americium and curium, unsatisfactory solutions exist (very tricky implementation and large secondary wastes) and interesting functionalities are under study (TPTZ and picolinamides). It is hoped that they will be suitable and, if so, the molecules will have to be tailored to the extraction conditions and the implementation made.

Table 2 gives the research status for the partitioning of actinides and fission products.

Concerning partitioning, once the feasibilities have been acquired, performances will be excellent and a preliminary calculation may be made by assuming losses of long-lived radionuclides from 0.1 % to 1 % in the high level effluent sent to vitrification.

2.2. Conclusion and prospects

Partitioning-transmutation can be considered from the technical point of view, in a dual approach at the medium and long-term.

At medium term, it is theoretically possible to design nuclear reactor parks that fission all the transuranic heavy nuclei they produce.

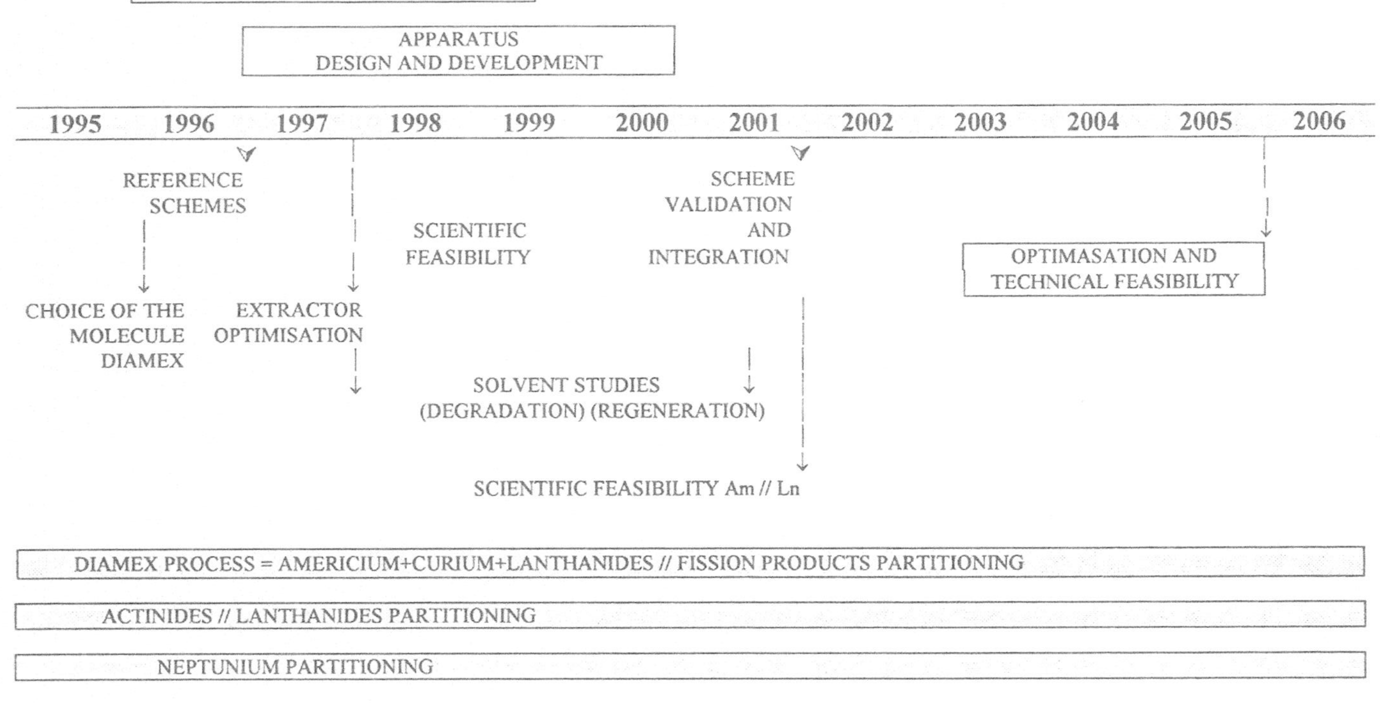
MAIN LINE 1 : PARTITIONING PROCESSES
BUILDING OF ATALANTE LABS.
APPARATUS
DESIGN AND DEVELOPMENT
1995 1996 1997 1998 1999 2000 2001 2002 2003 2004 2005 2006
REFERENCE
SCHEMES
SCIENTIFIC
FEASIBILITY
SCHEME
VALIDATION
AND
INTEGRATION
OPTIMASATION AND
TECHNICAL FEASIBILITY
CHOICE OF THE
MOLECULE
DIAMEX
EXTRACTOR
OPTIMISATION
SOLVENT STUDIES
(DEGRADATION) (REGENERATION)
SCIENTIFIC FEASIBILITY Am // Ln
DIAMEX PROCESS = AMERICIUM+CURIUM+LANTHANIDES // FISSION PRODUCTS PARTITIONING
ACTINIDES // LANTHANIDES PARTITIONING
NEPTUNIUM PARTITIONING
FUEL CYCLE DIVISION
The atom, from research to industry

Figure 2

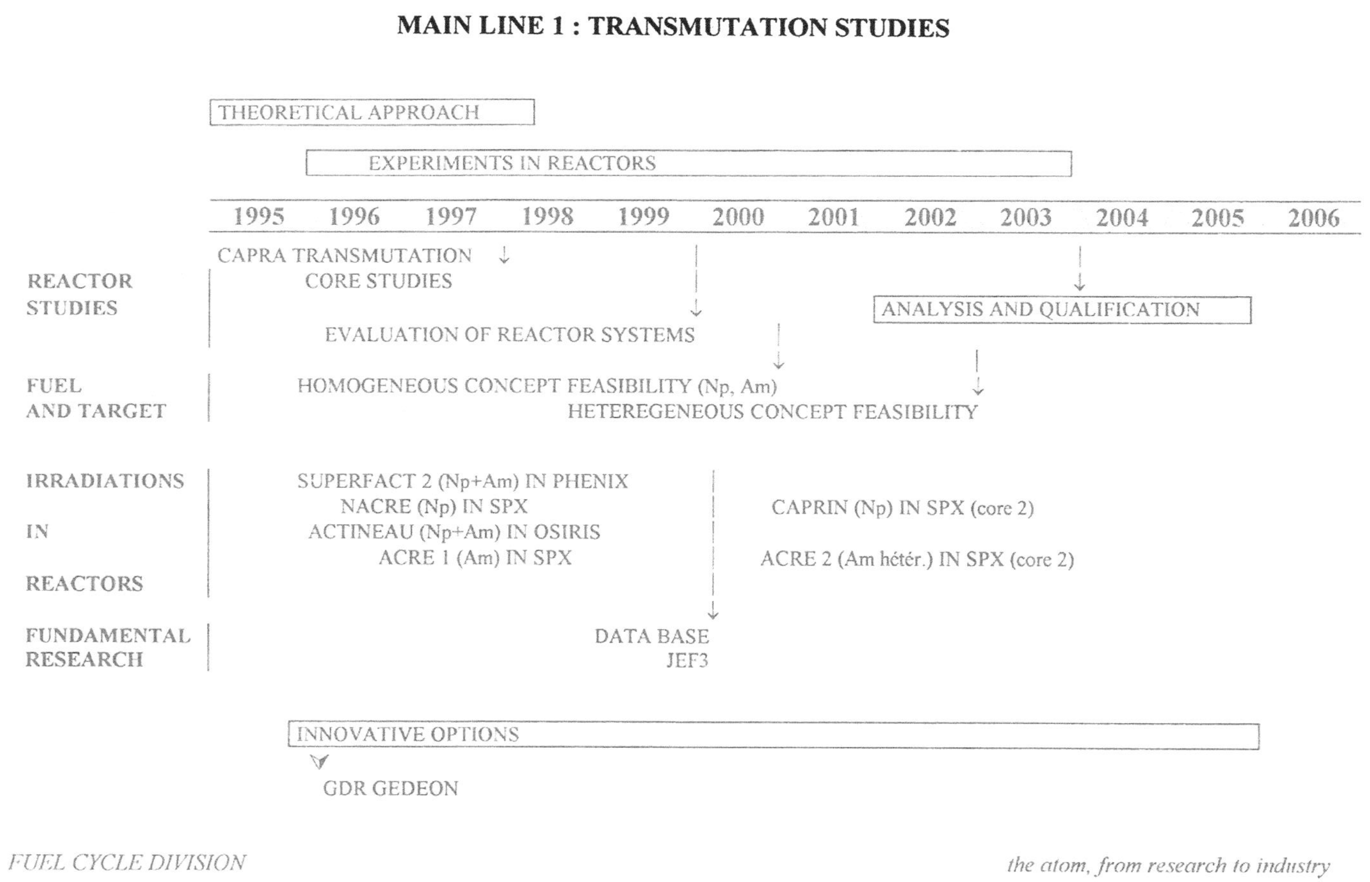

MAIN LINE 1 : TRANSMUTATION STUDIES

THEORETICAL APPROACH

EXPERIMENTS IN REACTORS

1995 1996 1997 1998 1999 2000 2001 2002 2003 2004 2005 2006

REACTOR STUDIES

CAPRA TRANSMUTATION CORE STUDIES

EVALUATION OF REACTOR SYSTEMS

ANALYSIS AND QUALIFICATION

FUEL AND TARGET

HOMOGENEOUS CONCEPT FEASIBILITY (Np, Am)

HETEREGENEOUS CONCEPT FEASIBILITY

IRRADIATIONS IN REACTORS

SUPERFACT 2 (Np+Am) IN PHENIX
NACRE (Np) IN SPX
ACTINEAU (Np+Am) IN OSIRIS
ACRE 1 (Am) IN SPX

CAPRIN (Np) IN SPX (core 2)
ACRE 2 (Am hétér.) IN SPX (core 2)

FUNDAMENTAL RESEARCH

DATA BASE
JEF3

INNOVATIVE OPTIONS

GDR GEDEON

FUEL CYCLE DIVISION *the atom, from research to industry*

Figure 3

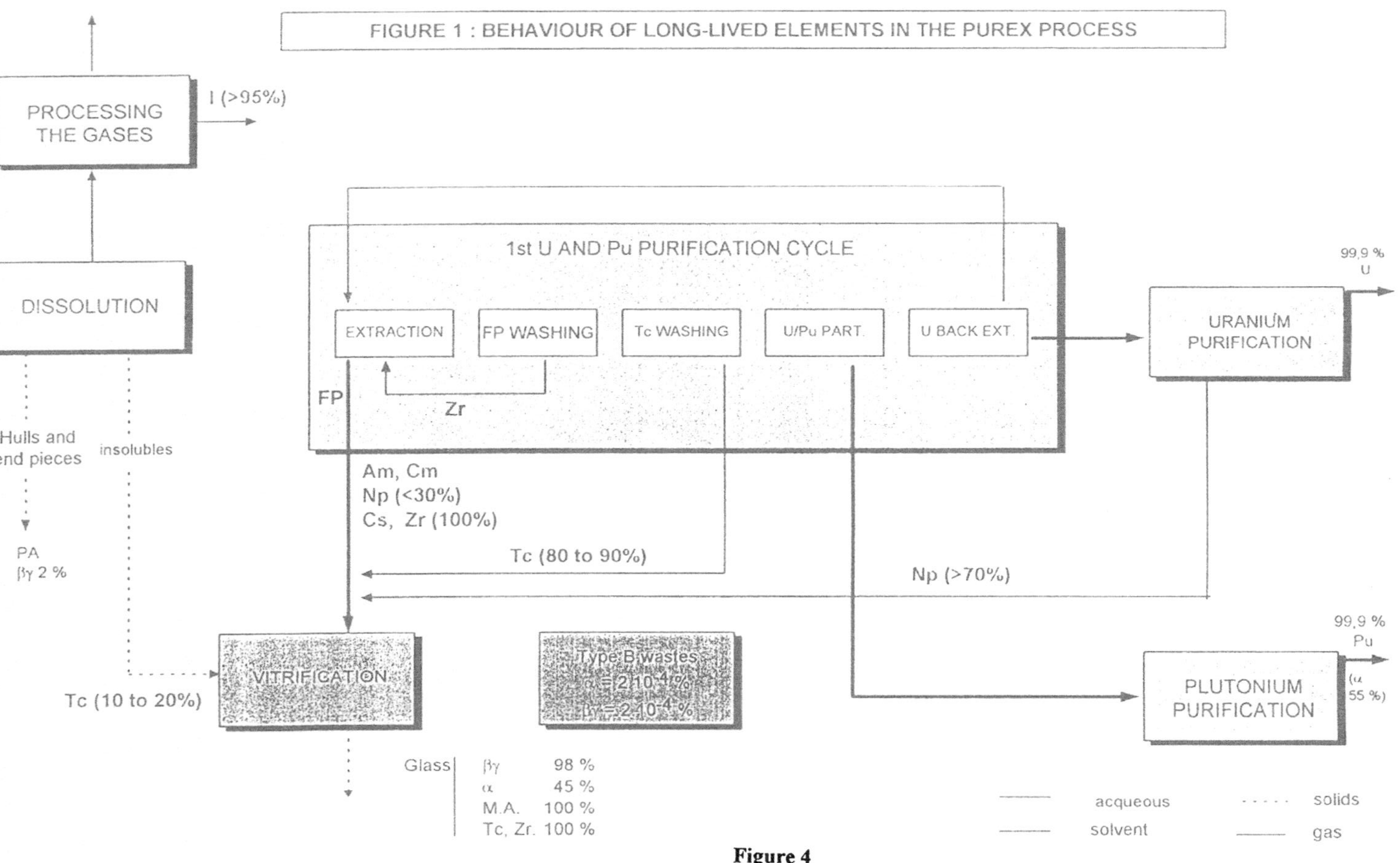

FIGURE 1 : BEHAVIOUR OF LONG-LIVED ELEMENTS IN THE PUREX PROCESS
PROCESSING THE GASES
I (>95%)
DISSOLUTION
Hulls and end pieces
insolubles
PA
βγ 2 %
1st U AND Pu PURIFICATION CYCLE
EXTRACTION
FP WASHING
Tc WASHING
U/Pu PART.
U BACK EXT.
FP
Zr
Am, Cm
Np (<30%)
Cs, Zr (100%)
Tc (80 to 90%)
Np (>70%)
URANIUM PURIFICATION
99,9 % U
PLUTONIUM PURIFICATION
99,9 % Pu
(α 55 %)
Tc (10 to 20%)
VITRIFICATION
Type B wastes
α = 2 10-4 %
βγ = 2 10-4 %
Glass
βγ 98 %
α 45 %
M.A. 100 %
Tc, Zr. 100 %
acqueous
solvent
solids
gas

Figure 4

LONG LIVED NUCLIDE INVENTORY IN 2010, ACCORDING TO EDF'S PLAN (GLOBAL 1995)

FUELS :			WASTES :		
PWR operation buffer (UOX)	=	8 500 mtHM	GLASSES	=	2 800 m^3
PWR in storage (UOX)	=	3 440 mtHM	Type B	=	70 000 m^3
PWR in storage (MOX)	=	1 900 mtHM			

ACTINIDES (metric tons)

	Pu	Np	Am	Cm
In glass and type B waste	0.15	7.6	7.1	0.9
In stored fuels and operation buffer	223(1)	6.9	17	2.3

LONG LIVED FISSION PRODUCTS (metric tons)

	^{79}Se	^{93}Zr	^{99}Tc	^{107}Pd	^{126}Sn	^{129}I	^{135}Cs
In glass and type B waste	0.01	9.7	13.7	3.7	0.4	---	6.3(2)
In stored fuels and operation buffer	0.01	11	13.8	4.4	0.4	3	7.2

(1) Including 100 tons of Pu in MOX fuel assemblies

(2) ^{135}Cs = 6.3 tons for Cs total = 44 tons

Table 1

RESEARCH STATUS ON ACTINIDE AND F.P. PARTITIONING

	Basic R & D	Process Development	Process Industrialisation
U/Pu partitioning			
Np partitioning		☐ > 99 %	☐ 95 %
Am/Cm partitioning			
Diamex		☐	
Ac/Ln	☐		
Oxidised Am partitioning	☐		
Tc partitioning	☐ insoluble franction		☐ soluble fraction
I partitioning			☐ 95 %
Zr partitioning		☐	
Cs partitioning	☐		
Pd, Se, Sn partitioning			
☐ : Present status of research			

Once the feasibility is demonstrated, performances will be excellent and 99 to 99,9 % of the radionuclides should be separated.

Table 2

The main sensitive points identified today are actinide-lanthanide separation, making americium targets and managing curium.

Major technical changes are necessary, in reactors, reprocessing and fuel fabrication.

These points make up the essential part of the technical research programme.

This management of materials should be performed on a large scale and on an extended timescale. It will not eliminate the «high level and long-lived» wastes as the long-lived fission products and the actinide losses from the cycle (~ 0.1 %) will remain in the glass. Type B wastes could be increased in proportion to the flows processed. Furthermore, the controlled phasing out of nuclear must be taken into account.

Several studies of complete scenarios over time, inventory of material, wastes, risks and costs according to the service rendered will be useful to the 2006 decisions and are included in the research programme.

For the long-term, innovative techniques may be required. They could include new hybrid reactor systems allowing to incinerate the capturing only isotopes as well. Their potentiality and the whole of the associated cycel remain to be studied.

A rather fundamental assessment-oriented research programme has been launched.

In any case, type B wastes will remain. Endeavour to rationalise and reduce the volumes are being made in the framework of current reprocessing (Pu partitioning).

The Puretex research programme deals with this subject.

2.3. Can partitioning-transmutation complement geological disposal ?

Projects for disposing of radioactive wastes in geological formations aim at isolating the radionuclides from man and the environment for a long enough time for their impact to have decreased below the regulatory limit of 0.25 mSv/year. The performance of the multibarrier confinement system (package, engineered barrier and geological formation) is assessed by means of safety models taking into account :

- the behaviour of the waste packages,
- the transfer rate of the various radionuclides through the engineered barrier,
- the transfer rate of the various radionuclides in the geosphere,
- the transfer paths and times of the various radionuclides within the biosphere,
- the effects of the various radionuclides on man.

The different confinement barriers allow to limit the flux of radionuclides at the outlet and hence their impact on man. The safety assessment allows this impact to be assessed for each radionuclide versus time, and hence to identify and classify the radionuclides that are important for safety. In particular, it allows to identify those that would be at the origin of a relatively greater impact, linked either to their properties, or to the uncertainty concerning their behaviour. These calculations are performed right from the design phase, taking the conditions offered by the site into account.

In the event the impact of some radionuclides from disposal in a given site were too great in view of regulatory requirements, means must be found to reduce it. If their management involved technical conditions that are difficult to achieve, other processes must be considered. Several paths may then be envisaged, including :

- a reduction at the source of the wastes disposed of, corresponding to a reduction in the radionuclide inventory. This is the partitioning followed by transmutation system,

- partitioning for special conditioning,

- adapting the performance of the engineered barrier to meet the safety requirement,

- adding an extra barrier, of the over-packing type or similar.

Each of these systems must be examined and assessed with respect to the safety objectives and the technical means to reach them.

Some orientations may already be provided based on generic type exercices, including the international exercices. They indicate that, in a normal evolution scenario :

- the actinides are well contained by the multibarrier system. Owing to their very low solubility in a reducing medium, they are immobilised by precipitation either in the barriers or within the natural system,

- the only significantly predominant impact could be due to mobile fission products. Iodine is often mentioned, an to a lesser extent caesium. Elements for which the data on behaviour in a natural medium are still uncertain are also mentioned : technetium, palladium, niobium and selenium.

A specific request concerning partitioning and transmutation originating from disposal studies can therefore not be considered before the end of 1997, namely in the framework of the first choice of concept, once the inventory per radionuclide has been specified, the long-term behaviour of the packages described and the specifications laid down.

2.4. What does partitioning-transmutation contribute as a way to reduce the inventory of long-lived radionuclides in waste ?

It may first be considered that P&T fits into a simple rationale to reduce the inventory of long-lived radionuclides sent to waste to potentially reduce the dangers that future generations may be subjected to. In that case, it results only from a subjective need to reduce «potential» dangers independently of any waste management scenario.

The analysis of the concentration of radioisotopes in a spent fuel, of their specific radioactivity and of their dose factor when ingested or inhaled has shown that the «radiotoxic inventory» after 1000 years was linked only to the actinides a thousandfold to ten thousandfold more than to the long-lived fission products. Plutonium represents 90 % of the inventory, minor actinides 10 %. Plutonium and minor actinides are therefore the priority challenge.

Let us recall that the «radiotoxic inventory» is only a global indicator. It is obtained by weighing each element in the inventory (its concentration) by a coefficient which is strictly valid only in the event of inhalation or ingestion in small doses.

Reducing the inventory of plutonium and minor actinides in wastes obliges global scenarios to be taken into account when they are recycled in a reactor park.

These theoretical scenarios compared with the once-through scenario where the spent fuel is considered a waste, brings out the following conclusions :

In the case of multirecycling plutonium alone, the radiotoxic inventory of the wastes is only reduced by a factor less than 10 (from 2 to 5 depending on the reactor type), since 10 % of the plutonium is transmuted into americium and curium.

In the case of multirecycling of plutonium and minor actinides, the inventory gain in wastes will then practically depend only on the partitioning performances.

A P&T strategy in a reactor park really makes sense only if it is mastered from begining to end : reaching a steady state in the park, operating this balanced park over the desired period of time, programmed phasing out necessary to eliminate the inventory in the cycle and putting an end to the nuclear programme. Any brusk interruption in the sequencing of these phases would heavily penalise, or even annihilate the expected gains in the tonnage to be disposed of.

The establishment, operating and programmed phase-out times for nuclear in a P&T strategy is counted in centuries. The problem of the availability of the foreseable uranium resource in a PWR park will then be acutely felt for carrying out such a strategy. In this regard, the position of FBRs is much more favourable in all respects : sufficient uranium resources, proven physical feasibility of multirecycling, lower production of minor actinides, etc.

The inventory of fission products in wastes can hardly be reduced using fission reactors and delicated, potentially more efficient, innovative systems must be considered. However their feasibility remains to be proven.

2.5. Conclusion

Partitioning-transmutation applied to the management of materials is an operation that could modify the composition of the wastes and act on the long-lived radionuclide inventory (espacially actinides). We do not know today how to assess the gain of such operations, or the disadvantages resulting from as service rendered, namely the TWhe of electricity produced.

This step will have to be included in nuclear power evolution and waste nanagement scenarios to bring out more significant parameters of the pros and cons, for example the short-term radiologic impact on the workers, the impact that could be produced in the future by geological disposal, the costs, the consumption of raw materials resources, etc.

Typical scenarios could perhaps suffice to bring out easier to use assessment critera by showing «classes» of radionuclides (fissile materials, γ n emitters, mobile elements, very low radioactive β emitters, etc.).

However, the difficulty in short-term and long-term risk intercomparisons will remain.

3. AXIS 2 - WASTE GEOLOGICAL DISPOSAL PROGRAM

This part is out of the scope of this present seminar, and, then presented very briefly.

As it was explained in the introduction, four new sites were selected and are at the present time investigated. Files deposits were made recently for the public enquiries until about the end of 1997. At this date it will be decided the construction of two or three underground laboratories.

The general schedule is showed on Figure 1. The four sites are in fact reduced at three because the two sites in Meuse and Haute-Marne departments are joint on one common site. The three hosts rocks are : clay for East site, granite for site in Vienne department and silt for site in Gard department.

4. AXIS 3 - WASTE CONDITIONING, PACKAGING AND LONG TERM INTERIM STORAGE

It is not precise in the law the signification of «long term» interim storage, it is clear that the goal of this third research axis is to answer to an eventual delay in the implementation of final geological disposal site.

In the case of long term interim storage it is necessary to have avalable special conditioning materials and packaging systems.

The National Assessment Committe suggest also to extend this area of researchs to particular conditioning for mobile long-lived radionuclides isolated by partitioning processes.

This strategy, named Partioning-Conditioning (P&C) is a complementary one of the Partitioning-transmutation (P&T) strategy.

Globally, we can say the P&T concerned chiefly the actinides and P&C the long-lived fission products for which transmutation is particularly difficult.

The general schedule concerning axis 3 researches is shown on **Figure 5**.

The program in this area is the following :

- for CEA research centers waste, chemically diversified, volume reduction techniques,

- studies of actual conditioning materials and improvements of conditioning processes.

  - Glass in the main industrial waste confinement material. In a chemically appropriate medium, only a few % of a glass cylinder are altered after several ten of thousands of years.

  - Basic research on phosphate materials with a cristalline structure adapted to chemical trapping of selected radionuclides.

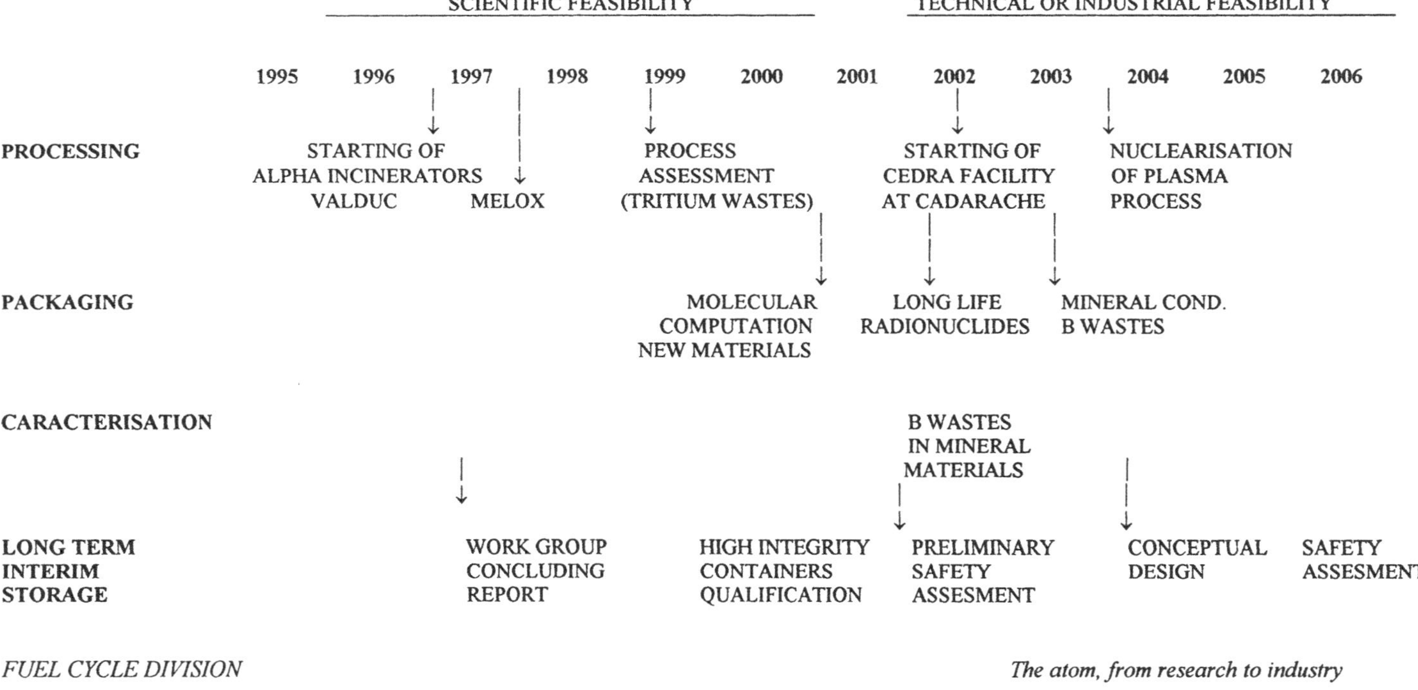
MAIN LINE 3 : WASTE PROCESSING-PACKAGING-CARACTERISATION-INTERIM STORAGE
SCIENTIFIC FEASIBILITY
TECHNICAL OR INDUSTRIAL FEASIBILITY
1995 1996 1997 1998 1999 2000 2001 2002 2003 2004 2005 2006
PROCESSING
STARTING OF ALPHA INCINERATORS VALDUC
MELOX
PROCESS ASSESSMENT (TRITIUM WASTES)
STARTING OF CEDRA FACILITY AT CADARACHE
NUCLEARISATION OF PLASMA PROCESS
PACKAGING
MOLECULAR COMPUTATION NEW MATERIALS
LONG LIFE RADIONUCLIDES
MINERAL COND. B WASTES
CARACTERISATION
B WASTES IN MINERAL MATERIALS
LONG TERM INTERIM STORAGE
WORK GROUP CONCLUDING REPORT
HIGH INTEGRITY CONTAINERS QUALIFICATION
PRELIMINARY SAFETY ASSESMENT
CONCEPTUAL DESIGN
SAFETY ASSESMENT
FUEL CYCLE DIVISION
The atom, from research to industry

Figure 5

- Above the ground, interim strorages.

- From a research point of view, extension of the capabilities of the facilities in terms of strorage duration for each category of waste.

- Measurements of physico-chemical properties of conditioning materials for medium activity waste.

5. FINAL CONCLUSION

As in many contries the difficulties met with public acceptance for the radioactive waste disposal site implementation have conducted the French Government to decide a plamed research program on 15 years to try to solve these difficulties.

Partitioning and Transmutation strategy find naturally its place in the french fuel cycle policy including the reprocessing step and MOX recycling. In this general scheme we can found all the steps necessary for actinides elimination and in particular for plutonium treatment and of course also for weapon-grade plutonium.

Different phases ar even existing at the industrial stage for a long time, as the fuel reprocessing, MOX fuel fabrication, fast breeder reactors and corresponding waste treatment.

The goals of the December 30, 1991 law reinforce R&D in all these fields and, furthermore, are the opportunity to develop new techniques.

GLOSSARY

CEA : Commissariat à l'Energie Atomique.

ANDRA : Agence Nationale pour la gestion des Déchets Radioactifs.

EDF : Electricité de France.

COGEMA : Compagnie Générale des Matières Nucléaires.

CNRS : Centre Nationale pour la Recherche Scientifique.

BRGM : Bureau de Recherches Géologique et Minière.

Self - Regulating Processes in Nuclear Reactor

V. Ya. Gol'din
Institute for Mathematical Modeling RAS

1. An analysis performed in several papers [1,2 and others] shows a necessity of nuclear power development. However, the nuclear power is admissible only provided security maintenance of the whole cycle from fuel obtaining till burial of radioactive wastes (RW). The crucial points are elimination of large reactor accidents and non- proliferation of nuclear weapons. Plutonium fate is tightly connected to the latter.

2. Nuclear power development convinces that the engineering methods are not able to provide necessary security. Technics with natural security [1,2] is needed. An important role in its development can be played by self- regulating processes. Let's show that a lot of conditions providing security can be realized with the help of two self- regulating processes:

a) self- regulating neutron- nuclear process (SNNP) [1,3-6],

b) self- regulating processes with reactiveness drop at temperature growth (negative temperature coefficient of reactiveness) [7,8] which ensure chain reaction switch- off at reactor over- heating.

3. A SNNP is proposed in [1,3,4]. The SNNP is studied with the help of mathematical modeling in [5,6] and a new variant SNNP2 is found. If in the active reactor zone U^{238} (or Th^{232}) is present together with U^{235} (or Pu^{239} or U^{233}), then Pu^{239} (or U^{233}) reproduces itself along with U^{235} burning according to reaction

$$U^{238}+n \rightarrow U^{239} \xrightarrow{\beta} Np^{239} \xrightarrow{\beta} Pu^{239}$$

with delay T = 2,4 days (or $Th^{232}+n \rightarrow Th^{233} \xrightarrow{\beta} Pa^{233} \xrightarrow{\beta} U^{233}$ with 27 days delay [3]).

There is an equilibrian concentration of Plutonium N_{eq}^{Pu9} at which these processes are in equilibrium [1,3,4]

$$N_{eq}^{Pu9} \cong \frac{\sigma_{n\gamma}^{U8}}{(\sigma_{n\gamma} + \sigma_{n,2n} + \sigma_f)^{Pu9}} N^{U8} \ldots^{1}$$

If $N^{Pu9} > N_{eq}^{Pu9}$, then N^{Pu9} drops, and reactor becomes sub critical. If $N^{Pu9} < N_{eq}^{Pu9}$, then N^{Pu9} grows. At exceeding of Plutonium critical concentration N_{cr}^{Pu9}, the growth of neutron flow during $\tau \sim 10^{-6}$ sec burns out Plutonium surplus and additional Plutonium is generated about 2,4 days later. It determines a SNNP existence. In fact, a cheap U^{238} burns there.

This process leads to a weak super- criticality, reactiveness is $\rho < 10^{-7}$. At increase of neutron losses due to fragments and neutron flow from the active zone, it leads to a weak sub- criticality $-\rho < 10^{-7}$ [6]. That is why the flow changes by less than 8% during twenty-four hours. A similar process, but with larger periods of time, arises in Thorium variant. For reactor start up, any of active elements can be used: U^{235}, Pu^{239} or U^{233}. At right chosen concentrations, reactor reaches SNNP.

4. A variation of considered SNNP2 makes it possible to decrease a deviation from criticality without control [6]. Let's divide the active zone at sub-zones "a" and "b". Let's set in sub- zone "a" $N^{Pu9} > N_{eq}^{Pu9}$, and in sub-zones "b" $N^{Pu9} < N_{eq}^{Pu9}$. In this case, the growth of N^{Pu9} in "b" will be compensated by the drop of N^{Pu9} in "a". The choice of concentrations and parameters is possible which minimizes the deviation from criticality in non-controlling state. So the correction can be minimized.

As in sub-zones "a" $N^{Pu9} > N_{eq}^{Pu9}$, then Plutonium of high concentration can be burned there. And the mass of initial Plutonium will decrease. In the sub- zones, it is possible to burn a mixture of

---

[1] **More accurate value of N_{eq} is determined in [6]. The calculations show that N_{eq}^{Pu9} depends on a place in reactor and on burning, i.e. average cross-sections depend on spectrum.**

E. R. Merz and C. E. Walter (eds.), Advanced Nuclear Systems Consuming Excess Plutonium, 331–333.
© 1997 *Kluwer Academic Publishers.*

Pu^{239} , Pu^{240} , Pu^{241} , Pu^{242} and other actinides which contains some percentage of fragments [6] obtaining after separation of burned fuel at heavy (A > 230) and light (A < 170) fractions [1,6]. It was announced in reports [9,10] that the technology of the separation was developed. That is why SNNP2 let us pass to a closed cycle. SNNP enables high burning out of fuel(Pu + U^{238}): to 30-40%, and practically, only U^{238} burns. However, a stability of fuel pin shells enables considerably smaller burning out. A reduction of fuel prices allows a decrease of output power density which will possibly help to raise the stability of fuel pin shell.

5. Inside the reproduction zone, Plutonium also accumulates. A running wave arises in the zone [1,3-6]. At the disposition of reproduction zone outside the active zone, Plutonium mass increases, and at the disposition of this zone inside the active zone, whole Plutonium mass drops while reactor works. At the accumulation of Plutonium in fuel pin of reproduction zone, they can be exchanged with the burned out fuel pin [6]. The reactor life time can be increased due to it.

6. A fraction of elder actinides increases in closed cycle [1]. It demands development of radiation secure technology at reloading and repairs. From the other side, the increase of actinide fraction together with technology of fuel separation at heavy and light fractions are important for non- proliferation of nuclear weapons. Here a prohibition or strong restriction of heavy isotope separation technology is possible which is not needed for nuclear power industry. Let's point out that the similar difficulties arise in Thorium cycle. When using Th^{232} in closed cycle, the elder actinides will also appear. Besides, U^{233} generation is accompanied with U^{232} generation due to reaction n-2n. The products of its radioactive decay are strong radiators with E= 2,61 MeV which demands a high radiation shelter.

In SNNP variant, U^{233} and U^{232} are generated inside reactor having strong radiation shelter. However these problems will arise at shut down, repair etc. These problems have to be solved before production. Let's notice that, at U^{233} variant, a RW problem is slightly softened because among it's fission products there are no Kr^{85} , Sr^{90} , Sb^{125} , Pm^{147} , Sm^{151} , Eu^{155} and five short- living nuclei which arise for U^{235} , U^{238} and Pu^{239}.

7. The most important point of security is the ceasing of chain reaction at heat removal breach. A good solution is to use heat expansion of active zone with transition to sub- critical state at dangerous temperature growth [8]. As a limit variant, a melting link connecting two parts of active zone can be used [7].

The reactiveness change at SNNP leads to decaying oscillations with reconstruction of critical state [6]. That is why, for reactor shut down, the time of reactiveness drop has to be less than semi-period of oscillations.

8. The SNNP2 variant is useful in hybrid systems. A (heat) power W of sub- critical reactor is determined by the power of neutron source q n/sec and K_f

$$\cong \frac{q}{\nu_f}\frac{K_f}{1-K_f}3{,}2\cdot 10^{-7}\ \mathrm{MW}$$

For security, it is proposed to use $K_f \leq 0{,}95$ that leads to tough, now hardly feasible demands to a source. K_f changes during reactor work. In SNNP2, it is easier to realize maintaining a constancy of K_f , which will enable to draw it nearer to 0,99.

REFERENCES

1. L.P.Feoktistov. Security as a Key Point of Nuclear Power Revival. Uspekhi Phys. Nauk. 1993 N8, p.89-102 (in Russian); Phys.Uspekhi v.36, N8, 1993, p.733 (in English).
2. V.V.Orlov, E.I. Avrorin, E.O.Adamov and others. Non- traditional Concepts of APS with Natural Security. Atomnaya Energya 1992,v.72, p.317- 328 (in Russian).
3. L.P.Feoktistov. Analysis of A Concept of Physically Secure Reactor. M. 1988, preprint of I.V.Kurchatov IAE, N 4605/4 (in Russian).
4. L.P.Feoktistov. A Variant of Secure Reactor. Priroda, 1989, N1, p.11- 15 (in Russian).
5. V.Ya.Gol'din and D.Yu.Anistratov. A Self-Adjusting Neutron-Nuclide Regime in Fast Neutron Reactors. TOPSAFE' 95. International East-West TOPical Meeting Safety at Operating Nuclear Power Plants. 1995, v.II, p.179-182.

6. V.Ya.Gol'din, D.Yu.Anistratov. Reactor at Fast Neutrons in Self- Regulating Neutron- Nuclear State. Mathematical Modeling, 1995, v.7, N10, p.12-32 (in Russian).

7. Holms and others. Experimental Studies in Reactors at Fast Neutrons. In: Experimental Reactors and Reactor Physics. M. GNTTL, 1956, p.230-254.

8. D.R.Pedersen and B.R.Seidel. The Safety Basis of Integral East. Nuclear Safety 1990, v.31 N4, p.443-457.

9. V.Hannum and others. Advantages of Perspective Fuel Cycle with Application of Fast Reactors for Plutonium Utilization. Argon National Laboratory, USA. Report at International Scientific Seminar "Perspective Nuclear Systems Using Plutonium".

10. V.Ivanov. Experience of GNC RF HIAR in Using Uranium- Plutonium Oxide Fuel in Nuclear Reactors. Ibid.

SUMMARIES of the PAPERS PRESENTED at the WORKSHOP

A major component of global efforts to prevent proliferation is the efficient control of plutonium. The growing world surplus of separated plutonium from power generation and excess weapon plutonium from disarmament under the reciprocal arms reduction agreed upon by Russia and the U. S. require urgent attention. Suitable methods are being sought for safe disposition of plutonium, particularly weapon plutonium.

Unless all plutonium is fissioned, a large amount will remain in some form for a long time as separated plutonium, in spent fuel, or fixed in a matrix such as high-level waste glass or inert ceramics. Plutonium contained in spent fuel or, together with high-level waste fixed in a matrix, makes the storage form self-protecting to a significant degree due to its intense gamma radiation.

An advanced approach is required to go beyond this "spent fuel standard" and to further reduce the proliferation risk. Advanced options include burning plutonium more or less totally by utilizing special fuel management with repeated fuel reprocessing in reactors or in accelerator-driven systems. Elimination of plutonium is more costly, complex, time-consuming, and risky than minimizing accessibility to meet the spent fuel standard. Nevertheless, relevant aspects with of the potential and promise of various elimination approaches were considered in this NATO Advanced Research Workshop. Summaries of each paper presented at the Workshop are given below. Acronyms are defined in the Glossary.

Country: Russia
Presenter: Alexander Chebeskov
Title (1): Problems of Excess Plutonium Utilization in Reactor

The large amount of fissile material being released under nuclear disarmament has drawn the attention of nuclear specialists, politicians, and the public to the problem of making mankind secure against the negative consequences of possible unauthorized diversion of plutonium. In view of Russia's strategy of a closed nuclear fuel cycle, separated plutonium will be utilized in nuclear reactors for electricity production.

Three scenarios for use of this material are under detailed consideration by specialists at Russian research centers, both under Russian initiative, and under broad international cooperation with Western Europe, United States, and Japan. Problems related to introducing plutonium to Russian reactors are identified. Nonproliferation issues of plutonium utilization in nuclear reactors in Russia are discussed in context of the strategic importance of plutonium.

Country: Germany
Presenter: Cornelius Broeders
Title (2): Recent Investigations on Plutonium Utilization.

Early strategic considerations of nuclear power from fission envisioned plutonium as the main fuel in fast breeder reactors in order to achieve high utilization of uranium resources. Startup of fast reactors would be improved by enhanced plutonium production rates in the

E. R. Merz and C. E. Walter (eds.), Advanced Nuclear Systems Consuming Excess Plutonium, 335–351.
© 1997 *Kluwer Academic Publishers.*

generally used LWRs. However, political and economical aspects have lead to a significant delay in the broad use of fast reactors. Instead of enhanced plutonium production in LWRs, the incineration of plutonium that they generate and the considerable amount of weapon plutonium, is now the main objective of strategic studies.

Recent investigations on plutonium incineration in three types of nuclear systems, PWRs, fast reactors, and ADS are discussed. All systems need a closed fuel cycle with sufficient capabilitiy. The use of MOX in PWRs is proven technology in several countries. Strategic investigations for plutonium multi-recycling in PWRs is presented for pools of PWRs with full UOX and full MOX core-loadings. The investigations show that for a ratio of UOX/MOX loadings of ~5/3 a nearly equilibrium plutonium inventory is reached after about 80 y. First investigations for the use of weapon plutonium in PWRs is also discussed.

Plutonium incineration in fast reactors is studied within the common European CAPRA project. Consequences for the buildup of Np and Am are discussed in some detail. Calculational procedures have been established for ADSs. First results of studies for plutonium incineration with a fast neutron spectrum are presented.

Country: United States
Presenter: William Hannum
Title (3): The Benefits of an Advanced Fast Reactor Fuel Cycle for Plutonium Management

The U.S. has no program to investigate nuclear fuel cycles for the large-scale consumption of plutonium. The official U.S. position is to focus on means to bury spent fuel from nuclear reactors and to achieve the spent fuel standard for excess weapon plutonium (considered by policy makers to be an urgent international priority). Recently, the National Research Council published a long-awaited report on its study of potential separations technology and transmutation systems (STATS), which concluded that in the nuclear energy phase-out scenario that they evaluated, transmutation of plutonium and long-lived radioisotopes was not cost effective. However, the STATS committee endorsed further study of partitioning to achieve superior waste forms for burial, and suggested that further consideration of transmutation should be in the context of energy production, not waste management.

The Department of Energy is planning for disposition of excess fissile material in the near term, but has no current program for fast reactor development. Nevertheless, sufficient data exist to identify the potential advantages of an advanced fast reactor metallic fuel cycle for long-term management of plutonium. Some of the key advantages are: (1) several tons of plutonium secured in a single reactor system, (2) metal alloy fuel allows economic fuel recycling to match energy production, (3) all actinides remain in the fuel cycle, out of the waste stream, (4) throughout the fuel cycle plutonium exists in a highly radioactive environment equivalent to the spent fuel standard, (5) the net rate of plutonium consumption can be controlled to meet future energy requirements, (6) because all actinides fission in the fast spectrum, transuranic isotopes would not build up as they do in a thermal spectrum, (7) specific fission products would be partitioned into the waste forms in which they would be most stable for disposal.

Country: Russia
Presenter: Oleg Skiba
Title (4): Experience of RINR on Usage of Uranium-Plutonium Oxide Fuel in Nuclear Reactors

The basic elements of the closed nuclear fuel cycle have been formulated and experimentally substantiated at RINR. These elements utilize (1) non-aqueous pyroelectrochemical methods of uranium-plutonium fuel reprocessing and waste utilization, (2) a polydispersed fractional composition granulate with high particle density, (3) the vibropack method to produce fuel elements from granulated fuel, and (4) remotely-controlled automated equipment for fuel reprocessing and production of fuel elements and fuel assemblies.

The initial materials in such a fuel cycle can be uranium and plutonium in any chemical form (from any origin including weapon plutonium) and spent fuel. Fuel granules are produced in a media of molten salts of alkaline metals chlorides— radiation resistant ion liquids without moderators. This permits high-productivity, compact processes that result in only a small amount of solid radioactive waste. The final product (granulated UO_2, PuO_2 or $UPuO_2$) has a particle density >10.7 g/cm^3. Chemical impurities in the fuel obtained in this way are acceptable from the performance viewpoint but hinder usage of the material for military purposes.

About 4500 kg of granulated fuel of different compositions have been produced at RINR. This product was used without additional preparation to fabricate >28,000 vibropack fuel elements. The vibropack fuel elements using uranium and uranium-plutonium fuel, tested in the SM-2, MIR, BOR-60, BN-350, and BN-600 reactors, showed high reliability. Burnup of 26% h.a. was achieved in the BOR-60 reactor on standard design fuel elements. Irradiation of experimental fuel elements with burnup >30% h.a. is under way. Burnup of 9.6% h.a. was achieved in the BN-600 reactor while testing vibropack fuel elements. Post-reactor investigations showed that the service life of these fuel elements was not exhausted. Burnup >40 MWd/kg U was achieved in both basic operating mode and in a power cycling mode under thermal neutron reactor conditions.

Apart from tests of relatively standard fuel element design, RINR is conducting an extensive investigation of fuel element performance with fuel material having increased content (<40%) of Am, Pu, and Np. The program includes investigation of transmutation processes of these isotopes.

On the basis of these investigations, facility flow sheets for converting weapon plutonium to both pellet and vibropack fuel elements were developed. The productivity of BN-600 fuel in the shielded facility at RINR is: granulated fuel, 1500 kg/y; fuel elements, 10 fuel elements/h; and fuel assemblies, 200 pieces/y.

The combined scientific-investigation, experimental-design, and fabrication efforts at RINR on the problem of a closed fuel cycle and plutonium utilization allows the technology of dry reprocessing of fuel and vibropack fuel elements to be recommended as a basic technology of the nuclear fuel cycle.

Country: Germany
Presenter: Enno Hicken
Title (5): Safety Aspects of Advanced Nuclear Systems Consuming Plutonium

The future of nuclear power is strongly influenced by safety aspects, because safety of nuclear systems is the main public concern. From the beginning, proposed designs must be evaluated with regard to safety. Following the core degradation at Three Mile Island and the accident at Chernobyl, new safety requirements have been developed in some countries. The stiffest requirements, in Germany and France, require that in case of an accident sequence with core melting it not be necessary to evacuate or relocate persons outside the fence of a nuclear power station. In principle, the safety of the complete fuel cycle including fuel fabrication, interim storage, power production, reprocessing (if applicable), and final disposal of waste has to be reconciled with existing licensing rules or the regulations modified.

Country: Japan
Presenter: Sadao Hattori
Title (6): Energy Source for Human Demands

The dramatic population growth beginning in the 19th century brings three fundamental problems: (1) lack of energy and food resources, (2) loss of comfortable climate and suitable environment, and (3) long-term depression of the world economy. These problems can be managed by aggressive industrial development as practiced in many countries of Asia. The most fundamental problem is to find an abundant clean energy source.

Water vapor and, to a lesser extent, carbon dioxide affect global warming. A world-wide water temperature increase of 1-2° C on the surface of the sea greatly increases vaporization of sea water. High temperature above the desert causes less clouds there and, more sea water vaporization causes heavier rain in usually highly-populated wet areas. Higher sea-surface temperature produces larger typhoons and hurricanes. A dramatic increase of persons encountering disaster together with an increasing shortage of food will cause a serious societal catastrophe. Nuclear energy is essential for human survival.

A fast reactor can produce as much as 100 times the energy from uranium as does a thermal reactor without recycling its fuel. A strategy is introduced for production of numerous very small reactors into which weapon plutonium could be quickly packed for burning under excellent IAEA control. An example of the design concept, called Super-Safe, Small, and Simple (4S) is introduced. This is an 80-MWt plutonium burner, refueled only once after 10 y of operation. Specific features of the reactor design include: (1) Reactor power follows load changes by inherent characteristics of the metallic fuel without use of control rods (no operator needed). (2) Refueling is not needed for ten years, then the nuclear fuel is packed in the reactor vessel (IAEA control of the nuclear materials is greatly simplified). (3) Isolation of the reactor from the public is not necessary because of the inherent passive safety achieved by virtue of its small size and the fundamental safety feature of metal fuel. (4) The dry pyro-process of spent fuel developed in the U. S. has inherent proliferation resistance and because of its simplicity is economical.

Country: Russia
Presenter: Victor Orlov
Title (7): Plutonium-Based Nuclear Power and Nonproliferation

First generation nuclear power, based on relatively low-cost U-235, can solve the problems of some fuel-deficient countries. However, its share of global fuel consumption will not grow far beyond the present level of ~5%. The anticipated growth of global energy demand is problematical in face of the call for stabilizing consumption of fossil fuels without a realistic large-scale alternative other than nuclear fission. The rise in fossil fuel consumption could be reduced by increasing the world nuclear power capacity by an order of magnitude, to above 3000 GWe, by the middle of 2000. This is a reasonable objective for the next stage in nuclear power development.

Such a scale can be attained with fast reactors operating in a closed fuel cycle with a conversion ratio of ~1 using ~10^4 t of plutonium to be produced by first-generation reactors and also plutonium from disarmaments. Such a system could further develop at a rate of ~1-2% per year, without constraint of cheap fuel resources.

Experience with sodium-cooled fast reactors and naval reactors can be drawn upon to create, within a limited time, economical fast reactors with inherent physical and chemical properties that rule out accidents of the TMI and Chernobyl types. Effective transmutation of actinides and long-lived fission products in fast reactors offers a promise of an acceptable solution for disposal of high-level waste.

The nuclear fuel cycle for the next stage is predetermined by the accumulation of large quantities of plutonium from the first generation of reactors and by the advantages of the U-Pu cycle with fast reactors over the Th-U cycle. At a later time, U-233 production in fast reactors would make it possible to use the Th-U cycle in thermal reactors, which are preferable for small plants designed to supply local electricity and heat needs.

A nuclear solution to the global energy problem would be unacceptable if it added to the risk of nuclear weapon proliferation. This will not happen if a new technology for nuclear fuel processing is developed, that neither requires separation of fuel components nor is capable of doing so. Another requirement is the means for physical protection against theft.

Fast reactors operating at a conversion ratio ~1, pave the way for employment of a simplified recycling technology that amounts merely to achieving a reasonable degree of removal of fission products. Development of such fuel cycle technology could be viewed as an immediate task of a long-term nuclear program as an effort more productive than creation of special burners of plutonium. Illegal production and extraction of plutonium from such a cycle by modern techniques is a risk comparable to illegal uranium enrichment. This is a risk that already exists and is bound to persist in the future, irrespective of the course taken by nuclear power development. The only way to overcome this risk appears to lie in improving the security measures and the political nonproliferation regime.

Country: United Kingdom
Presenter: Kevin Hesketh
Title (8): The Costs of Weapons Plutonium Disposition Through MOX Utilisation in Existing Commercial LWRs

The urgent need to convert excess weapon plutonium to at least the "spent fuel standard" has led to considerable research activity into potential plutonium consuming systems. Advanced thermal and fast reactors and accelerator based systems potentially offer advantages over MOX fuel in the current generation of thermal reactors. But the advantages of such advanced systems may not be sufficient reason for holding up disposition of plutonium for the considerable time that will be needed to get them established. Resources spent on supporting plutonium utilization in existing commercial reactors may be more effective.

The rationale for pursuing advanced plutonium systems needs to considered very carefully in view of the fact that their development times will be extended. New nuclear systems will need a considerable time from initial conception to deployment on a significant scale. Initial design and development work, followed by construction and demonstration of a prototype will take >10 y. Subsequent construction of full-scale plants will take several more years. The cost of development and the cost of building full-scale plants will run into tens of billions of dollars for any new system. Despite this considerable cost, advanced systems cannot be relied on to contribute significantly to weapon plutonium disposition for at least 20 y.

A solution is required that can be applied on a much shorter timescale, preferably with shorter and less extensive research and development requirements. Use of LWRs, for which MOX fuel is already available commerically is such a solution. Even allowing for such factors as age constraints and technical constraints, the world's 400 LWRs would be able to remove ~20 t/y of weapon plutonium given sufficient MOX fabrication capacity. This rate well exceeds that which would be needed to remove the 100 to 200 t of excess weapon plutonium from stocks in a reasonable period of time, say 10 to 15 y. The approach of encouraging existing LWR operators to support plutonium utilization through technical and financial support could well be cheaper than developing advanced systems and a more rapid means of achieving the end goal.

Disposition of weapon plutonium utilization is a global issue that requires an international solution. Harnessing the existing infrastructure of LWR plants with existing MOX technology is likely to be the most effective means of actually achieving the spent fuel standard. An international solution implies recognition of all involved that existing means of safeguarding the plutonium, perhaps with some enhancements, are politically acceptable. NATO may have an important role to play in facilitating the necessary international understanding. This paper is intended as a positive contribution to help the Workshop achieve a balanced view of the options

Country: Switzerland
Presenter: Carlo Rubbia
Title (9): A Realistic Plutonium Elimination Scheme with Fast Energy Amplifiers and Thorium-Plutonium Fuel

The presentation explores the possibility of using an ADS, previously proposed as a conceptual design for energy amplification, for incineration of unwanted actinides from PWRs and from the disassembly of weapons. The key idea is to use a thorium-plutonium mixture instead of the more conventional uranium-plutonium mixture in a sub-critical device. The ADS operates as a converter of plutonium to U-233. The latter can be later mixed with U-238 and used as fuel for PWRs. The ADS is preferred over a fast reactor from a safety standpoint. A cluster of ADSs operated in conjunction with PWRs is an effective and realistic solution to the ultimately complete elimination of existing plutonium and minor actinide stockpiles and facilitates definitive geologic disposal. Preliminary economic considerations show that plutonium incineration, compared to direct geologic disposal, is not only environmentally more acceptable but is also a more profitable alternative.

Country: United States
Presenter: David Alberstein
Title (10): Weapons Grade Plutonium Destruction in the Gas Turbine Modular Helium Reactor (GT-MHR)

The GT-MHR, when fueled with surplus weapon plutonium, has the unique capability to destroy 90% of the initially charged Pu-239 and 65% of all the intitially charged plutonium in a once-through reactor cycle while generating electricity at ~ 50% efficiency. Plutonium content and quality in the spent fuel is so low that there is little or no military or commercial incentive for reprocessing and recycle. The spent fuel is well suited for disposal as whole elements in a geologic repository. The unique inherent passive safety characteristics of the GT-MHR result in a design that is meltdown proof and insensitive to operator errors. Its high efficiency results in minimal environmental impact and substantial advantages in plant economics.

Since the summer of 1994, MINATOM in Russia, GA in the U.S., and more recently Framatome in France have been participating in a cooperative program to develop the GT-MHR for disposition of weapon plutonium in Russia. The near term objective of this program is to construct a GT-MHR plant at Seversk to burn weapon plutonium and replace the power currently provided at that site. Fuel development and conceptual design activities are underway at several Russian institutes and are scheduled to be completed in October 1997. International support for this effort has been increasing in recent months. Besides Framatome, organizations in several other major nations with nuclear power capability have expressed interest in participating.

Russia is interested in the GT-MHR for plutonium destruction because it offers a uniquely high level of reactor safety, it is the only nuclear reactor capable of using the direct cycle gas turbine for production of electricity at high efficiency. The uranium fueled version of the GT-MHR has high commercial potential as an export commodity that will address balance of payments issues. It achieves a higher level of plutonium destruction without recycle than any other reactor technology and its fuel cycle offers superior diversion and proliferation resistance.

A description of the GT-MHR and its plutonium disposition characteristics, including diversion and proliferation resistance characteristics of the fuel cycle. Information is also presented on deployment cost and schedule, on the gas-cooled reactor experience base, and on the status of the GT-MHR development.

Country: Italy
Presenter: Herbert Rief
Title (11): Some Safety and Fuel Cycle Considerations in Accelerator Driven Systems

A fast spectrum ADS may become a versatile nuclear industry tool and allow for a number of new applications. ADS may serve as a means of transmuting large quantities of minor actinides and long-lived fission products into short-lived waste. Takahashi (BNL) has analyzed the cross-progeny fuel cycle that burns plutonium in a thorium matrix and generating U-233 that can be used to refuel LWRs. In Rubbia (CERN) states that the thorium cycle results in reduced actinide production and leads to a lower radiotoxicity of the waste.

The focus here is on the use of thorium-hosted weapon and reactor grade plutonium in a fast ADS from the viewpoint of neutron economy and plutonium annihilation rate. A sub-critical system (k_{eff} = 0.98 - 0.99) is chosen such that prompt critical excursions, as they may result from coolant voiding, fuel movement, positive temperature coefficients due to a large actinide inventory, etc., are not possible. ADS combines its intrinsic safety with the advantages of a reactor having a large delayed neutron fraction (imitated by the spallation neutrons) and a relatively flat power distribution providing nearly uniform burnup in the fuel.

In a system of this kind a proton beam power of 10-20 MW produces enough spallation neutrons for a 1 GWe power plant. An inexpensive, compact "multistage" cyclotron might produce sufficient proton beam power. Initially, a series of safety studies of near-critical systems was conducted using a simplified kinetics code (Takahashi et al.) that accounts only for negative Doppler feedback in an adiabatic system to calculate the effect of fast transients. Later, the modified European Accident Code was used. This code includes additional feedback due to axial fuel expansion, sodium voiding, clad melting, and fuel motion. The analysis shows that for fast, or medium to fast reactivity ramps, ADS has a major advantage in coping with serious reactivity insertions when compared to conventional reactors. On the other hand, "slow" core melting may occur, if in a loss-of-heat -ink or coolant-flow accident the accelerator beam is not switched off. Therefore, passive means for shutting-off the proton beam are of primary importance.

Country: Canada
Presenter: Peter Boczar
Title (12): Advanced CANDU Systems for Plutonium Destruction

High neutron economy, on-line refueling, and a simple fuel-bundle design of the CANDU reactor provide a high degree of versatility for the disposition of weapon plutonium. A study by AECL, concluded that 50 t of weapon plutonium could be dispositioned in 25 y with two reactors within the current operating and safety envelopes for natural-uranium fuel. MOX fuel would be fabricated in the U.S. and/or Russia and MOX bundles would be transported to Canada. A full core of MOX fuel can be accommodated, with no reactor design changes. To suppress excess reactivity, a burnable absorber would be mixed with depleted uranium in the central elements of each bundle, avoiding performance issues associated with integral burnable absorbers. Void reactivity in this design is negative, which eliminates the LOCA power pulse.

A fuel qualification program, currently underway in the U.S. and Russia, involves the fabrication of MOX pellets (tested since the late 1970s) from weapon plutonium for irradiation testing in Canada. These new tests (endorsed by G-7 leaders at the nuclear summit meeting in Moscow in April) will confirm the behavior of weapon plutonium. This conventional MOX option is readily available in the near-term. More advanced fuel options would result in near-complete destruction of the plutonium. Two such options use a different carrier for weapon plutonium: (1) an inert non-fertile material and (2) thorium.

A study of an inert matrix fuel examined actinide annihilation in CANDU reactors, and these results can be extrapolated to weapon plutonium. The existing 37-element bundle uses SiC as the carrier for ~0.25 kg of plutonium. To suppress excess reactivity the seven central elements consist of only Gd and SiC. Void reactivity with this fuel is negative, resulting in a fairly strong negative power coefficient. Realistic fuel-management simulations were performed, confirming that the power envelopes are within the limits for natural-uranium fuel. The very high thermal conductivity of the SiC fuel would result in extremely low fuel centreline temperatures, with significant benefits in performance and safety. About 0.9 t/y of weapon plutonium would be disposed of in a Bruce A reactor with an estimated refueling rate of 12 bundles each full-power day Over 95% of the initial Pu-239 would be destroyed, compared with >50% in the conventional MOX option.

Using thorium as the carrier for plutonium offers another means of achieving high efficiency in plutonium destruction. A burnup of 30 MWd/kg HM can be achieved with ~2.5% weapon plutonium in ThO_2 while destroying >90% of the fissile plutonium. The U-233 that is produced, (some fissions insitu), is not attractive for weapons because of contamination with U-232. Irradiated Pu/ThO_2 can be disposed of directly, with the spent fuel having a much lower content of minor actinides (Np, Am, and Cm) compared to spent uranium fuel. Alternatively, the spent fuel can be reprocessed to recycle U-233. The high neutron economy of CANDU reactors may open up new options in this regard.

Country: France
Presenter: Massimo Salvatores
Title (13): Advanced Systems for Pu Utilisation

Plutonium utilization is presently underway in LWRs as part of irradiated fuel reprocessing strategies, for example in France. Standard PWRs with MOX fuel are used. However, plutonium recycling can have limitations beyond one or two recycles. In this respect, fast reactors offer a unique opportunity to use plutonium (of whatever quality) to regulate and control plutonium inventories, in mid-term strategies.

Cores with high plutonium content in MOX fuel and in an inert matrix have been studied in the CAPRA program and several feasible examples are described. Advanced concepts, in principle, show Pu consumption features. Some studies performed at CEA are mentioned. In recent years, ADSs have been proposed for various applications, such as energy production, waste transmutation, and plutonium consumption.

The understanding of the physical principles which are at the basis of these concepts helps to put into perspective their potential and facilitates their comparison to conventional reactors. This topic is discussed to some extent on the basis of a physical approach that allows emphasizing the features of interest of these hybrid systems, their transmutation potentials, and the trade-offs (e.g. on the overall energy balance of the system) that should be expected. French programs in these areas are related to the back-end of the fuel cycle and to the waste management and involve the use of large experimental installations (like Super Phenix).

Country: Russia
Presenter: I. Levina, E. Glushkov
Title (14): Weapon-Grade Pu Destruction in VVER and HTGR Reactors

Among various options for the disposition of weapon plutonium Russia favors burning it in power reactors. This choice is explained by the desire to use the high energy value of plutonium and the capabilities of the existing nuclear-industrial complex. Several variants of burning weapon plutonium in power reactors, VVERs, BNs, HTGRs, are under development in Russia. Russian specialists, in collaboration with AECL, are also examining the burning of plutonium in CANDU reactors. This paper only considers the use of VVERs and HTGRs.

Utilization of plutonium in VVER-1000 reactors is one of the most realistic options, because reactors of this type are in operation in Russia and abroad. Physical features of the core with MOX fuel (reduction in the fraction of delayed neutrons and in the efficiency of absorbers used) restrict the fraction of MOX fuel assemblies that can be used in the existing design of VVER-1000 to 1/3 in most of the cycles considered.

Determination of this fractional limit of MOX fuel in the VVER-1000 core requires a complex set of calculations and experiments, including, critical experiments with MOX fuel; upgrading and verification of the codes for the calculation of the neutronic characteristics of the VVER-1000 cores with plutonium fuel (comparison of the results from the precisie calculations made with Russian and foreign codes and of results of Russian and foreign critical experiments with plutonium fuel); complex calculations for the substantiation of safety with particular emphasis on reactivity accidents; and trial operation of MOX fuel assemblies in operating VVER-1000s.

In HTGRs, plutonium is used in the form of undiluted plutonium dioxide. The employment of microspheres with multilayer coatings of pyrocarbon and silicon carbide allows a high burnup of plutonium (up to 90% of the initial Pu-239), to be reached in a once-through reactor cycle. Plutonium remaining in the spent fuel is of no interest for weapon production. The ceramic structure of the spent fuel and the multilayer spherical particle coatings make possible its long-term disposal in geological formations without reprocessing. At present a joint Russia-U.S. development project of a conceptual GT-MHR is underway.

Country: Russia
Presenter: Stanislav Subbotin
Title (15): Utilization of Excess Pu in Molten Salt Reactors

Technical options for resolving the plutonium issue cannot be limited to ways of incineration or denaturing, if continued use of nuclear power is considered. Plutonium must be creatively used in the development of new reactor components of the future nuclear power infrastructure. One such component, the MSR, can be used alongside solid-fuel thermal and fast reactors to achieve cost effectiveness, resource sufficiency, safety, acceptable environmental impact, and nonproliferation of weapon grade materials.

Closing the fuel cycle not only for uranium and plutonium but also for minor actinides should be considered. Transmutation of long-lived fission products is awaiting a solution as well. A more sophisticated infrastructure is needed, that includes additional technological steps in fuel reprocessing and new reactor designs. A three-component infrastructure consisting of thermal and fast reactors and the burner-MSR provides the most expedient approach for utilization of excess plutonium, incineration of minor actinides, and transmutation of some fission products.

The fuel cycle for the burner-MSR, in such an infrstructure would have the following features: (1) spent fuel goes through dry gas-fluoride processing; (2) uranium and some of the plutonium are recycled as a fuel in solid-fuel thermal and fast reactors; and (3) the remainder of the plutonium together with all minor actinides and some fission products are incinerated in a burner-MSR. Stable and short-lived fission products are removed from fuel of any composition by the separation systems inherent in the MSR. Stable and short-lived fission products would be placed in interim storage awaiting application in processes or medicine.

Introduction of MSR in the nuclear power infrastructure will stimulate consideration of (1) the gas-fluoride technique for reprocessing spent fuel efficiently from power reactors, including RBMKs, (2) the use of plutonium from RBMKs for denaturating weapon plutonium (not suitable for the thermal reactors), and (3) the use of a fluoride molten salt system for destruction of dangerous chemical waste. Eventual sole use of thorium will permit an almost non-transuranium fuel cycle.

Country: Sweden
Presenter: Waclaw Gudowski
Title (16): Accelerator-Driven Systems— A New Perspective on Fission Energy

Intense, external neutron sources combined with nuclear reactors into hybrid systems, called ADS, can significantly improve the neutron economy of conventional nuclear reactors. Improved neutron economy provides an excess of neutrons that may be usefully utilized to fission most of actinide isotopes including plutonium as well as for transmutation of some radioactive isotopes into short-lived or stable isotopes. Breeding U-233 in the thorium fuel cycle, uranium enrichment with U-233, and possibilities for extended burnup are also attractive options for these hybrids. Moreover, the subcritical mode of operation offers unique safety advantages and provides the opportunity to use "exotic" nuclear fuel that would not be acceptable in critical systems. The accelerator provides a convenient control mechanism for subcritical systems and can, in principle, work without safe-shutdown mechanisms, such as control rods.

In the last few years ADS have been proposed for different purposes: (1) destruction of radioactive waste from conventional nuclear reactors, (2) production of nuclear energy without conventional nuclear reactors or in synergistic system with conventional reactors, (3) destruction/denaturating of excess weapon plutonium, and (4) rapid production of weapon materials like tritium, U-233, and Pu-239. Proposed designs for ADSs are reviewed. The national perspective and the prospects for nuclear energy development in specific countries places constraints on the design of a particular ADS. There is no single design which can optimally address different objectives for the transmutation systems. Moreover, proliferation concerns can place severe additional constraints on ADS designs.

Today, the most important objectives for ADSs seem to be incineration of weapon (and reactor) plutonium and destruction of radioactive waste from conventional nuclear reactors. Production of nuclear energy with ADSs should be a long-term future objective. These ADS designs will naturally emerge from ADS designs for plutonium and waste transmutation.

Country: Russia
Presenter: Yuri Trutnev
Title (17): Plutonium Problem and One of the Solution Techniques

Nuclear energy is a side product of the industrial production of weapon plutonium. Both science and engineering solutions for this technology were achieved through evolution of the U-Pu fuel cycle. The principal purpose of this cycle, the production of weapon plutonium in special reactors, converted naturally to production of plutonium in power reactors. Therefore, the plutonium "problem" is not a problem associated with nuclear energy but with that of the U-Pu cycle. Its solution is to convert thermal reactors to use the closed U-Th fuel cycle.

Most weapon plutonium and uranium and power plutonium can be used in these converted reactors. This would allow a significant removal of plutonium isotopes that would be fissioned by thermal neutrons and technologically prohibit use of the remaining plutonium for manufacturing explosive devices.

Currently, the unique situation posed by excess plutonium has given rise to the belief that the next evolutionary phase of nuclear energy will be conversion to modern thermal reactors operating on the closed U-Th fuel cycle. These thermal reactors would take full advantage of the accumulated science and engineering achievements, validated technologies, and operating experience of the most advanced reactors.

The closed U-Th fuel cycle has clear advantages over the U-Pu fuel cycle in the solution of the challenging problems in modern nuclear energy such as: handling of actinides in spent fuel and high-level radioactive waste; nonproliferation of fissionable materials and nuclear technologies; reasonable use of weapon uranium and plutonium; transmutation of generated long-lived isotopes including power plutonium; increase in the efficiency of use of nuclear fuel resources; and increase in the safety of reactor operation.

Country: Russia
Presenter: Vasilii Marshalkin
Title (18): Solution for the Nonproliferation of Fissionable Materials in the Closed U-Th Fuel Cycle

The basic solution for restricting access to fissile materials that could be used for manufacturing weapons has two parts: first, these materials must be kept out of storage facilities and, second, use of the material should be unacceptable for manufacturing explosive devices. These conditions are implemented through operation of thermal reactors on a closed U-Th fuel cycle.

In converting from presently used reactors to this closed cycle, weapon uranium and plutonium as well as reactor plutonium can and must be used as startup fuel to be fissioned by thermal neutrons. Their burnup through fission would be followed by generation of the U-233 isotope that is subsequently fissioned by thermal neutrons. Estimates were made of the initial loading requirements and the required refueling rates for six scenarios. The estimates show that the scenarios considered are feasible.

Country: United States
Presenter: Francesco Venneri
Title (19): Destruction of Plutonium and Other Nuclear Waste Materials Using the Accelerator-Driven Transmutation of Waste Concept

The ADS concept can provide for nuclear material destruction. The performance of nuclear systems aimed at destruction of plutonium and other long-lived radionuclides is a function of available excess neutrons and ability to maintain safe operating conditions. The choice of an accelerator/fluid-fuel combination reduces the amount of fuel in the system by eliminating the need to maintain unassisted criticality and by providing a fuel form where neutron utilization for fission is optimized and neutron absorber removal is simplified. Important parameters in the evaluation of subcritical ADSs are: increased safety for burning actinides; increased flexibility of operation; impact of fluid fuel on burnup scenarios for plutonium; and impact of fluid fuel on destruction of key isotopes.

ADSs can burnup existing inventories of nuclear waste while producing power and little new actinide waste. ADS has *four unique attributes*, that are the result of its subcriticality and the use of liquid fuel: process robustness, absorber insensitivity, burn efficiency, and completeness. ***Robustness*** of the ADS allows it to accept a wide range of nuclear waste with significant variability in isotopic assay and chemical contamination, from defense spent fuel and scrap plutonium to commercial spent fuel. As there is no need to reach and maintain criticality, the ADS is relatively ***insensitive*** to neutron absorbers and fuel burnup. ADS should be able to burn plutonium and most components of nuclear waste ***efficiently***, i.e. with low standing inventories, and to a high degree of ***completeness***, so that waste destined for permanent storage in a repository might not only be reduced, but as it contains little TRU or mobile fission products, the repository becomes technically feasible and politically acceptable.

ADSs can accept weapon plutonium, but also a variety of other less pure and well characterized waste forms, including spent fuel plutonium, higher actinides and fission products. If weapon plutonium is destroyed in MOX systems or deep-burn reactors,

ADSs can accept and destroy the residual inventory of higher actinides and long-lived fission products. In the management of plutonium and nuclear waste, ADS can help achieve material inventory reduction factors on the order of 100 to 1000.

Country: Russia
Presenter: Leonid Bolshov
Title (20): Environmental & Safety Problems in Pu Utilization & Power Generation

RAS performed an assessment of the possible radiological consequences of a severe accident involving melting of a core fueled with weapon plutonium. The assessment included calculations: of the possible activity of Pu-239 released to the atmosphere during an accident of a VVER-1000; landscape contamination by plutonium isotopes resulting from such an accident; and population and staff exposure levels resulting from accidental release of Pu-239.

The informational database used in the assessment included: actual data on the levels of the environment contamination by plutonium isotopes that resulted from the Chernobyl accident; experimental data on plutonium isotope content in the human body in the regions around Chernobyl and around plutonium-processing plants at the Mayak factory in the Chelyabinsk region; modern dosimetry data on assessment of plutonium impact on the human body.

The consequences of severe accidents at VVER-type reactors were determined using a modified computer code that allows for specific design features of these reactors. Results of the asssesment are that following a severe accident at a reactor using weapon plutonium fuel, contamination of the landscape at 5 km from the plant would not exceed 1.3 kBq/m^2. The average effective accumulated exposure dose resulting from Pu-239 release even after 70 y of residence in the contaminated zone would not exceed 0.85 mSv, i.e. 1.2% of the natural background for the same time period. The doses for the plant personnel working in the zones of intensive dust production might be 5–10 times higher, but even in this case they are not hazardous. Thus, using weapon plutonium fuel in VVER-1000s does not aggravate the consequences of an accident for the population residing near the reactor site.

Country: Germany
Presenter: Peter Phlippen
Title (21): Accelerator-Driven Transmutation System with Fluid Fuel and Liquid Lead Cooling

Effective transmutation of plutonium as well as the minor actinides (Np, Am, Cm) can be achieved by subcritical neutronic systems. Desirable criteria for such systems include: low pressure, low actinide inventory, low amount of secondary waste, low beam power at maximum power rating, extended safety, and easy accessibility for fuel operations. General requirements for transmutation systems are discussed followed by a description of a thermal system, operated with liquid lead that serves as coolant, fuel carrier, and the spallation target. Target and blanket fluid circuits are separate and thus can be dealt with independently with respect to safety evaluation and mechanical layout.

Investigations have mostly concentrated on neutronic behavior. Graphite is chosen as the moderator. Therefore, low actinide concentrations in the fuel (1-2%) dispersed in liquid lead are sufficient to operate the system at a multiplication ratio of ~ 0.95, resulting in low beam power and high transmutation rates. The actinides are assumed to be dispersed as oxides having nearly the same density as liquid lead at 400° C and thus not likely to experience settling. To optimize resource utilization, a nearly constant radial power rating in the fuel must be obtained. This goal can be achieved by increasing the moderation ratio with radius resulting in a power density peaking in the fuel of less than 1.5 for a blanket outer radius of 2.5 m. The service life of the graphite structures, which is dominated by the fast neutron fluence, is thus homogenized over the whole system as well.

A special layout is presented for a system to transmute plutonium and minor actinides as taken from a PWR. A proton beam of 1.6 GeV, 30 mA (mean) is assumed to strike a lead target 0.8 m in diameter. TRU-oxides, at low concentration, are dispersed in the lead, resulting in ~ 4200 kg TRU in the blanket. Annual TRU consumption is ~530 kg (including 480 kg Pu). For a ~1360 MW_t system (165 MW/m^3 in the fuel, 26.5 MW/m^3 in the blanket) operating at 40% thermal efficiency and a beam power efficiency of 50% in the accelerator, a net power of ~440 MW can be provided to the grid.

Country: United States
Presenter: Carl Walter
Title (22): The Utilitarian Approach to the Disposition of Weapon Plutonium

The NAS report on the management and disposition of weapon-grade plutonium observes that: (1) weapon plutonium presents a "clear and present danger" and (2) there is little advantage to annihilating weapon plutonium in view of the large amount of plutonium contained in spent fuel resulting from operation of today's nuclear power plants. These observations led to a two-part recommendation that begins with conversion of weapon plutonium into a form that is similar to that of plutonium in spent fuel. NAS refers to this step as meeting the "spent-fuel standard" and considers it to be a sufficient <u>first step</u> in disposition of excess weapon plutonium. The second part of the recommendation is that attention must subsequently be given to the proliferation risks associated with the much greater, and growing, amount of plutonium contained in the world's nuclear spent fuel. Attention to the second step also resolves the first step.

NAS considers disposition of weapon plutonium to be of paramount concern, tand conflicting priorities should noit delay its implementation. To do so would extend the period of clear and present danger. Without presenting explicit cost and timing data, DOE considers that use of ALMRs for disposition of weapon plutonium would be costly and not timely. DOE has dismissed the ALMR as a candidate for disposition of weapon plutonium, and in addition suspended the successful development effort on the ALMR/ Fuel Recycle System that had been ongoing. The latter action is based on a narrowly held opinion that recycling plutonium is counter to nonproliferation objectives. In view of advances in reactor design and pyro-metallurgical processes, as well as the sizable inventory of LWR spent fuel, there is little basis to support such an opinion today.

A utilitarian approach for deployment of an advanced nuclear-electric power sector that is ultimately fueled only by depleted uranium is described. The sector is optimized on a system basis to meet several objectives in the context of international safeguards against diversion of plutonium and proliferation of nuclear weapons. These objectives include (1) generation of electric power efficiently and economically; (2) performance with utmost predictable safety; (3) minimization of environmental impacts through conservation of natural resources, consumption of actinides and long-lived fission products with responsible disposal of unavoidable waste; and (4) consumption of spent fuel from currently used reactors.

Country: Japan
Presenter: Hiroshi Sekimoto
Title (23): Neutron Spectrum Concerning Ex-Weapon Plutonium Degradation and Plutonium Utilization in the Future

The reactor burn option seems to be the optimal alternative for the disposition of excess weapon plutonium compared to the other alternatives such as vitrification and emplacement in deep boreholes. It is the only option able to reduce excess weapon plutonium inventories, and the only option that returns equity for the huge capital investment made in the past.

Among the burn options, transmutation of weapon-grade plutonium to reactor-grade plutonium is much easier than total incineration of the plutonium in the reactor. For this purpose the thermal reactor is superior to the fast reactor, as the former incinerates Pu-239 more rapidly as well as obtains the higher concentration of higher plutonium isotopes.

The amount of reactor-grade plutonium produced is about half of the original weapon-grade plutonium when utilized in commercial reactors. In Japan it is intended to start using MOX fuel in a few LWRs in the second half of the nineties and to increase the number of such reactors in a planned fashion to about ten by the year 2000 and ten more by 2010.

High-level waste disposal from LWRs is an important problem in Japan. In 1988, Japan published a report entitled "Long-Term Program for Research, Development on Nuclide Partitioning and Transmutation Technology". In this program, called OMEGA, fast reactors and accelerators are studied in depth, as their harder neutron spectrum reduces the production of higher actinides. This approach produces opposite results than that of burning weapon-grade plutonium in thermal reactors as mentioned above.

The following system can be considered: Weapon-grade plutonium is transmuted to reactor-grade plutonium in a thermal reactor designed, built and operated for this purpose. Subsequently, the transmuted plutonium is utilized in commercial reactors and the long-life radioactive waste is incinerated in special purpose reactors. Continued use of the nuclear fission system is considered to be inevitable, even in the far future. The future nuclear equilibrium system is being studied. In such a system, since more neutrons are necessary to change fertile materials to fissile materials and to incinerate long-life fission products, fast reactors are generally superior to thermal reactors.

Country: Russia
Presenter: Boris Zakharkin
Title (24): Radiochemical Reprocessing Problems Allied to Plutonium Utilization

Plutonium is a highly efficient nuclear energy ingredient to be used as a promising alternative to natural uranium. Currently there are two sources of plutonium: reactor-grade plutonium from nuclear spent fuel, and weapon-grade plutonium that is declared excess. In the context of gradual progress of the nuclear fuel cycle, a radiochemical plant was commissioned in 1976-1977 for reprocessing spent fuel from power, ship, and research reactors to obtain pure plutonium. As much as 30 t of plutonium (as dioxide) suitable for secondary fuel is in stock at present. The challenge is to build up MOX fuel production for primary fast reactors and thermal reactors. The plant applies equally to the utilization of weapon-grade plutonium but requires an optional facility to dissolve plutonium metal and alloys. Results are presented from scientific analyses and pilot-plant substantiations of the problems that have been encountered.

Country: France
Presenter: J. Lefevre
Title (25): French Law on Waste Management Research— General View on Partitioning Program and Current State of Partitioning Studies

Because ANDRA met with opposition in 1989 during investigations of HLW disposal sites, the French government decided to temporarily halt the investigations. After subsequent discussions, the Government passed a law at the end of 1991, to relaunch the investigations in the future, in a more favorable public and political acceptance climate. This law consists of a long-term research program extending to 2006, along three research axes: (1) partitioning and transmutation of long-lived wastes in order to minimize their long-term toxicity, (2) assessment of the capacity of geological structures to confine radioactive wastes through research on performance in underground laboratories under ANDRA leadership, (3) and improvement of waste packaging techniques and prolonged surface storage.

Periodic evaluations of the R&D program will be made until 2006 by 12 members of a national evaluation commission. Flowsheets with general R&D planning for these three research axes are presented and details are given on the partitioning and transmutation R&D program concerning plutonium cycling and minor actinide management. Examples of long-lived radionuclides partitioning are given.

WORKSHOP SUMMARY and EVALUATION

The growing world surplus of separated plutonium from power generation and excess weapon plutonium from disarmament require urgent attention to avoid proliferation of atomic weapons. Unless all plutonium is fissioned, a large amount will remain in some form for a long time in spent fuel, as separated plutonium, or fixed in a matrix such as high-level waste glass or inert ceramics. Plutonium contained in spent fuel or, together with high-level waste fixed in a matrix, makes the storage form self-protecting to a significant degree due to its intense gamma radiation. An advanced approach is required to go beyond this "spent fuel standard" and to further reduce the proliferation risk. Advanced options include almost total burning of plutonium by utilizing special fuel management in reactors with repeated fuel processing or in ADS with continuous processing. The 25 papers presented at the Workshop can be grouped into five topical areas as noted below, although the boundaries separating these areas are not well defined.

Topic	No. of Papers
Plutonium Management	9
Advanced Reactor Concepts	7
Accelerator Driven Systems	5
Safety	2
Waste Management/Processing	2

A summary of the information presented in each of these topical areas together with some assessment by the editors follows. These assessments appear in bold italics.

Plutonium Management

Russia's strategy of achieving a closed nuclear fuel cycle requires that separated plutonium will be utilized in nuclear reactors for electricity production. Several scenarios for its use are being considered in Russian and joint international studies. Nonproliferation issues of plutonium utilization in nuclear reactors in Russia are considered in context of the strategic energy value of plutonium. ***We believe that this is the most practical approach to plutonium management.***

Utilization of plutonium in VVER-1000 reactors is a realistic near-term option, because these reactors are in general use in Russia. Physical features of the core with MOX fuel restrict the fraction of MOX fuel assemblies that can be used in the existing design of VVER-1000, ***however we believe that near-term action in this manner should be undertaken.***

The basic elements of the closed nuclear fuel cycle have been formulated and experimentally substantiated in Russia. These elements utilize (1) non-aqueous pyroelectrochemical methods of uranium-plutonium fuel reprocessing and waste utilization, (2) a polydispersed oxide granulate with high particle density, (3) the vibropack method to produce fuel elements from granulated fuel, and (4) remotely-controlled automated equipment for fuel reprocessing and production of fuel elements and fuel assemblies. The processing provides adequate purity from the reactor performance viewpoint, but hinders usage of the material in weapons. Vibropack MOX fuel elements tested in several reactors demonstrated high burnup and reliability. Facility flow sheets

E. R. Merz and C. E. Walter (eds.), Advanced Nuclear Systems Consuming Excess Plutonium, 353–362.
© 1997 *Kluwer Academic Publishers.*

have been developed and there is capability to produce reasonably large quantities of fuel assemblies from vibropack elements. This experience supports recommending utilization of dry reprocessing and vibropack fuel elements as a basic technology of the nuclear fuel cycle.

Anticipated growth of global energy demand coupled with restrictions on the use of fossil fuels can not be satisfied by any large-scale alternative other than nuclear fission. The rise in fossil fuel consumption could be reduced by increasing world nuclear power capacity by an order of magnitude, to above 3000 GWe, by the middle of 2000. A nuclear solution to the global energy problem would be acceptable if a new technology for nuclear fuel processing is developed that neither requires separation of fuel components nor is capable of doing so. Increased nuclear capacity could be attained with fast reactors operating in a closed fuel cycle with a conversion ratio of ~1 using plutonium to be produced by first-generation reactors and also weapon plutonium. Such a system could further develop at a rate of ~1-2% per year, without constraint of cheap fuel resources. The plutonium inventory would be of the order of 10^4 t. ***We do not find this amount of plutonium to be of concern in nuclear power infrastructures that operate under international surveillance.***

Experience with sodium-cooled fast reactors and naval reactors can be drawn upon to create, within a limited time, economical fast reactors with inherently safe characteristics. Effective transmutation of actinides and long-lived fission products in fast reactors promises an acceptable solution for disposal of high-level waste. Fast reactors operating at a conversion ratio ~1, pave the way for employment of a simplified recycling technology that amounts merely to achieving a reasonable degree of removal of fission products. Developing such a fuel cycle is considered to be more productive than creation of special burners of plutonium. Clandestine production and extraction of plutonium from such a cycle is a risk comparable to the existing risk of clandestine uranium enrichment. ***Denaturing of U-233 is not a clear solution to atomic weapon proliferation, as enrichment of U-233 is simpler than enrichment of U-235.***

The United States has no program to investigate nuclear fuel cycles for large-scale consumption of plutonium. The official U.S. position is to focus on means to bury spent fuel from nuclear reactors and to achieve the spent fuel standard for excess weapon plutonium. ***We believe that this position is erroneous.***

There is sufficient information that identifies the potential advantages of an advanced fast reactor metallic fuel cycle for long-term management of plutonium. Some of the key advantages are: (1) several tons of plutonium secured in a single reactor system, (2) metal alloy fuel allows economic fuel recycling to match energy production, (3) all actinides remain in the fuel cycle, out of the waste stream, (4) throughout the fuel cycle plutonium exists in a highly radioactive environment equivalent to the spent fuel standard, (5) the net rate of plutonium consumption can be controlled to meet future energy requirements, (6) because all actinides fission in the fast spectrum, transuranic isotopes would not build up as they do in a thermal spectrum, (7) specific fission products would be partitioned into the waste forms in which they would be most stable for disposal.

Germany has recently investigated plutonium incineration in three types of nuclear systems: PWRs, fast reactors, and ADS. All systems ***need a closed fuel cycle.*** The use of MOX in PWRs is proven technology in several countries. Plutonium incineration in fast reactors is being studied within the common European CAPRA project. Procedures have been established for assessing the performance of ADSs and some results are available.

In France, plutonium utilization is presently underway in LWRs as part of spent fuel recycling strategies. Standard PWRs with MOX fuel are used. However, plutonium recycling can have limitations beyond one or two recycles. In this respect, fast reactors offer a unique opportunity to use plutonium (of whatever quality) to regulate and control plutonium inventories.

The United Kingdom argues that the rationale for pursuing advanced plutonium incineration systems needs to considered very carefully in view of the fact that their development times will be extended. Despite their considerable cost, advanced systems cannot be relied on to contribute significantly to weapon plutonium disposition for at least 20 y. ***We believe that this is probably correct and that achieving the spent fuel standard at an early date is of prime importance.*** Existing LWRs fueled with MOX could be used on a much shorter time scale with less extensive R & D. Existing LWRs would be able to convert ~20 t/y of weapon plutonium to the spent fuel standard, given sufficient MOX fabrication capacity.

Japan considers the reactor burn option to be the optimal alternative for the disposition of excess weapon plutonium compared to other alternatives such as vitrification and emplacement in deep boreholes. The reactor burn option is the only option able to reduce excess weapon plutonium inventories, and the only option that returns equity for the huge capital investment made in the past. Transmutation of weapon-grade plutonium to reactor-grade plutonium is much easier than total incineration of the plutonium. The amount of reactor-grade plutonium produced is about half of the original weapon-grade plutonium when utilized in commercial reactors. Further, the thermal reactor is superior to the fast reactor, as the former incinerates Pu-239 more rapidly as well as obtains the higher concentration of higher plutonium isotopes.

Japan intends to start using MOX fuel (from reprocessed spent fuel) in a few LWRs in the second half of the nineties and to increase the number of such reactors in a planned fashion to about ten by the year 2000 and ten more by 2010. Waste management is an important issue in Japan. The OMEGA program is studying the use of fast reactors and accelerators as a means to reduce production of higher actinides. In the fast neutron spectrum plutonium transmutation produces opposite results than in a thermal reactor.

In the near term, the following plutonium management system can be considered: Weapon-grade plutonium is transmuted to reactor-grade plutonium in thermal reactors. Subsequently, the transmuted plutonium is utilized in fast reactors and the long-life radioactive waste is incinerated in special reactors. Japan considers that continued use of nuclear fission into the far future is inevitable and is studying the future nuclear equilibrium system. In such a system, since more neutrons are necessary to change fertile material to fissile material and to incinerate long-life fission products, fast reactors are generally superior to thermal reactors. ***Japan appears to have appropriate and focussed long range plans for insuring the availability of nuclear power in their country.***

Advanced Reactor Concepts

Russia believes that the plutonium "problem" is not a problem associated with nuclear energy but with that of the Pu-U cycle. The problem is solved by converting thermal reactors to use the closed U-Th fuel cycle. Most weapon plutonium and uranium and power plutonium could be used in these converted reactors. This would allow a significant removal of plutonium isotopes that would be fissioned by thermal neutrons and technologically prohibit use of the remaining plutonium for manufacturing explosive devices.

The closed U-Th fuel cycle is said to have clear advantages over the Pu-U fuel cycle in areas such as: handling of actinides in spent fuel and high-level radioactive waste, nonproliferation of fissionable materials and nuclear technologies, reasonable use of weapon uranium and plutonium, transmutation of long-lived isotopes including power plutonium, increase in the efficiency of use of nuclear fuel resources, and increase in the safety of reactor operation. Technical options for resolving the plutonium issue cannot be limited to ways of incineration or denaturing, if future development of nuclear power is considered. ***We agree with this latter position. but are concerned that the weapon proliferation aspects of the U-Th cycle are not well understood, and therefore not well evaluated by those who propose its use.***

Complete closure of the fuel cycle (uranium, plutonium, minor actinides, long-lived fission products) should be sought. ***In fact, we believe that complete optimization of the closed fuel cycle (energy resources, environment, economics uranium, plutonium, minor actinides, long-lived fission products, byproducts, waste , reactors, processing, etc.) should be performed for the nuclear power system of each country.*** A three-component infrastructure consisting of a synergistic combination of thermal and fast reactors and the burner-MSR may provide an expedient approach. Such an infrastructure would have the following features: (1) spent fuel goes through dry gas-fluoride processing, (2) uranium and some plutonium are recycled in solid-fuel thermal and fast reactors, and (3) the remainder of the plutonium together with all minor actinides and some fission products are incinerated in a burner-MSR. Stable and short-lived fission products would be removed by the separation systems inherent in the MSR.

Advantages of the burner MSR include the application of the gas-fluoride technique for fuel processing which produces only a small quantity of waste and its capability to use fuel of various compositions. In the future, MSR would become a necessary reactor component for wide-scale, long-term use of thorium.

An international cooperative program to develop the GT-MHR for disposition of weapon plutonium in Russia is well under way. The near term objective of this program is to construct a GT-MHR plant at Seversk to burn weapon plutonium and replace the power currently provided at that site. Fuel development and conceptual design activities are underway at several Russian institutes and are scheduled to be completed in October 1997.

The GT-MHR, when fueled with surplus weapon plutonium, has the unique capability to destroy 90% of the initially charged plutonium-239 and 65% of all the initially charged plutonium in a once-through reactor cycle while generating electricity at ~ 50% efficiency. The unique inherent passive safety characteristics of the GT-MHR result in a design that is meltdown proof and insensitive to operator errors. Its high efficiency results in minimal environmental impact and substantial advantages in plant economics. ***We believe that the GT-MHR has great potential in future power production, and its development at this time to assist with disposition of excess weapon plutonium should be undertaken at high priority.***

The NAS report on the management and disposition of weapon-grade plutonium observes that: (1) weapon plutonium presents a "clear and present danger" and (2) there is little advantage to annihilating weapon plutonium in view of the large amount of plutonium contained in spent fuel resulting from operation of today's nuclear power plants. ***In this Workshop we explored some of the ways to go beyond the spent fuel standard. These ways do not result in the elimination of all fissile material inventories in an ongoing nuclear power era. At the same time, it was shown that the amount of residual fissile material could be essentially eliminated if there is a need to do that in the future.***

A utilitarian approach for deployment of an advanced nuclear-electric power sector that is ultimately fueled only by depleted uranium appears to satisfactorily address the issues of concern. The sector is optimized on a system basis to meet several objectives in the context of international safeguards against diversion of plutonium and proliferation of nuclear weapons. These objectives include (1) generation of electric power efficiently and economically; (2) performance with utmost predictable safety; (3) minimization of environmental impacts through conservation of natural resources, consumption of actinides and long-lived fission products with responsible disposal of unavoidable waste; and (4) consumption of spent fuel from currently used reactors.

Japan has studied the global interactions of population growth, energy resources, and the environment. The dramatic population growth beginning in the 19th century brings three fundamental problems: (1) lack of energy and food resources, (2) loss of comfortable climate and suitable environment, and (3) long-term depression of the world economy. The most fundamental problem is to find an abundant clean energy source. Nuclear energy is essential for human survival.

Japan proposes a strategy for production of numerous very small reactors into which weapon plutonium could be quickly loaded for burning under excellent IAEA control. The design concept, called Super-Safe, Small, and Simple (4S) is an 80-MWt plutonium burner, refueled only once after 10 y of operation. ***We believe that a small reactor with these type of characteristics should be developed for use in developing countries.***

High neutron economy, on-line refueling, and a simple fuel-bundle design of the CANDU reactor provide a high degree of versatility for the disposition of weapon plutonium. A Canadian study concludes that 50 t of weapon plutonium could be dispositioned in 25 y with two reactors within the current operating and safety envelopes for natural-uranium fuel. MOX fuel would be fabricated in the U.S. and/or Russia and MOX bundles would be transported to Canada.

A fuel qualification program, currently underway in the U.S. and Russia, involves the fabrication of MOX pellets (tested since the late 1970s) from weapon plutonium for irradiation testing in Canada. These new tests will confirm the behavior of weapon plutonium. A major advantage of the CANDU option is participation by a third country, thus enhancing security and safeguards in a balanced, simultaneous draw down of both

U.S. and Russian excess weapon plutonium. A joint Canada-Russia feasibility study sponsored by the Canadian government builds upon previous DOE studies. Its aim is to establish the viability of a CANDU MOX fuel fabrication plant in Russia, addressing related safeguards and security issues. Work has begun earlier this year, and the first interim report is scheduled for this fall.

This conventional MOX option is readily available in the near-term. More advanced fuel options would result in near-complete destruction of the plutonium. Two such options being studied use a different carrier for weapon plutonium: (1) inert non-fertile material and (2) thorium.

Using an inert matrix fuel, about 0.9 t/y of weapon plutonium could be processed through a CANDU reactor. Over 95% of the initial Pu-239 would be destroyed, compared with ~50% in the conventional MOX option. Using thorium as the carrier for plutonium, a burnup of 30 MWd/kg can be achieved while destroying >90% of the fissile plutonium. The U-233 that is produced, (some fission insitu), is not attractive for weapons because of contamination with U-232. Irradiated Pu/ThO_2 can be disposed of directly, with the spent fuel having a much lower content of minor actinides (Np, Am, and Cm) compared to spent uranium fuel.

Accelerator Systems

The United States, Germany, Italy, Switzerland, and Sweden have studied the use of ADSs for one objective or another. We believe that there appear to be no technical feasibility issues at the outset, but enough work has not been completed to determine whether ADSs are an economically viable concept. The United States is developing ADS to destroy certain nuclear materials. ADS performance is a function of available excess neutrons and ability to maintain safe operating conditions. The choice of an accelerator/fluid-fuel combination reduces the amount of fuel in the system by eliminating the need to maintain unassisted criticality and by providing a fuel form where neutron utilization for fission is optimized and neutron absorber removal is simplified. Important parameters in the evaluation of ADS are: increased safety of subcritical operation for burning actinides; increased flexibility of operation; impact of subcritical operation and fluid fuel on burnup scenarios for plutonium; and impact of subcritical operation and fluid fuel on the destruction of key radionuclides.

ADS can burn existing inventories of nuclear waste while producing power and little new actinide waste. ADS has four unique attributes, that are the result of its subcriticality and the use of liquid fuel: (1) process robustness, allowing a wide range of nuclear waste with significant variability in isotopic and chemical assay, (2) insensitivity to neutron absorbers and fuel burnup, (3) efficient burning with low contained plutonium inventory, and (4) Completeness, such that waste destined for a repository is minimized and contains little TRU or mobile fission products.

If weapon plutonium is destroyed in MOX fueled or deep-burn reactors, ADS can accept and destroy the residual inventory of higher actinides and long-lived fission products. In the management of plutonium and nuclear waste, ADS can help achieve material inventory reduction factors on the order of 100 to 1000.

German studies indicate that effective transmutation of plutonium as well as of the minor actinides can be achieved by subcritical neutronic systems. Desirable criteria for such systems include: low pressure, low actinide inventory, low amount of secondary waste, low beam power at maximum power rating, extended safety, and easy accessibility for fuel operations. A thermal transmutation system is proposed that uses liquid lead as coolant, fuel carrier, and spallation target.

These investigations have concentrated on neutronic behavior. Graphite is chosen as the moderator. Therefore, low actinide concentrations in the fuel (1-2 wt-%) dispersed in liquid lead are sufficient to operate the system at a multiplication ratio of ~ 0.95, resulting in low beam power and high transmutation rates. The actinides are assumed to be dispersed as oxides having nearly the same density as liquid lead at 400 °C and thus not likely to experience settling. To optimize resource utilization, a nearly constant radial power rating in the fuel must be obtained. This goal can be achieved by increasing the moderation ratio with radius resulting in low power density peaking in the fuel. The service life of the graphite structures, which is dominated by the fast neutron fluence, is thus homogenized over the whole system as well.

Italy is also studying the performance of a fast spectrum ADS as a potential versatile tool for the nuclear industry that could allow a number of new applications. ADS may serve as a means of transmuting large quantities of minor actinides and long-lived fission products into short-lived waste. A focus of the study is on the use of thorium-hosted weapon and reactor grade plutonium in a fast ADS from the viewpoint of neutron economy and plutonium annihilation rate. A subcritical system ($k_{eff} = 0.98 - 0.99$) is chosen such that prompt critical excursions, as they may result from coolant voiding, fuel movement, positive temperature coefficients due to a large actinide inventory, etc., are not possible. ADS combines its intrinsic safety with the advantages of a reactor having a large delayed neutron fraction (imitated by the spallation neutrons) and a relatively flat power distribution providing nearly uniform burnup in the fuel.

In a system of this kind a proton beam power of 10-20 MW produces enough spallation neutrons for a 1 GWe power plant. Initial safety studies of near-critical systems show that for fast, or medium to fast reactivity ramps, ADS has a major advantage in coping with serious reactivity insertions when compared to conventional reactors. Core melting may occur, however, in a loss-of-coolant-flow accident if the accelerator is not switched off thus passive means for shutting-off the proton beam must be provided.

ADS can significantly improve the neutron economy of conventional nuclear reactors. Improved neutron economy provides an excess of neutrons that may be beneficially utilized to fission most of actinide isotopes including plutonium as well as for transmutation of some radioactive isotopes into short-lived or stable isotopes. Swedish studies indicate that breeding U-233 in the thorium fuel cycle, uranium enrichment with U-233, and possibilities for extended burnup are also attractive options for these hybrids. Moreover, the subcritical mode of operation offers unique safety advantages and provides the opportunity to use nuclear fuels that would not be acceptable in critical systems. The accelerator provides a convenient control mechanism for subcritical systems and can, in principle, work without control rods.

ADS has been proposed for different purposes: (1) destruction of radioactive waste from conventional nuclear reactors, (2) production of nuclear energy without conventional nuclear reactors or in a synergistic system with conventional reactors, (3) destruction/denaturating of excess weapon plutonium, and (4) rapid production of weapon materials like tritium, U-233, and Pu-239. No single design optimally addresses these different objectives. Moreover, proliferation concerns can place severe additional constraints on ADS designs.

Unlike the liquid-fuel ADS design concepts, Switzerland proposes a hybrid system comprised of an ADS and a PWR. The ADS using solid fuel consisting of uranium, plutonium, and minor actinides, would transform plutonium to U-233 and the PWR fuel (containing U-238) would burn the U-233. This system would operate indefinitely in a closed fuel cycle. The cycle would be fed with additional plutonium (the objective being to incinerate plutonium) and smaller amounts of uranium and thorium to makeup the amounts that are burned or transmuted. Fission products, devoid of actinides, would be disposed in a geologic repository that would have lower radiotoxicity by four orders of magnitude in 600 years, than one that produced the same power in PWRs alone that operated on a once-through cycle. The ADS would be lead cooled. Lead would also serve as the spallation target for 1 GeV protons from a cyclotron. ***We believe that toxicity issues associated with the large amounts of lead being considered in these systems should be carefully assessed. Also, the weapon nonproliferation advantage of the U-Th cycles is not obvious to us, and should be examined more closely.***

Safety

Because safety of nuclear systems is the main public concern, continuous attention must be given to the safety of new reactor designs from their inception. New safety requirements have been developed in some countries. The stiffest requirements, in Germany and France, require that in case of an accident sequence with core melting it not be necessary to evacuate or relocate persons residing beyond the power station fence. In principle, the safety of the complete fuel cycle including fuel fabrication, interim storage, power production, reprocessing (if applicable), and final disposal of waste has to be assessed and reconciled with licensing regulations. ***Safety is paramount, but we believe that safety parameters should be based on a realistic understanding of the hazards and risks in the context of other hazards and risks that are acceptable to society.***

Russia has assessed the possible radiological consequences of a severe accident involving melting of a core fueled with weapon plutonium. The activity of Pu-239 released to the atmosphere during the postulated accident of a VVER-1000, landscape contamination by plutonium isotopes resulting from such an accident, and population and staff exposure levels resulting from accidental release of Pu-239 were calculated. The results indicate that using weapon plutonium fuel in VVER-1000s does not aggravate the consequences of an accident for the population residing near the reactor site.

The informational database used in the assessment included data on the: environmental contamination by plutonium isotopes that resulted from the Chernobyl accident, plutonium isotope content in the human body in the regions around Chernobyl and around plutonium-processing plants in the Chelyabinsk region, and impact of plutonium on the human body based on modern dosimetry. Following the postulated severe accident, contamination of the landscape at 5 km from the plant would not exceed 1.3

kBq/m^2. The average effective accumulated exposure dose resulting from Pu-239 release even after 70 y of residence in the contaminated zone would not exceed 0.85 mSv, i.e. 1.2% of the natural background for the same time period. The doses for the plant personnel working in the zones of intensive dust production might be 5–10 times higher, but even in this case they are not hazardous. ***We believe that the activities discussed at this Workshop can be carried out with utmost safety.***

Waste Management/Processing

In the context of gradual progress of the nuclear fuel cycle, Russia commissioned a radiochemical plant in 1976-1977 for reprocessing spent fuel from power, ship, and research reactors to obtain pure plutonium. As much as 30 t of plutonium (as dioxide) suitable for secondary fuel is in stock at present. The challenge is to build up MOX fuel fabrication facilities for thermal and fast reactors.

ANDRA met with opposition in 1989 during investigations of radioactive waste disposal sites in France. After subsequent discussions, a law was passed at the end of 1991, to restart the investigations in the future, in a more favorable public and political acceptance climate. This law provides for a long-term research program continuing to 2006. Research will be conducted in three areas: (1) partitioning and transmutation of long-lived wastes in order to minimize long-term toxicity, (2) assessment of the capacity of geologic structures to confine radioactive wastes, and (3) improvement of waste packaging techniques and prolonged surface storage. ***It is indeed fortuneate that nuclear power generates so little waste compared to waste from other types of power plants. Solving the waste disposal problem, therefore, has not been an urgent priority. However we believe that it is time to proceed diligently to a sound long-range solution to this annoying issue. Only then can the nuclear fuel cycle be deemed invulnerable to diversion of fissile material for atomic weapons.***

GLOSSARY

ADS	Accelerator Driven System
AECL	Atomic Energy of Canada, Limited
ANDRA	(French: National Agency for Disposal of RadioActive-waste)
BN	Russian design fast breeder reactor
BNL	Brookhaven National Laboratory
CANDU	Canadian Deuterium Uranium
CAPRA	Project on advanced plutonium(actinide) burner cores,U-free fue
CEA	(French: Atomic Energy Commission)
CERN	(Swiss: European Center for Nuclear Research)
DOE	Department of Energy
GT-MHR	Gas Turbine-Modular Helium Reactor
HLW	High-Level Waste
HM	Heavy Metal
HTGR	High Temperature Gas-cooled Reactor
IAEA	International Atomic Energy Agency
LOCA	Loss of Cooling Accident
LWR	Light Water Reactor
MINATOM	Ministry of Atomic Energy
MOX	Mixed (Uranium/Plutonium) OXide
MSR	Molten Salt Reactor
NAS	National Academy of Sciences
NATO	North Atlantic Treaty Organization
OMEGA	Option Making Extra Gains from Actinides/Fission Products
PWR	Pressurized Water Reactor
RAS	Russian Academy of Sciences
RINR	Russian Institute of Nuclear Reactors
RBMK	Russian design boiling-ater type reactor;graphire-moderated,wat
STATS	Separations Technology and Transmutation Systems
TRU	TransUranic
UOX	Uranium OXide
VVER	(Russian: Water-cooled/moderated Energy Reactor)

The manufacturer's authorised representative in the EU is Springer Nature Customer Service Centre GmbH, Europaplatz 3, 69115 Heidelberg, Germany. If you have any concerns regarding our products, please contact ProductSafety@springernature.com

Printed and bound by CPI Group (UK) Ltd, Croydon, CR0 4YY

29/04/2026

02099472-0006